ISNM
International Series of Numerical Mathematics
Vol. 142

Advanced Problems in Constructive Approximation

3rd International Dortmund Meeting on Approximation Theory (IDoMAT) 2001

Edited by

M.D. Buhmann
D.H. Mache

Birkhäuser Verlag
Basel · Boston · Berlin

Editors:

Martin D. Buhmann
Justus-Liebig University
Mathematical Institute
Arndtstr. 2
D-35392 Giessen
Germany
e-mail: martin.buhmann@math.uni-giessen.de

Detlef H. Mache
University of Dortmund
Institute of Applied Mathematics
Vogelpothsweg 87
D-44221 Dortmund
Germany
e-mail: mache@math.uni-dortmund.de

2000 Mathematics Subject Classification 30E10, 33F05, 41A05, 42A10, 42A15, 42A82, 47A58, 47B65, 65D05, 65D07, 65D15, 65D20, 65D30, 65T60

A CIP catalogue record for this book is available from the Library of Congress, Washington D.C., USA

Bibliographic information published by Die Deutsche Bibliothek
Die Deutsche Bibliothek lists this publication in the Deutsche Nationalbibliografie; detailed bibliographic data is available in the Internet at <http://dnb.ddb.de>.

ISBN 978-3-0348-7602-5 ISBN 978-3-0348-7600-1 (eBook)
DOI 10.1007/978-3-0348-7600-1

Softcover reprint of the hardcover 1st edition 2003
Member of the BertelsmannSpringer Publishing Group
Printed on acid-free paper produced of chlorine-free pulp. TCF ∞
Cover design: Heinz Hiltbrunner, Basel

ISBN 978-3-0348-7602-5

9 8 7 6 5 4 3 2 1

www.birkhauser.ch

Table of Contents

Preface

The current form of modern approximation theory is shaped by many new developments which are the subject of this series of conferences. The International Meetings on Approximation Theory attempt to keep track in particular of fundamental advances in the theory of function approximation, for example by (orthogonal) polynomials, (weighted) interpolation, multivariate quasi-interpolation, splines, radial basis functions and several others. This includes both approximation order and error estimates, as well as constructions of function systems for approximation of functions on Euclidean spaces and spheres.

It is a piece of very good fortune that at all of the IDoMAT meetings, colleagues and friends from all over Europe, and indeed some countries outside Europe and as far away as China, New Zealand, South Africa and U.S.A. came and discussed mathematics at IDoMAT conference facility in Witten-Bommerholz. The conference was, as always, held in a friendly and congenial atmosphere.

After each meeting, the delegates were invited to contribute to the proceeding's volume, the previous one being published in the same Birkhäuser series as this one. The editors were pleased about the quality of the contributions which could be solicited for the book. They are refereed and we should mention our gratitude to the referees and their work.

The recent meeting in August 2001 was particular in that it was held at the time of Professor Manfred Müller's retirement. It was therefore both a celebration of approximation theory and of the many mathematical contributions Professor Müller made to approximation theory, as well as his friendship with many of the meeting's delegates, including, of course, the editors of this volume. We are grateful for the fine contributions that were delivered at the time of the conference, several of which are now included in this book. It is meant not only as a proceedings of the IDoMAT meeting, but also as a *Festschrift* in honour of Manfred Müller. This volume is therefore in its entirety dedicated to him.

At this point we also thank the Deutsche Forschungsgemeinschaft (Bonn) for providing the majority of the financial support of this conference and the publisher for accepting the proceedings into its International Series of Numerical Mathematics.

Also we would like to thank all participants for their efforts towards making this a successful meeting.

Leading experts and colleagues in approximation theory and quite a number of young researchers made the conference a stimulating event, with interesting discussions and scientific interactions to support and initiate future research. In this sense the success of the IDoMAT conferences in the years 1995, 1998 & 2001 and the positive resonance will encourage us to continue in future this series of International Meetings with new developments in approximation theory and applied mathematics in Witten-Bommerholz.

Witten-Bommerholz, August 2002

Martin D. Buhmann
Detlef H. Mache

3rd International Meeting on Approximation Theory
Witten – Haus Bommerholz (Germany)
August 20–24, 2001

List of Participants

1. **Bavinck, Herman**, Delft University of Technology,
Faculty of Information Technology and Systems, Mekelweg 4
NL-2628 CD Delft, The Netherlands;
e-mail: bavinck@its.tudelft.nl

2. **Beatson, Rick**, University of Canterbury,
Department of Mathematics and Statistics, Private Bag 4800
Christchurch, New Zealand;
e-mail: R.Beatson@math.canterbury.ac.nz

3. **Berdysheva, Elena E.**, Universität Erlangen-Nürnberg,
Mathematisches Institut, Bismarckstr. 1 1/2
D-91054 Erlangen, Germany;
e-mail: berdyshe@mi.uni-erlangen.de

4. **Berens, Hubert**, Universität Erlangen-Nürnberg,
Mathematisches Institut, Bismarckstr. 1 1/2
D-91054 Erlangen, Germany;
e-mail: berens@mi.uni-erlangen.de

5. **Blum, Heribert**, Universität Dortmund,
Institut für Angewandte Mathematik, Lehrstuhl X, Vogelpothsweg 87
D-44221 Dortmund, Germany;
heribert.blum@math.uni-dortmund.de

6. **Braess, Dietrich**, Ruhr-Universität Bochum,
Fakultät für Mathematik, Universitätsstr. 150
D-44780 Bochum, Germany;
e-mail: braess@num.ruhr-uni-bochum.de

7. **Bruin, Marcel G. de**, Delft University of Technology,
Faculty of Information Technology and Systems, Department of Applied Mathematical Analysis, Mekelweg 4
2628 CD Delft, The Netherlands;
e-mail: m.g.debruin@its.tudelft.nl

8. **Buhmann, Martin D.**, Justus-Liebig Universität Giessen,
Mathematisches Institut, Arndtstr. 2
D-35392 Giessen, Germany
e-mail: Martin.Buhmann@math.uni-giessen.de

9. **Butzer, Paul L.**, RWTH Aachen,
Lehrstuhl A für Mathematik, Templergraben 55
D-52056 Aachen, Germany;
e-mail: butzer@rwth-aachen.de

10. **Davydov, Oleg**, Universität Giessen,
Mathematisches Institut, Arndtstr. 2
D-35392 Giessen, Germany
e-mail: Oleg.Davydov@math.uni-giessen.de

11. **Delvos, Franz J.**, Universität-Gesamthochschule-Siegen,
Fachbereich Mathematik, Hölderlinstr. 3
D-57068 Siegen, Germany;
delvos@mathematik.uni-siegen.de

12. **Felten, Michael**, FernUniversität-GHS-Hagen,
LG Numerische Mathematik, Lützowstr. 125
D-58084 Hagen, Germany;
e-mail: michael.felten@fernuni-hagen.de

13. **Fernández, Noemi Lain**, Medizinische Universität Lübeck,
Institut für Mathematik, Wallstr. 40
D-23560 Lübeck, Germany;
fernande@math.mu-luebeck.de

14. **Filbir, Frank**, Medizinische Universität Lübeck,
Institut für Mathematik, Wallstr. 40
D-23560 Lübeck, Germany;
filbir@math.mu-luebeck.de

15. **Fredebeul, Christoph**, Universität Dortmund,
Institut für Angewandte Mathematik, Lehrstuhl VIII, Vogelpothsweg 87
D-44221 Dortmund, Germany;
e-mail: Christoph.Fredebeul@math.uni-dortmund.de

16. **Grajewski, Matthias**, Universität Dortmund,
Institut für Angewandte Mathematik, Vogelpothsweg 87
D-44221 Dortmund, Germany;

17. **Heilmann, Margaretha**, Universität-Gesamthochschule-Wuppertal,
Gaußstr. 20, D-42097 Wuppertal;
e-mail: heilmann@math.uni-wuppertal.de

18. **Hollenhorst, Manfred**, Universität Giessen,
HRZ, Heinrich-Buff-Ring 44
D-35392 Giessen, Germany;
e-mail: hollenhorst@hrz.uni-giessen.de

19. **Impens, Chris**, University of Gent,
Department of Pure Mathematics and Computer Algebra, Galglaan 2
B-9000 Gent (Belgium);
e-mail: ci@cage.rug.ac.be

20. **Köster, Michael**, Universität Dortmund,
Institut für Angewandte Mathematik
D-44221 Dortmund, Germany;

21. **Kroo, Andras**, Hungarian Academy of Sciences,
Mathematical Institute, Reáltanoda u. 13–15
H-1053 Budapest, Hungary;
e-mail: kroo@renyi.hu

22. **Leviatan, Dany**, Tel Aviv University,
School of Mathematical Sciences
69978 Tel Aviv, Israel;
e-mail: leviatan@math.tau.ac.il

23. **Lubinsky, Doron S.**, Witwatersrand University Johannesburg,
Mathematics Department
Wits 2050, South Africa;
and: Georgia Institute of Technology,
Atlanta, Georgia, 30332-1060, U.S.A.
e-mail: lubinsky@math.gatech.edu

24. **Luh, Lin-Tian**, Providence University,
Dept. of Mathematics, Shalu Town,
Taichung County, Taiwan;
e-mail: ltluh@pu.edu.tw

25. **Mache, Detlef H.**, Universität Dortmund,
Institut für Angewandte Mathematik (Lehrstuhl VIII), Vogelpothsweg 87
D-44221 Dortmund, Germany;
e-mail: Detlef.Mache@math.uni-dortmund.de

26. **Mache, Petra**, FernUniversität-GHS-Hagen,
Lehrgebiet Numerische Mathematik, Lützowstr. 125,
D-58084 Hagen, Germany;
e-mail: Petra.Mache@fernuni-hagen.de

27. **Martin, Uwe**, Universität Dortmund,
Institut für Angewandte Mathematik, Vogelpothsweg 87,
D-44221 Dortmund, Germany;

28. **Mazure, Marie L.**, Université Joseph Fourier,
LMC – IMAG, BP 53
38041 Grenoble Cedex 9, France;
e-mail: mazure@imag.fr

29. **Meckbach, Sabine**, Universität – GHS – Kassel,
Fachbereich 17 – Mathematik / Informatik, Holländische Str. 36,
D-34109 Kassel, Germany;
e-mail: meckbach@hrz.uni-kassel.de

30. **Méhauté, Alain Le**, Université de Nantes,
Department de Mathématiques, 2 rue de la Houssinière,
F-44072 Nantes Cedex, France;
e-mail: alm@math.univ-nantes.fr

31. **Mhaskar, Hrushikesh**, California State University,
Department of Mathematics
Los Angeles, CA 90032, U.S.A.;
e-mail: hmhaska@calstatela.edu

32. **Müller, Manfred W.**, Universität Dortmund,
Institut für Angewandte Mathematik (Lehrstuhl VIII), Vogelpothsweg 87
D-44221 Dortmund, Germany;
e-mail: mueller@math.uni-dortmund.de

33. **Obermaier, Josef**, GSF-Forschungszentrum
für Umwelt und Gesundheit,
Institut für Biomathematik und Biometrie, Ingolstädter Landstr. 1
D-85764 Neuherberg, Germany;
e-mail: josef.obermaier@gsf.de

34. **Pinkus, Allan**, Technion,
Department of Mathematics,
32000 Haifa, Israel;
e-mail: pinkus@tx.technion.ac.il

35. **Plonka, Gerlind**, G. Mercator-Universität-GHS-Duisburg,
Fachbereich Mathematik, Lotharstr. 65,
D-47057 Duisburg, Germany;
e-mail: plonka@math.uni-duisburg.de

36. **Prünte, Ludger**, Universität Dortmund,
Institut für Angewandte Mathematik, Vogelpothsweg 87,
D-44221 Dortmund, Germany;

37. **Rasa, Ioan**, Technical University Cluj-Napoca,
RO-3400 Cluj-Napoca, Romania;
e-mail: Ioan.Rasa@math.utcluj.ro

38. **Reimer, Manfred**, Universität Dortmund,
Institut für Angewandte Mathematik (Lehrstuhl III), Vogelpothsweg 87
D-44221 Dortmund, Germany;
e-mail: reimer@math.uni-dortmund.de

39. **Revers, Michael**, Universität Salzburg,
Mathematisches Institut, Hellbrunnerstr. 34
A-5020 Salzburg, Austria;
e-mail: Michael.Revers@sbg.ac.at

40. **Rohwer, Carl**, University of Stellenbosch,
Department of Mathematics, University of Stellenbosch, Private Bag X1
Matieland, 7602, South Africa
e-mail: chr@maties.sun.ac.za

41. **Sablonnière, Paul**, INSA de Rennes,
20 Avenue des Buttes de Coesmes
35043 Rennes cédex, France;
e-mail: Paul.sablonniere@insa-rennes.fr

42. **Schaback, Robert**, Universität Göttingen,
Institut für Numerische und Angewandte Mathematik, Lotzestr. 16–18
D-37083 Göttingen, Germany;
e-mail: schaback@math.uni-goettingen.de

43. **Schempp, Walter**, Universität-GHS-Siegen,
Fachbereich Mathematik, Lehrstuhl I, Hölderlinstr. 3
D-57068 Siegen, Germany;
schempp@mathematik.uni-siegen.de

44. **Skrzipek, Michael R.**, FernUniversität-GHS- Hagen,
Fachbereich Mathematik, Lützowstr. 125
D-58084 Hagen, Germany;
e-mail: michael.skrzipek@fernuni-hagen.de

45. **Stöckler, Joachim**, Universität Dortmund,
Institut für Angewandte Mathematik, Lehrstuhl VIII
D-44221 Dortmund, Germany;
e-mail: joachim.stoeckler@math.uni-dortmund.de

46. **Strauß, Hans**, Universität Erlangen-Nürnberg,
Institut für Angewandte Mathematik, Martensstr. 3
D-91058 Erlangen, Germany;
e-mail: strauss@am.uni-erlangen.de

47. **Szabados, József**, Hungarian Academy of Sciences,
Alfred Renyi Institute of Mathematics, Reáltanoda u. 13–15
H-1053 Budapest, Hungary;
e-mail: szabados@renyi.hu

48. **Szwarc, Ryszard**, Wroclaw University,
Institute of Mathematics, pl. Grunwaldzki 2/4
PL-50-384 Wroclaw, Poland;
e-mail: szwarc@math.uni.wroc.pl

49. **Totik, Vilmos**, University of Szeged,
Bolyai Institute, Aradi v. tere 1
H-6720 Szeged, Hungary;
e-mail: totik@math.u-szeged.hu or totik@math.usf.edu

50. **Vértesi, Peter**, Hungarian Academy of Sciences,
Alfred Renyi Institute of Mathematics, Reáltanoda u. 13–15
H-1053 Budapest, Hungary;
e-mail: veter@renyi.hu

51. **Wang, Renhong**, Dalian University of Technology,
Institute of Mathematical Sciences
Dalian 116024, P.R. China;
e-mail: renhong@dlut.edu.cn

Advanced Problems in Constructive Approximation
(Eds.) M.D. Buhmann and D.H. Mache
International Series of Numerical Mathematics Vol. 142, 1–19

Linear Perturbations of the Classical Orthogonal Polynomials which are Eigenfunctions of Linear Differential Operators

H. Bavinck

Abstract

In this paper we consider polynomials orthogonal with respect to an inner product which consists of the inner product of the classical orthogonal polynomials combined with some perturbation and we give a survey of the work done to derive linear differential operators having these orthogonal polynomials as eigenfunctions.

1 Introduction

In his paper [24] S. Bochner classified the sequences of real or complex polynomials $\{P_n(x)\}_{n=0}^{\infty}$ of a real variable x with $\deg(P_n(x)) = n$, which are eigenfunctions of a second-order linear differential operator. He showed that, up to a complex linear change of variables, the only systems of polynomials with this property are the well-known polynomials of Jacobi, Laguerre and Hermite, the Bessel polynomials and the polynomials $\{x^n\}_{n=0}^{\infty}$. For certain values of the parameters the first three are orthogonal with respect to a real weight function. Recently Kwon and Littlejohn [44] followed Bochner's work showing that, up to a real change of variable, there are six distinct orthogonal polynomial systems (Jacobi, Laguerre, Hermite, Bessel, twisted Jacobi and twisted Hermite) that arise as eigenfunctions of a linear differential operator. H.L. Krall tried to classify all differential operators of the form

$$\mathbf{A(x,D)} := \sum_{\mathbf{i=0}}^{\mathbf{r}} \mathbf{a_i(x)D^i}, \tag{1}$$

having orthogonal polynomials, (polynomials orthogonal with respect to a real weight function) as eigenfunctions, where $\mathbf{D} = \frac{\mathbf{d}}{\mathbf{dx}}$, r is an integer ≥ 3, $\{a_i(x)\}_{i=0}^{r}$ are real continuous functions on $\mathbb{R}$. It is not difficult to see that if (1) has real polynomials as eigenfunctions, then $a_i(x)$ has to be a real polynomial of degree $\leq i$ for all $i = 0, 1, 2, \ldots, r$. Thus the differential operator must have the form

$$\mathbf{A}(\mathbf{x}, \mathbf{D}) = \sum_{\mathbf{i}=1}^{\mathbf{r}} \left(\sum_{\mathbf{j}=0}^{\mathbf{i}} \mathbf{a}_{\mathbf{ij}} \mathbf{x}^{\mathbf{j}} \right) \mathbf{D}^{\mathbf{i}} \tag{2}$$

and the eigenvalues$\{\lambda_n\}_{n=0}^{\infty}$ such that

$$\mathbf{A}(\mathbf{x}, \mathbf{D}) \mathbf{P}_{\mathbf{n}}(\mathbf{x}) = \lambda_{\mathbf{n}} \mathbf{P}_{\mathbf{n}}(\mathbf{x}), \tag{3}$$

with $\lambda_0 = 0$ and λ_n does not vanish for all $n \in \mathbb{N} \backslash \{0\}$. In [42] H.L. Krall showed, that if r is the smallest order of a differential operator of the form (2) having certain orthogonal polynomials as eigenfunctions, then r must be even and he gave an example of a fourth-order operator having nonclassical orthogonal polynomials as eigenfunctions. In another paper [43] he classified all fourth-order linear differential operators having orthogonal polynomials as eigenfunctions and he discovered two more of such operators. More than forty years later his son A.M. Krall [39] (see also [40]) studied the orthogonal polynomials which are eigenfunctions of these new operators, using the technique of distributional weight functions. A.M. Krall found weight functions and the explicit representations for the polynomials and he derived several properties of them including the appropriate boundary value problems. Because of their similarity to the corresponding classical polynomials A.M. Krall called these polynomials Laguerre type, Jacobi type and Legendre type polynomials.

2 Koornwinder's representation

In 1984 T.H. Koornwinder [38] considered the polynomials $\{P_n^{\alpha,\beta,M,N}(x)\}_{n=0}^{\infty}$, now usually called Jacobi type polynomials, orthogonal with respect to the inner product

$$\begin{aligned}(f, g) \quad &= \quad \frac{\Gamma(\alpha+\beta+2)}{2^{\alpha+\beta+1}\Gamma(\alpha+1)\Gamma(\beta+1)} \int_{-1}^{1} f(x)g(x)(1-x)^{\alpha}(1+x)^{\beta}dx + \\ &\quad +Mf(-1)g(-1) + Nf(1)g(1),\end{aligned}$$

$M \geq 0, N \geq 0, \alpha > -1, \beta > -1$. He gave an explicit representation of these polynomials in terms of the classical Jacobi polynomials $\left\{P_n^{(\alpha,\beta)}(x)\right\}_{n=0}^{\infty}$ and their derivatives. In the cases $\beta = 0, N = 0$ and $\alpha = \beta = 0, M = N$ they correspond to polynomials studied earlier by A.M. Krall [39]. Furthermore, using the limit relation

$$L_n^{\alpha,N}(x) := \lim_{\beta\to\infty} P_n^{\alpha,\beta,0,N}(1 - 2\beta^{-1}x)$$

Koornwinder introduced the polynomials $\{L_n^{\alpha,N}(x)\}_{n=0}^{\infty}$, now usually called Laguerre type polynomials, given by

$$L_n^{\alpha,N}(x) = \left[1 + N\binom{n+\alpha}{n-1}\right] L_n^{(\alpha)}(x) + N\binom{n+\alpha}{n}\mathbf{D}\mathbf{L}_\mathbf{n}^{(\alpha)}(\mathbf{x}), \tag{4}$$

which are orthogonal with respect to the inner product

$$(f,g) = \frac{1}{\Gamma(\alpha+1)}\int_0^\infty f(x)g(x)x^\alpha e^{-x}dx + Nf(0)g(0), \quad N \geq 0, \alpha > -1. \tag{5}$$

For $\alpha \in \{0,1,2\}$ these polynomials in have been considered in [39], [48] and [41].

3 Differential operators

After Koornwinder's paper [38] it became a challenge to find a differential operator having the Laguerre type polynomials $\{L_n^{\alpha,N}(x)\}_{n=0}^{\infty}$ as eigenfunctions and a differential operator having the Jacobi type polynomials $\{P_n^{\alpha,\beta,M,N}(x)\}_{n=0}^{\infty}$ as eigenfunctions. In the case of Laguerre type polynomials this would generalize the operators found by H.L. Krall [43], L.L. Littlejohn and A.M. Krall [48], and L.L. Littlejohn [41] for $\alpha \in \{0,1,2\}$ respectively and the problem was solved by J. and R. Koekoek [29] (see also [35]). Since for the classical Laguerre polynomials $\left\{L_n^{(\alpha)}(x)\right\}_{n=0}^{\infty}$ it is known that

$$\mathbf{L}^{(\alpha)}\mathbf{L}_\mathbf{n}^{(\alpha)}(\mathbf{x}) = \mathbf{n}\mathbf{L}_\mathbf{n}^{(\alpha)}(\mathbf{x}) \tag{6}$$

with

$$\mathbf{L}^{(\alpha)} := -\mathbf{x}\mathbf{D}^\mathbf{2} - (\alpha + \mathbf{1} - \mathbf{x})\mathbf{D}, \tag{7}$$

they looked for an operator of the form

$$\mathbf{L}^{(\alpha)} + \mathbf{N}\mathbf{A}^{(\alpha)}, \tag{8}$$

where

$$\mathbf{A}^{(\alpha)} := \sum_{\mathbf{i}=\mathbf{1}}^{\infty} \mathbf{a}_\mathbf{i}(\mathbf{x};\alpha)\mathbf{D}^\mathbf{i},$$

and for numbers $\left\{\alpha_n^{(\alpha)}\right\}_{n=0}^{\infty}$ such that

$$\left[\left(\mathbf{L}^{(\alpha)} - \mathbf{n}\mathbf{I}\right) + N\left(\mathbf{A}^{(\alpha)} - \alpha_\mathbf{n}^{(\alpha)}\mathbf{I}\right)\right] L_n^{\alpha,N}(x) = 0 \tag{9}$$

for $n \in \mathbb{N}$. Here $\mathbf{I}$ denotes the identity operator. By substituting (4) into (9) and by equating the coefficients of N and N^2 on both sides, J. and R. Koekoek obtained two systems of equations and they showed that the two systems of equations for the unknown constants $\left\{\alpha_n^{(\alpha)}\right\}_{n=1}^{\infty}$ and the unknown functions $\{a_i(x;\alpha)\}_{i=1}^{\infty}$, have a unique solution given by

$$\alpha_n^{(\alpha)} = \binom{n+\alpha+1}{n-1}, n \in \{1,2,3,\dots\} \tag{10}$$

and

$$a_i(x;\alpha) = \frac{1}{i!}\sum_{j=1}^{i}(-1)^{i+j}\binom{\alpha+1}{j-1}\binom{\alpha+2}{i-j}(\alpha+3)_{i-j}x^j, \quad i \in \{1,2,3,\dots\}. \tag{11}$$

Actually they guessed the result and proved afterwards that this solution satisfies both systems of equations. From (11) it follows at once that in the case $N > 0$ the differential operator $\mathbf{L}^{(\alpha)} + N\mathbf{A}^{(\alpha)}$ is of order $2\alpha+4$ if $\alpha \in \mathbb{N}$ and of infinite order if $\alpha \notin \mathbb{N}$.

4 Sobolev type orthogonal polynomials

Using the method of [38] H.G. Meijer and H. Bavinck [52], [22] introduced polynomials orthogonal with respect to the inner product

$$\begin{aligned}(f,g) &= \frac{\Gamma(2\alpha+2)}{2^{2\alpha+1}\Gamma(\alpha+1)^2}\int_{-1}^{1} f(x)g(x)(1-x^2)^{\alpha}dx + \\ &\quad + M[f(-1)g(-1)+f(1)g(1)] + N[f'(-1)g'(-1)+f'(1)g'(1)],\end{aligned}$$

$M \geq 0, N \geq 0, \alpha > -1$. Since $(x,x) \neq (1,x^2)$, these polynomials are no longer orthogonal to a weight function and in [23] are dealt with properties of the zeros and recurrence relations. Such a kind of inner product, involving derivatives evaluated at certain points, has been called Sobolev type. In [34], [35] R. Koekoek studied the polynomials $\left\{L_n^{\alpha,M_0,M_1,\dots,M_l}(x)\right\}_{n=0}^{\infty}$, polynomials which are orthogonal with respect to the inner product

$$(f,g) = \frac{1}{\Gamma(\alpha+1)}\int_0^{\infty} f(x)g(x)x^{\alpha}e^{-x}dx + \sum_{k=0}^{l} M_k f^{(k)}(0)g^{(k)}(0), \tag{12}$$

$\alpha > -1, M_k \geq 0$ for $k \in \{0,1,2,\dots,l\}$ and $l \in \mathbb{N}$. If $l = 0$ they are Laguerre type polynomials, for $l \geq 1$ we call them Sobolev type Laguerre polynomials.

5 The inversion method and its applications

At a conference in Erice (June 1990) R. Askey [25] raised the problem of finding difference equations for generalizations of the Meixner polynomials, which are orthogonal with respect to a weight function obtained by adding a point-mass to the ordinary weight function for Meixner polynomials. The following limit relation connects the Meixner polynomials $\{M_n(x;\beta,c)\}_{n-0}^{\infty}$, defined by

$$M_n(x;\beta,c) = (-1)^n \sum_{k=0}^{n} \binom{x}{k}\binom{-x-\beta}{n-k} c^{-k}$$

with the Laguerre polynomials:

$$\lim_{c\to 1} M_n(\frac{cx}{1-c};\alpha+1,c) = L_n^{(\alpha)}(x), \;\; n=0,1,2,\ldots.$$

H. Bavinck and R. Koekoek decided to investigate the easier case of Charlier polynomials first and solved the problem [20] in that case using a new technique which I will call the inversion method. Later H. Bavinck and H. van Haeringen [18] treated the case of Meixner polynomials by the same method. In both discrete cases the difference operator turned out to be of infinite order for all relevant values of the parameters. In [4] this inversion method was used in the Laguerre case to retrieve the differential operator found by J. and R. Koekoek [29] in a direct way. For Laguerre polynomials the method is based on two well-known formulae

$$\mathbf{D}\mathbf{L}_{\mathbf{n}}^{(\alpha)}(\mathbf{x}) = -\mathbf{L}_{\mathbf{n-1}}^{(\alpha+\mathbf{1})}(\mathbf{x}), \;\; \mathbf{n} = \mathbf{1},\mathbf{2},\mathbf{3},\ldots \tag{13}$$

and the generating function

$$\sum_{n=0}^{\infty} L_n^{(\alpha)}(x)t^n = (1-t)^{-\alpha-1}\exp\left(\frac{xt}{t-1}\right). \tag{14}$$

It follows that

$$\begin{aligned}
(1-t)^{i-j-1} &= (1-t)^{\alpha+i}\exp\left(\frac{-xt}{t-1}\right)(1-t)^{-\alpha-j-1}\exp\left(\frac{xt}{t-1}\right)\\
&= \sum_{m=0}^{\infty} L_m^{(-\alpha-i-1)}(-x)t^m \sum_{k=0}^{\infty} L_k^{(\alpha+j)}(x)t^k\\
&= \sum_{n=0}^{\infty}\left(\sum_{k=0}^{n} L_{n-k}^{(-\alpha-i-1)}(-x)L_k^{(\alpha+j)}(x)\right)t^n.
\end{aligned}$$

By comparing the coefficient of t^{i-j} on both sides one obtains

$$\sum_{k=0}^{i-j} L_{i-j-k}^{(-\alpha-i-1)}(-x)L_k^{(\alpha+j)}(x) = \delta_{ij}, \quad j\le i, i,j\in\mathbb{N}$$

or

$$\sum_{k=j}^{i} L_{i-k}^{(-\alpha-i-1)}(-x)L_{k-j}^{(\alpha+j)}(x) = \delta_{ij}, \quad j\le i, i,j\in\mathbb{N}. \tag{15}$$

Formula (15) can be interpreted as follows. If we define the matrix $T = (t_{ij})_{i,j=0}^n$ with entries

$$t_{ij} = \begin{cases} L_{i-j}^{(\alpha+j)}(x), & j \leq i, \\ 0, & j > i, \end{cases}$$

then this matrix T is a triangular matrix with determinant 1 and the inverse U of this matrix is given by $T^{-1} = U = (u_{ij})_{i,j=0}^n$ with entries

$$u_{ij} = \begin{cases} L_{i-j}^{(-\alpha-i-1)}(-x), & j \leq i, \\ 0, & j > i. \end{cases}$$

This leads to the following lemma (see [33] Lemma 5)

Lemma 5.1 *Suppose that for a certain $k \in \mathbb{N}$ we have the system of equations*

$$\sum_{i=1}^{\infty} A_i(x) \mathbf{D}^{\mathbf{i}+\mathbf{k}} \mathbf{L}_{\mathbf{n}}^{(\alpha)}(\mathbf{x}) = \mathbf{F}_{\mathbf{n}}(\mathbf{x}), \quad \mathbf{n} = \mathbf{k}+\mathbf{1}, \mathbf{k}+\mathbf{2}, \mathbf{k}+\mathbf{3}, \ldots,$$

where $\{A_i(x)\}_{i=1}^{\infty}$ are independent of n. Then this system has a unique solution given by

$$A_i(x) = (-1)^{i+k} \sum_{j=1}^{i} L_{i-j}^{(-\alpha-i-k-1)}(-x) F_{j+k}(x), \quad i = 1, 2, 3, \ldots.$$

In order to apply this lemma one has to cope with three problems:

Problems

1. Find an expression for the numbers $\left\{\alpha_n^{(\alpha)}\right\}_{n=1}^{\infty}$.
2. Show the equivalence of the two systems or at least show that all the solutions of one of the systems are also solutions of the other one.
3. Write the coefficients, found by means of Lemma 1, in such a way that the finite order of the differential operator in the case of $\alpha \in \mathbb{N}$ can be seen.

These problems have been solved in [4], the last one by a tedious computation.

In [33] J. Koekoek, R. Koekoek and H. Bavinck looked for a linear differential operator of a special form having the polynomials $\left\{L_n^{\alpha,M,N}(x)\right\}_{n=0}^{\infty}$, orthogonal with respect to the inner product (of Sobolev type if $N > 0$)

$$(f, g) = \frac{1}{\Gamma(\alpha+1)} \int_0^{\infty} f(x) g(x) x^{\alpha} e^{-x} dx + M f(0) g(0) + N f'(0) g'(0),$$

with $M \geq 0, N \geq 0, \alpha > -1$, as eigenfunctions. The aim was to find operators $\mathbf{A}^{(\alpha)}, \mathbf{B}^{(\alpha)}, \mathbf{C}^{(\alpha)}$ and numbers $\left\{\alpha_n^{(\alpha)}\right\}_{n=0}^{\infty}, \left\{\beta_n^{(\alpha)}\right\}_{n=0}^{\infty}, \left\{\gamma_n^{(\alpha)}\right\}_{n=0}^{\infty}$ such that for $n = 0, 1, 2, \ldots$

$$[(\mathbf{L}^{(\alpha)} - \mathbf{nI}) + \mathbf{M}(\mathbf{A}^{(\alpha)} - \alpha_{\mathbf{n}}^{(\alpha)}\mathbf{I})$$
$$+N(\mathbf{B}^{(\alpha)} - \beta_{\mathbf{n}}^{(\alpha)}\mathbf{I}) + \mathbf{MN}(\mathbf{C}^{(\alpha)} - \gamma_{\mathbf{n}}^{(\alpha)}\mathbf{I})]\mathbf{L}_{\mathbf{n}}^{\alpha,\mathbf{M},\mathbf{N}}(\mathbf{x}) = \mathbf{0}. \tag{16}$$

Here $\mathbf{L}^{(\alpha)}$ is given by (7) and $\mathbf{A}^{(\alpha)} = \mathbf{A}^{(\alpha)}(\mathbf{x}), \mathbf{B}^{(\alpha)} = \mathbf{B}^{(\alpha)}(\mathbf{x}), \mathbf{C}^{(\alpha)} = \mathbf{C}^{(\alpha)}(\mathbf{x})$ are of the form

$$\mathbf{A}^{(\alpha)}(\mathbf{x}) = \sum_{\mathbf{i}=\mathbf{1}}^{\infty} \mathbf{a_i}(\mathbf{x};\alpha)\mathbf{D^i}, \mathbf{B}^{(\alpha)}(\mathbf{x}) = \sum_{\mathbf{i}=\mathbf{1}}^{\infty} \mathbf{b_i}(\mathbf{x};\alpha)\mathbf{D^i}, \mathbf{C}^{(\alpha)}(\mathbf{x}) = \sum_{\mathbf{i}=\mathbf{1}}^{\infty} \mathbf{c_i}(\mathbf{x};\alpha)\mathbf{D^i}.$$

Clearly we have to take $\alpha_0^{(\alpha)} = \beta_0^{(\alpha)} = \gamma_0^{(\alpha)} = 0$. It is clear that similar of problems as mentioned above for the operator $\mathbf{A}^{(\alpha)}$ and the numbers $\left\{\alpha_n^{(\alpha)}\right\}_{n=1}^{\infty}$, which occur here again, return here in a much more complicated way for the operators $\mathbf{B}^{(\alpha)}, \mathbf{C}^{(\alpha)}$ and the numbers $\left\{\beta_n^{(\alpha)}\right\}_{n=1}^{\infty}$ and $\left\{\gamma_n^{(\alpha)}\right\}_{n=1}^{\infty}$. A new phenomenon is the fact that the operators $\mathbf{B}^{(\alpha)}$ and $\mathbf{C}^{(\alpha)}$ are no longer determined uniquely. Lemma 1 plays an important role in deriving the unknown operators, but showing that the operators are of finite order if α is a nonnegative integer causes considerable difficulties. We state the main result of [33]:

Theorem 5.2 *For $\alpha > -1$ and $M^2 + N^2 > 0$ the only differential equations of the form (16) satisfied by the polynomials $\left\{L_n^{\alpha,M,N}(x)\right\}_{n=0}^{\infty}$ are those where the coefficients $\{a_i(x;\alpha)\}_{i=1}^{\infty}, \{b_i(x;\alpha)\}_{i=1}^{\infty}$ and $\{c_i(x;\alpha)\}_{i=1}^{\infty}$ and the numbers $\left\{\alpha_n^{(\alpha)}\right\}_{n=1}^{\infty}$, $\left\{\beta_n^{(\alpha)}\right\}_{n=2}^{\infty}$ and $\left\{\gamma_n^{(\alpha)}\right\}_{n=1}^{\infty}$ are determined explicitly in the paper, and $\beta_1^{(\alpha)}$ is arbitrary. Only if $N\beta_1^{(\alpha)} = 0$ and $\alpha \in \mathbb{N}$ the order of this differential equation is finite and equal to*

$$\begin{cases} 2\alpha + 4 & \textit{if } M > 0 \textit{ and } N = 0 \\ 2\alpha + 8 & \textit{if } M = 0 \textit{ and } N > 0 \\ 4\alpha + 10 & \textit{if } M > 0 \textit{ and } N > 0. \end{cases}$$

Otherwise the differential equation is of infinite order.

For the orthogonal polynomials which are orthogonal with respect to a discrete measure (Charlier, Meixner etc.) we mean by a Sobolev type inner product an inner product involving differences (see [2], [3]). The inversion method is also used to derive a difference equation of infinite order for Sobolev type Charlier polynomials [5] and in [21] Sobolev type Meixner polynomials are studied in such a normalization, that they can be compared to Sobolev type Laguerre polynomials. It turns out that these Sobolev type Meixner polynomials are eigenfunctions of a difference operator of infinite order for all values of the parameter β, whereas the corresponding Sobolev type Laguerre polynomials are eigenfunctions of a differential operator, which for nonnegative integer values of the parameter is of finite order. In [31] J. and R. Koekoek extended the inversion method to Jacobi polynomials and in [30] they used the new inversion formula to obtain a direct way to derive the differential equations for symmetric generalized ultraspherical polynomials, found earlier in [36] by ingenious guessing. A survey of the inversion formulas and some applications are given in [37]

6 The existence of differential and difference operators

One of the problems we met in the preceding section was that of showing the equivalence of some systems of equations and another was that of finding the numbers $\left\{\alpha_n^{(\alpha)}\right\}_{n=1}^{\infty}$. The first problem is due to the fact that it is not a priori sure that a differential or difference operator of the desired form exists. In two papers, written independently, [28] and [6] the existence problem is treated and also a construction for the numbers $\left\{\alpha_n^{(\alpha)}\right\}_{n=1}^{\infty}$ is found. We state a part of the results in [6].

6.1 Notations

Let $\{P_n(x)\}_{n=0}^{\infty}$ be polynomials with $\deg[P_n(x)] = n$ for each $n \in \mathbb{N}$ and let $\{\lambda_n\}_{n=0}^{\infty}$ be real numbers with $\lambda_0 = 0$ and $\{\lambda_n\}_{n=1}^{\infty}$ not all equal to zero such that $\{P_n(x)\}_{n=0}^{\infty}$ is a polynomial set of solutions of

$$\mathbf{L(x)y(x)} \equiv \sum_{\mathbf{i=1}}^{\infty} \mathbf{l_i(x)}\mathcal{D}_{\S}^{\rangle}\dagger(\S) = \lambda_{\backslash}\dagger(\S). \tag{17}$$

Here $\{l_i(x)\}_{i=1}^{\infty}$ are polynomials with $\deg[l_i(x)] \leq i$ for all $i \in \{1, 2, 3, \ \ldots\}$.
$\mathcal{D}_{\S}\dagger(\S)$ may be read as the derivative $\mathbf{Dy(x)} = \frac{dy(x)}{dx}$, the forward difference $\mathbf{\Delta y(x) = y(x+1) - y(x)}$ or the backward difference $\mathbf{\nabla y(x) = y(x) - y(x-1)}$.
$\mathcal{D}_{\S}^{\rangle}\dagger(\S) = \mathcal{D}_{\S}(\mathcal{D}_{\S}^{\rangle-\infty}\dagger(\S)), \rangle \in \{\infty, \in, \ni, \ldots\}$ and $\mathcal{D}_{\S}^{\prime}\dagger(\S) = \dagger(\S)$.
If $\mathbf{\Lambda^{(i)}}$, $i \in \mathbb{N}$, indicates the difference operator, defined by

$$\mathbf{\Lambda^{(i)}} := \mathbf{\Delta^{i-2\lfloor \frac{i}{2} \rfloor}} \, (\mathbf{\Delta\nabla})^{\lfloor \frac{i}{2} \rfloor}, \tag{18}$$

(see [11]) then we can also take $\mathcal{D}_{\S}^{\rangle}\dagger(\S) = \mathbf{\Lambda^{(i)}y(x)}$, $i \in \mathbb{N}$.

6.2 Linear perturbations

Let $\{Q_n(x)\}_{n=0}^{\infty}$ be polynomials with $\deg[Q_n(x)] \leq n$ for each $n = 0, 1, 2, \ldots$ with

$$Q_n(x) = \sum_{k=0}^{n} q_{n,k} P_k(x) \tag{19}$$

and $\{P_n^{\mu}(x)\}_{n=0}^{\infty}$ be the polynomials given by

$$P_n^{\mu}(x) = P_n(x) + \mu Q_n(x), \; n = 0, 1, 2, \ldots, \mu \in \mathbb{R}. \tag{20}$$

The aim is to find an operator $\mathbf{A}$ of the form

$$\mathbf{A(x)y(x)} \equiv \sum_{\mathbf{i=1}}^{\infty} \mathbf{a_i(x)}\mathcal{D}^{\rangle}_{\S}\dagger(\S), \tag{21}$$

where $\{a_i(x)\}_{i=1}^{\infty}$ are polynomials with $\deg[a_i(x)] \leq i$ for all $i = 1, 2, 3, \; \ldots,$ and real numbers $\{\alpha_n\}_{n=0}^{\infty}$ with $\alpha_0 = 0$ such that

$$(\mathbf{L} + \mu \mathbf{A})\mathbf{P_n^{\mu}(x)} = (\lambda_\mathbf{n} + \mu\alpha_\mathbf{n})\mathbf{P_n^{\mu}(x)}, \; \mathbf{n = 0, 1, 2, \ldots}. \tag{22}$$

Definition 6.1 *Let $\{P_n(x)\}_{n=0}^{\infty}$ and $\{Q_n(x)\}_{n=0}^{\infty}$ be as stated. We call the polynomials $\{P_n^{\mu}(x)\}_{n=0}^{\infty}$ given by (20)* `a linear perturbation of` *$\{P_n(x)\}_{n=0}^{\infty}$* `of the class` *$m$ $(m \in \mathbb{N})$ when the following conditions are satisfied:*

1. *if $n \leq m$ then $q_{n,k} = 0$ for all $k \in \{0, 1, 2, \ldots, n\}$*
2. *if $n > m$ then $q_{n,n} \neq 0, q_{n,n-1} \neq 0$ and $q_{n,k} = 0$ for all $k \in \{0, 1, 2, \ldots, m - 1\}$.*

We now state the main result.

Theorem 6.2 *Let $\{P_n^{\mu}(x)\}_{n=0}^{\infty}$ be a linear perturbation of $\{P_n(x)\}_{n=0}^{\infty}$ of the class m.*
Then a necessary and sufficient condition for the existence of an operator $\mathbf{A}$ of the form (21) and real numbers $\{\alpha_n\}_{n=1}^{\infty}$ such that (22) holds, is

$$q_{n,k} \sum_{j=k+1}^{n} (\lambda_j - \lambda_{j-1}) q_{j,j} = \sum_{j=k+1}^{n} (\lambda_j - \lambda_k) q_{n,j} q_{j,k}, \tag{23}$$

for all $n \in \{1, 2, 3, \ldots\}, k \in \{0, 1, 2, \ldots, n - 1\}$.
If $m = 0$, then the real numbers $\{\alpha_n\}_{n=1}^{\infty}$ and the operator $\mathbf{A}$ are uniquely determined. If $m > 0$, then $\alpha_1, \alpha_2, \alpha_3, \ldots, \alpha_m$ can be chosen arbitrarily and the operator $\mathbf{A}$ is uniquely determined when $\alpha_1, \alpha_2, \alpha_3, \ldots, \alpha_m$ are chosen.
The numbers $\{\alpha_n\}_{n=m+1}^{\infty}$ are given by

$$\alpha_n = \alpha_m + \sum_{j=m+1}^{n} (\lambda_j - \lambda_{j-1}) q_{j,j}. \tag{24}$$

6.3 Application to Sobolev type orthogonal polynomials

Let $\{P_n(x)\}_{n=0}^{\infty}$ be a system of orthogonal polynomials relative to a positive-definite real moment functional σ, which satisfy a differential or difference equation of the form (17). Let ϕ be the symmetric bilinear form (of Sobolev type if $l \geq 0$ and $\mu > 0$) defined by

$$\phi(p,q) = \langle \sigma, pq \rangle + \mu p^{(l)}(c) q^{(l)}(c),$$

where $\mu(\neq 0)$ and c are real constants, $l \in \{0,1,2,\dots\}$, p and q are any real polynomials and the notation

$$p^{(l)}(x) = \mathcal{D}_{\S}^{\updownarrow}\surd(\S),\ \updownarrow \in \mathbb{N}$$

is used. If ϕ is positive-definite then in the case $\mathcal{D}_{\S} = \frac{\lceil}{\lceil \S}$ it is shown in [28] that if $P_n^{(l)}(c) \neq 0$ for all $n = l, l+1, l+2, \dots$, then the corresponding orthogonal polynomials $\{P_n^{\mu}(x)\}_{n=0}^{\infty}$ satisfy a differential equation (possibly of infinite order) of the form (22), where $\alpha_1, \alpha_2, \alpha_3, \dots, \alpha_l$ can be chosen arbitrarily and the operator $\mathbf{A}$ of the form (21) is uniquely determined when $\alpha_1, \alpha_2, \alpha_3, \dots, \alpha_l$ are chosen. We derive this and the corresponding result for differences directly from Theorem 6.2. If we write

$$K_n^{(r,s)}(x,y) = \sum_{i=0}^{n} \frac{\mathcal{D}_{\S}^{\nabla}\mathcal{P}_{\rangle}(\S)\mathcal{D}_{\dagger}^{\int}\mathcal{P}_{\rangle}(\dagger)}{\langle \sigma, P_i^2(x) \rangle},\ n,r,s \in \mathbb{N},$$

then (see [51], [1], [3]) the polynomials $\{P_n^{\mu}(x)\}_{n=0}^{\infty}$ can be written as (20) with

$$Q_n(x) = K_{n-1}^{(l,l)}(c,c) P_n(x) - P_n^{(l)}(c) K_{n-1}^{(0,l)}(x,c),$$

hence

$$q_{n,k} = -\frac{P_n^{(l)}(c) P_k^{(l)}(c)}{\langle \sigma, P_k^2(x) \rangle},\ k \in \{0,1,2,\dots,n-1\} \tag{25}$$

and

$$q_{n,n} = K_{n-1}^{(l,l)}(c,c). \tag{26}$$

It follows that if $P_n^{(l)}(c) \neq 0$ for all $n = l, l+1, l+2, \dots$, then $\{P_n^{\mu}(x)\}_{n=0}^{\infty}$ is a linear perturbation of class l of $\{P_n(x)\}_{n=0}^{\infty}$ and by using summation-by-parts (23) easily follows.

6.4 Special and symmetric linear perturbations

In [6], [7] two other kinds of linear perturbations are introduced, meant for symmetric orthogonal polynomials (Hermite, Gegenbauer). The special linear perturbation corresponds to a symmetric bilinear form (of Sobolev type if $l \geq 0$ and $\mu > 0$) defined by

$$\phi(p,q) = \langle \sigma, pq \rangle + \mu p^{(l)}(0) q^{(l)}(0),$$

and the symmetric linear perturbation corresponds to a symmetric bilinear form (of Sobolev type if $l \geq 0$ and $\mu > 0$) defined by

$$\phi(p,q) = \langle \sigma, pq \rangle + \mu \left(p^{(l)}(c)q^{(l)}(c) + p^{(l)}(-c)q^{(l)}(-c) \right).$$

If the orthogonal polynomials with respect to σ are symmetric, then the orthogonal polynomials with respect to both these perturbed functionals are symmetric as well. For these two cases theorems similar to Theorem 1 are derived.

6.5 Conclusion

In the papers [5], [18], [20], [21], [36], [29] and [33] (in the case $M = 0, N > 0$) differential and difference operators (in some cases of finite order, in some other cases of infinite order) are constructed having certain systems of orthogonal polynomials as eigenfunctions. All these orthogonal polynomials are linear perturbations of the classical orthogonal polynomials. The classical orthogonal polynomials are eigenfunctions of a differential or difference operator $\mathbf{L}$ of the second order with eigenvalues λ_n. In the papers mentioned above tedious proofs were needed to show the existence of an operator of the form $\mathbf{L}+\mu\mathbf{A}$ having the linear perturbations of the classical orthogonal polynomials as eigenfunctions with eigenvalues of the form $\lambda_n + \mu\alpha_n$. By the results in [6], [7] in all these cases and in many more such proofs have become superfluous and moreover it is shown that for a certain value of $m \in \mathbb{N}$, depending on the situation, the numbers $\alpha_1, \alpha_2, \alpha_3, \ldots, \alpha_m$ are arbitrary and for the numbers $\{\alpha_n\}_{n=m+1}^{\infty}$ the explicit expression (24) is given.

6.6 Two linear perturbations

In [10] combinations of two linear perturbations are considered and in [19] for the symmetric polynomials combinations of two linear special and/or symmetric perturbations are studied. We state the main result of [10] with a small modification shown in [19]:

Theorem 6.3 *Let $\{P_n(x)\}_{n=0}^{\infty}$ be a system of orthogonal polynomials relative to a positive-definite moment functional σ and let $\{\lambda_n\}_{n=0}^{\infty}$ be a sequence of real numbers with $\lambda_0 = 0$ and $\{\lambda_n\}_{n=1}^{\infty}$ not all equal to zero such that $\{P_n(x)\}_{n=0}^{\infty}$ are eigenfunctions of a linear differential or difference operator $\mathbf{L}$ with eigenvalues $\{\lambda_n\}_{n=0}^{\infty}$. If c_1 and c_2 are real constants, l_1 and l_2 are nonnegative integers,*

$$P_n^{(l_1)}(c_1) \neq 0 \text{ for all } n \in \{l_1, l_1+1, l_1+2, \ldots\} \tag{27}$$

and

$$P_n^{(l_2)}(c_2) \neq 0 \text{ for all } n \in \{l_2, l_2+1, l_2+2, \ldots\}, \tag{28}$$

then the polynomials $\{P_n^{\mu,\nu}(x)\}_{n=0}^{\infty}$, orthogonal with respect to the bilinear form of Sobolev type defined by

$$\psi(p,q) = \langle \sigma, pq \rangle + \mu p^{(l_1)}(c_1)q^{(l_1)}(c_1) + \nu p^{(l_2)}(c_2)q^{(l_2)}(c_2), \tag{29}$$

where $\mu > 0, \nu > 0$, and p and q are any polynomials, are eigenfunctions of one (or more if $\min(l_1, l_2) > 0$) linear differential or difference operators of form

$$\mathbf{L} + \mu\mathbf{A} + \nu\mathbf{B} + \mu\nu\mathbf{C} \tag{30}$$

with eigenvalues

$$\{\lambda_n + \mu\alpha_n + \nu\beta_n + \mu\nu\gamma_n\}_{n=0}^{\infty}. \tag{31}$$

Here $\alpha_0 = \beta_0 = \gamma_0 = 0$ and the numbers $\{\alpha_n\}_{n=1}^{l_1}$ (if $l_1 > 0$), $\{\beta_n\}_{n=1}^{l_2}$ (if $l_2 > 0$) and $\{\gamma_n\}_{n=1}^{\min\{l_1,l_2\}}$ (if $\min\{l_1, l_2\} > 0$) can be chosen arbitrarily. The operators $\mathbf{A}, \mathbf{B}$ and $\mathbf{C}$ and the numbers $\{\alpha_n\}_{n=l_1+1}^{\infty}, \{\beta_n\}_{n=l_2+1}^{\infty}$ and $\{\gamma_n\}_{n=\min\{l_1,l_2\}+1}^{\infty}$ are uniquely determined, when $\{\alpha_n\}_{n=1}^{l_1}, \{\beta_n\}_{n=1}^{l_2}$ and $\{\gamma_n\}_{n=1}^{\min\{l_1,l_2\}}$ are fixed.

7 The finite order cases

Krall [43] found three fourth-order linear differential operators having orthogonal polynomials as eigenfunctions, which A.M. Krall [39] called Laguerre type, Jacobi type and Legendre type polynomials. They have all three become the fathers of a large family of finite order linear differential operators having (Sobolev type) orthogonal polynomials as eigenfunctions. For the non-Sobolev cases it has been conjectured by Magnus [25] that if orthogonal polynomials are eigenfunctions of a differential operator (3) then they must be orthogonal with respect to a classical weight function $\omega(x)$ plus possibly point masses at the endpoints of the support of $\omega(x)$. Strong support to Magnus' conjecture is given in [45], [46]. For the connection with spectral theory and a general survey of the field the reader is referred to [26], [27].

7.1 The Sobolev type Laguerre polynomials

We consider the Sobolev type Laguerre polynomials $\left\{L_n^{\alpha,M_1,M_2}(x, l_1, l_2)\right\}_{n=0}^{\infty}$, which are orthogonal with respect to the inner product

$$\langle p, q\rangle = \frac{1}{\Gamma(\alpha+1)}\int_0^{\infty} p(x)q(x)x^{\alpha}e^{-x}dx + M_1p^{(l_1)}(0)q^{(l_1)}(0) + M_2p^{(l_2)}(0)q^{(l_2)}(0), \tag{32}$$

where $\alpha > -1, M_1 \geq 0, M_2 \geq 0$ and $l_1, l_2 \in \mathbb{N}$, $l_1 < l_2$. These polynomials are generalizations of the Laguerre polynomials, which are known to be eigenfunctions of the second order linear differential operator

$$\mathbf{L}^{(\alpha)} = -\mathbf{x}\mathbf{D}^2 - (\alpha + \mathbf{1} - \mathbf{x})\mathbf{D}$$

with eigenvalues $\lambda_n = n, n \in \mathbb{N}$.

As an application in [10] it follows that there exist linear differential operators $\mathbf{A}^{(\alpha,\mathbf{l}_1)}$, $\mathbf{A}^{(\alpha,\mathbf{l}_2)}$,$\mathbf{C}^{(\alpha,\mathbf{l}_1,\mathbf{l}_2)}$ (usually of infinite order) and numbers $\left\{\alpha_n^{(\alpha,l_1)}\right\}_{n=0}^{\infty}$,

$\left\{\alpha_n^{(\alpha,l_2)}\right\}_{n=0}^{\infty}$, $\left\{\gamma_n^{(\alpha,l_1,l_2)}\right\}_{n=0}^{\infty}$ such that the polynomials $\left\{L_n^{\alpha,M_1,M_2}(x,l_1,l_2)\right\}_{n=0}^{\infty}$ are solutions of the differential equation

$$[\left(\mathbf{L}^{(\alpha)}-\mathbf{nI}\right)+M_1\left(\mathbf{A}^{(\alpha,\mathbf{l_1})}-\alpha_{\mathbf{n}}^{(\alpha,\mathbf{l_1})}\mathbf{I}\right)+M_2\left(\mathbf{A}^{(\alpha,\mathbf{l_2})}-\alpha_{\mathbf{n}}^{(\alpha,\mathbf{l_2})}\mathbf{I}\right)$$
$$+M_1M_2\left(\mathbf{C}^{(\alpha,\mathbf{l_1},\mathbf{l_2})}-\gamma_{\mathbf{n}}^{(\alpha,\mathbf{l_1},\mathbf{l_2})}\mathbf{I}\right)]y(x)=0. \tag{33}$$

Here $\mathbf{A}^{(\alpha,\mathbf{l})}=\sum_{\mathbf{i}=\mathbf{1}}^{\infty}\mathbf{a_i}(\mathbf{x};\alpha,\mathbf{l})\mathbf{D^i}$, $\mathbf{l}\in\{\mathbf{l_1},\mathbf{l_2}\}$, $\mathbf{C}^{(\alpha,\mathbf{l_1},\mathbf{l_2})}=\sum_{\mathbf{i}=\mathbf{1}}^{\infty}\mathbf{c_i}(\mathbf{x};\alpha,\mathbf{l_1},\mathbf{l_2})\mathbf{D^i}$ and $\mathbf{I}$ denotes the identity operator. Further we have to take $\alpha_0^{(\alpha,l_1)}=\alpha_0^{(\alpha,l_2)}=\gamma_0^{(\alpha,l_1,l_2)}=0$ and the values $\left\{\alpha_n^{(\alpha,l_1)}\right\}_{n=1}^{l_1}$ (if $l_1>0$), $\left\{\alpha_n^{(\alpha,l_2)}\right\}_{n=1}^{l_2}$, $\left\{\gamma_n^{(\alpha,l_1,l_2)}\right\}_{n=1}^{l_1}$ (if $l_1>0$) can be chosen arbitrarily; for the other values formulas are given. To each choice of the arbitrary values corresponds precisely one linear differential operator of the form (30), usually of infinite order. In [15] it is shown that if all the arbitrary values are chosen to be 0 and further $\alpha\in\mathbb{N}$, then the corresponding operators $\mathbf{A_0}^{(\alpha,\mathbf{l_1})}$, $\mathbf{A_0}^{(\alpha,\mathbf{l_2})}$ and $\mathbf{C_0}^{(\alpha,\mathbf{l_1},\mathbf{l_2})}$ are of finite order: $\mathbf{A_0}^{(\alpha,\mathbf{l})}$ is of order $2\alpha+4l+4$, $l\in\{l_1,l_2\}$ and $\mathbf{C_0}^{(\alpha,\mathbf{l_1},\mathbf{l_2})}$ is of order $4\alpha+4l_1+4l_2+6$. Further it was proved that any other choice of the arbitrary values will lead to an operator of infinite order and also that if $\alpha\notin\mathbb{N}$, then the operator is of infinite order for any choice of the arbitrary values. In a number of special cases this problem has been considered before (see [9] for a complete survey). Here we only mention the case $M_2=0$, which for $l_1=0$ has been treated in [29] (see also [4] and [12]) and in general in [8], and the case $l_1=0, l_2=1$, which was studied in [33].

7.2 The Sobolev type Jacobi polynomials

We consider the Sobolev type Jacobi polynomials $\left\{P_n^{\alpha,\beta,M_1,M_2}(x,l_1,l_2)\right\}_{n=0}^{\infty}$, which are orthogonal with respect to the inner product

$$\langle p,q\rangle = \frac{\Gamma(\alpha+\beta+2)}{2^{\alpha+\beta+1}\Gamma(\alpha+1)\Gamma(\beta+1)}\int_{-1}^{1}p(x)q(x)(1-x)^{\alpha}(1+x)^{\beta}dx \tag{34}$$
$$+M_1p^{(l_1)}(-1)q^{(l_1)}(-1)+M_2p^{(l_2)}(1)q^{(l_2)}(1),$$

$\alpha,\beta>-1$, $l_1,l_2\in\mathbb{N}$, $M_1,M_2\geq 0$. These polynomials are generalizations of the Jacobi polynomials, which are eigenfunctions of the linear differential operator

$$\mathbf{L}^{(\alpha,\beta)}:=(\mathbf{x^2}-\mathbf{1})\mathbf{D^2}+[\alpha-\beta+(\alpha+\beta+\mathbf{2})\mathbf{x}]\mathbf{D},$$

with eigenvalues

$$\lambda_n^{(\alpha,\beta)}=n(n+\alpha+\beta+1)\ ,n\in\mathbb{N}.$$

As a consequence of the perturbation theory [10] there exists a class of linear differential operators of the form

$$\mathbf{L}^{(\alpha,\beta)}+\mathbf{M_1}\mathbf{A}^{(\alpha,\beta,\mathbf{l_1})}+\mathbf{M_2}\mathbf{B}^{(\alpha,\beta,\mathbf{l_2})}+\mathbf{M_1}\mathbf{M_2}\mathbf{C}^{(\alpha,\beta,\mathbf{l_1},\mathbf{l_2})} \tag{35}$$

for which the Sobolev type Jacobi polynomials are eigenfunctions with eigenvalues of the form

$$\left\{\lambda_n^{(\alpha,\beta)} + M_1\alpha_n^{(\alpha,\beta,l_1)} + M_2\beta_n^{(\alpha,\beta,l_2)} + M_1M_2\gamma_n^{(\alpha,\beta,l_1,l_2)}\right\}_{n=0}^{\infty}.$$

Here $\alpha_0^{(\alpha,\beta,l_1)} = \beta_0^{(\alpha,\beta,l_2)} = \gamma_0^{(\alpha,\beta,l_1,l_2)} = 0$, the values of $\left\{\alpha_n^{(\alpha,\beta,l_1)}\right\}_{n=1}^{l_1}$ (if $l_1 > 0$), $\left\{\beta_n^{(\alpha,\beta,l_2)}\right\}_{n=1}^{l_2}$ (if $l_2 > 0$), $\left\{\gamma_n^{(\alpha,\beta,l_1,l_2)}\right\}_{n=1}^{\min\{l_1,l_2\}}$ (if $\min\{l_1,l_2\} > 0$) can be chosen arbitrarily and for the other values formulas are given. To each choice of the arbitrary values corresponds a linear differential operator of the form (35), usually of infinite order.

Further it is shown in [16] that, if all the arbitrary values are chosen to be 0, then for the corresponding operators $\mathbf{A}_0^{(\alpha,\beta,\mathbf{l_1})}$, $\mathbf{B}_0^{(\alpha,\beta,\mathbf{l_2})}$ and $\mathbf{C}_0^{(\alpha,\beta,\mathbf{l_1},\mathbf{l_2})}$ we have: $\mathbf{A}_0^{(\alpha,\beta,\mathbf{l_1})}$ is of order $2\beta + 4l_1 + 4$, if $\beta \in \mathbb{N}$, $\mathbf{B}_0^{(\alpha,\beta,\mathbf{l_2})}$ is of order $2\alpha + 4l_2 + 4$, if $\alpha \in \mathbb{N}$, and $\mathbf{C}_0^{(\alpha,\beta,\mathbf{l_1},\mathbf{l_2})}$ is of order $2\alpha + 2\beta + 4l_1 + 4l_2 + 6$, if $\alpha, \beta \in \mathbb{N}$. Any other choice of the arbitrary values will lead to one or more operators of infinite order.

In the case $l_1 = l_2 = 0$ this was proved by J. Koekoek and R. Koekoek [32] (see also [13]). An important tool in their work was the inversion formula for Jacobi polynomials, introduced in [31]. They also showed that the operator $\mathbf{A}_0^{(\alpha,\beta,\mathbf{0})}$ is always of infinite order if $\beta \notin \mathbb{N}$, that $\mathbf{B}_0^{(\alpha,\beta,\mathbf{0})}$ is always of infinite order if $\alpha \notin \mathbb{N}$ and $\mathbf{C}_0^{(\alpha,\beta,\mathbf{0},\mathbf{0})}$ is always of infinite order if $\alpha, \beta \notin \mathbb{N}$. Such a result is likely to be true for the operators $\mathbf{A}_0^{(\alpha,\beta,\mathbf{l_1})}$, $\mathbf{B}_0^{(\alpha,\beta,\mathbf{l_2})}$ and $\mathbf{C}_0^{(\alpha,\beta,\mathbf{l_1},\mathbf{l_2})}$ in general.

7.3 The Sobolev type Gegenbauer polynomials

We consider the Sobolev type Gegenbauer polynomials $\left\{P_n^{\alpha,M_1,M_2,l_1,l_2}(x)\right\}_{n=0}^{\infty}$, orthogonal with respect to the inner product

$$\begin{aligned}(f,g) &= \frac{\Gamma(2\alpha+2)}{2^{2\alpha+1}\Gamma(\alpha+1)^2}\int_{-1}^{1} f(x)g(x)(1-x^2)^{\alpha}dx \\ &\quad + M_1[f^{(l_1)}(-1)g^{(l_1)}(-1) + f^{(l_1)}(1)g^{(l_1)}(1)] \\ &\quad + M_2[f^{(l_2)}(-1)g^{(l_2)}(-1) + f^{(l_2)}(1)g^{(l_2)}(1)], \end{aligned} \tag{36}$$

$M_1 \geq 0, M_2 \geq 0, \alpha > -1, l_1, l_2 \in \mathbb{N}, l_1 < l_2$. They are generalizations of the Gegenbauer or ultraspherical polynomials, which are eigenfunctions of the linear differential operator

$$\mathbf{L}^{(\alpha)} :=(\mathbf{x^2}-\mathbf{1})\mathbf{D^2} + \mathbf{2}(\alpha+\mathbf{1})\mathbf{xD},$$

with eigenvalues $\lambda_n^{(\alpha)} = n(n+2\alpha+1)$, $n \in \mathbb{N}$. As a consequence of the theory of two symmetric linear perturbations treated in [19] there exist linear differential operators $\mathbf{A}^{(\alpha,\mathbf{l})}$, $l \in \{l_1, l_2\}$and $\mathbf{C}^{(\alpha,\mathbf{l_1},\mathbf{l_2})}$ and numbers $\left\{\alpha_n^{(\alpha,l)}\right\}_{n=0}^{\infty}$, $l \in \{l_1, l_2\}$ and $\left\{\gamma_n^{(\alpha,l_1,l_2)}\right\}_{n=0}^{\infty}$ such that

$$\Big[\left(\mathbf{L}^{(\alpha)} - \lambda_{\mathbf{n}}^{(\alpha)}\mathbf{I}\right) + M_1\left(\mathbf{A}^{(\alpha,\mathbf{l_1})} - \alpha_{\mathbf{n}}^{(\alpha,\mathbf{l_1})}\mathbf{I}\right) + \mathbf{M_2}\left(\mathbf{A}^{(\alpha,\mathbf{l_2})} - \alpha_{\mathbf{n}}^{(\alpha,\mathbf{l_2})}\mathbf{I}\right)$$
$$+ M_1 M_2\left(\mathbf{C}^{(\alpha,\mathbf{l_1},\mathbf{l_2})} - \gamma_{\mathbf{n}}^{(\alpha,\mathbf{l_1},\mathbf{l_2})}\mathbf{I}\right)\Big] P_n^{\alpha,M_1,M_2,l_1,l_2}(x) = 0 \tag{37}$$

for $n \in \mathbb{N}$. Here $\alpha_0^{(\alpha,l_1)} = \alpha_0^{(\alpha,l_2)} = \gamma_0^{(\alpha,l_1,l_2)} = 0$, the values of $\left\{\alpha_n^{(\alpha,l_1)}\right\}_{n=1}^{l_1}$(if $l_1 > 0$), $\left\{\alpha_n^{(\alpha,l_2)}\right\}_{n=1}^{l_2}$, $\left\{\gamma_n^{(\alpha,l_1,l_2)}\right\}_{n=1}^{l_1}$ (if $\min(l_1, l_2) > 0$) can be chosen arbitrarily and for the other values formulas are given in [19]. To each choice of the arbitrary values corresponds a linear differential operator such that (37) holds, usually of infinite order.

Further it is shown in [17] that, if all the arbitrary values are chosen to be 0 and $\alpha \in \mathbb{N}$, then for the corresponding operators $\mathbf{A}_{\mathbf{0}}^{(\alpha,\mathbf{l})}$ ($\mathbf{l} \in \{\mathbf{l_1}, \mathbf{l_2}\}$) and $\mathbf{C}_{\mathbf{0}}^{(\alpha,\mathbf{l_1},\mathbf{l_2})}$ we have: $\mathbf{A}_{\mathbf{0}}^{(\alpha,\mathbf{l})}$ is of order $2\alpha+4l+4$ and $\mathbf{C}_{\mathbf{0}}^{(\alpha,\mathbf{l_1},\mathbf{l_2})}$ is of order $4\alpha+4l_1+4l_2+6$. Any other choice of the arbitrary values will lead to one or more operators of infinite order. In the case $M_2 = 0, l_1 = 0$, this was shown by R. Koekoek [36], (see also [30]). The case $l_1 = 0, l_2 = 1$ has been dealt with in [14].

References

[1] M. Alfaro, F. Marcellán, M.L. Rezola and A. Ronveaux: *On orthogonal polynomials of Sobolev type: algebraic properties and zeros.* SIAM J. Math. Anal. **23** (1992), 737–757.

[2] H. Bavinck: *On polynomials orthogonal with respect to an inner product involving differences.* J. Comp.Appl. Math. **57** (1995) 17–27.

[3] H. Bavinck: *On polynomials orthogonal with respect to an inner product involving differences (The general case).* Applicable Analysis **59** (1995) 233–240.

[4] H. Bavinck: *A direct approach to Koekoek's differential equation for generalized Laguerre polynomials.* Acta Math. Hungar. **66 (3)** (1995) 247–253.

[5] H. Bavinck: *A difference operator of infinite order with Sobolev-type Charlier polynomials as eigenfunctions.* Indag. Math. N.S. **7** (1996) 281–291.

[6] H. Bavinck: *Linear perturbations of differential or difference operators with polynomials as eigenfunctions.* J. Comp. Appl. Math. **78** (1997), 179–195.

[7] H. Bavinck: *Addenda and Errata to "Linear perturbations of differential or difference operators with polynomials as eigenfunctions".* J. Comp. Appl. Math. **83** (1997) 145–146.

[8] H. Bavinck: *Differential operators having Sobolev type Laguerre polynomials as eigenfunctions.* Proc. Amer. Math. Soc. **125** (1997) 3561–3567.

[9] H. Bavinck: *Differential operators having Laguerre type and Sobolev type Laguerre polynomials as eigenfunctions: a survey.* In: "Special Functions and Differential Equations" Proceedings Workshop Madras, Edited by K. Srinivasa Rao et al. Allied Publishers Private Limited 1998.

[10] H. Bavinck: *Differential and difference operators having orthogonal polynomials with two linear perturbations as eigenfunctions.* J. Comp. Appl. Math. **92** (1998) 85–95.

[11] H. Bavinck: *Generalizations of Meixner polynomials which are eigenfunctions of a difference operator.* J. Differ. Equations Appl. **5** (1999) 143–153.

[12] H. Bavinck: *On a linear perturbation of the Laguerre operator.* J. Comp. Appl. Math. **106** (1999) 197–202.

[13] H. Bavinck: *A note on the Koekoeks' differential equation for generalized Jacobi polynomials.* J. Comp. Appl. Math. **115** (2000) 87–92.

[14] H. Bavinck: *Differential operators having Sobolev type Gegenbauer polynomials as eigenfunctions.* J. Comp. Appl. Math. **118** (2000) 23–42.

[15] H. Bavinck: *Differential operators having Sobolev type Laguerre polynomials as eigenfunctions. New developments.* J. Comp. Appl. Math. **133** (2001) 183–193.

[16] H. Bavinck: *Differential operators having Sobolev type Jacobi polynomials as eigenfunctions.* Submitted for publication.

[17] H. Bavinck: *Differential operators having Sobolev type Gegenbauer polynomials as eigenfunctions (the general case with two symmetric perturbations at* -1 *and* 1). In preparation.

[18] H. Bavinck and H. van Haeringen: *Difference equations for generalized Meixner polynomials.* J. Math. Anal. Appl. **184** (1994) 453–463.

[19] H. Bavinck and J. Koekoek: *Differential operators having symmetric orthogonal polynomials as eigenfunctions.* J. Comp. Appl. Math. **106** (1999) 369–393.

[20] H. Bavinck and R. Koekoek: *On a difference equation for generalizations of Charlier polynomials.* J. Approx. Theory **81** (1995) 195–206.

[21] H. Bavinck and R. Koekoek: *Difference operators with Sobolev-type Meixner polynomials as eigenfunctions.* In: Advances in Difference Equations II. Comput. Math. Appl. **36** (1998) 163–177.

[22] H. Bavinck and H.G. Meijer: *Orthogonal polynomials with respect to a symmetric inner product involving derivatives.* Applicable Analysis **33** (1989) 103–117.

[23] H. Bavinck and H.G. Meijer: *On orthogonal polynomials with respect to a symmetric inner product involving derivatives: zeros and recurrence relations.* Indag. Math. N.S. **1** (1990) 7–14.

[24] S. Bochner: *Über Sturm-Liouvillesche Polynomsysteme.* Math. Z. **89** (1929) 730–736.

[25] C. Brezinski, L. Gori and A. Ronveaux (Eds.), *Open problems.* In: Orthogonal polynomials and their applications. IMACS Annals on Computing and Applied Mathematics Vol. 9, J.C. Baltzer AG, Basel 1991, pp. 417–419.

[26] W.N. Everitt and L.L. Littlejohn: *Orthogonal polynomials and spectral theory, a survey.* In: C. Brezinski, L. Gori and A. Ronveaux (Eds.), Orthogonal polynomials and their applications. IMACS Annals on Computing and Applied Mathematics Vol. 9, J.C. Baltzer AG, Basel 1991, pp. 21–55.

[27] W.N. Everitt, K.H. Kwon, L.L. Littlejohn and R. Wellman: *Orthogonal polynomial solutions of linear ordinary differential equations.* J. Comp. Appl. Math. **133** (2001) 85–109

[28] I.H. Jung, K.H. Kwon and G.J. Yoon: *Differential equations of infinite order for Sobolev-type orthogonal polynomials.* J. Comp. Appl. Math. **78** (1997) 277–293.

[29] J. Koekoek and R. Koekoek: *On a differential equation for Koornwinder's generalized Laguerre polynomials.* Proc. Amer. Math. Soc. **112** (1991) 1045–1054.

[30] J. Koekoek and R. Koekoek: *Finding differential equations for symmetric generalized ultraspherical polynomials by using inversion methods,* International Workshop on Orthogonal Polynomials in Mathematical Physics (Leganés, 1996), Univ. Carlos III, Leganés, 1997, pp. 103–111.

[31] J. Koekoek and R. Koekoek: *The Jacobi inversion formula. Complex Variables Theory Appl.* 39 (1999) 1–18.

[32] J. Koekoek and R. Koekoek: *Differential equations for generalized Jacobi polynomials.* J. Comp. Appl. Math.126 (2000) 1–31.

[33] J. Koekoek , R. Koekoek and H. Bavinck: *On differential equations for Sobolev-type Laguerre polynomials.* Trans. Amer. Math. Soc. **350** (1998) 347–393.

[34] R. Koekoek: *Generalizations of the classical Laguerre polynomials.* J. Math. Anal. Appl. **153** (1990) 576–590.

[35] R. Koekoek: *Generalizations of the classical Laguerre polynomials and some q-analogues.* Thesis Delft University of Technology 1990.

[36] R. Koekoek: *Differential equations for symmetric generalized ultraspherical polynomials.* Trans. Amer. Math. Soc. **345** (1994) 47–72.

[37] R. Koekoek: *Inversion formulas involving orthogonal polynomials and some of their applications.* In: Proceedings of the International Workshop on Special Functions. Hong Kong 1999. World Scientific. Singapore, New Jersey, London, Hong Kong, pp. 166–180.

[38] T.H. Koornwinder: *Orthogonal polynomials with weight function* $(1-x)^\alpha (1+x)^\beta + M\delta(x+1) + N\delta(x-1)$. Canad. Math. Bull. **27**(2) (1984) 205–214.

[39] A.M. Krall: *Orthogonal polynomials satisfying fourth order differential equations.* Proc. Roy. Soc. Edinburgh, **87A**(1981) 271–288.

[40] A.M. Krall: *A review of orthogonal polynomials satisfying boundary value problems.* In: Proceedings of an International Symposium on Orthogonal Polynomials and their Applications. Segovia, Spain 1986. Lecture Notes in Mathematics, No. 1329, 1988, Springer-Verlag, New York, pp. 73–97.

[41] A.M. Krall and L.L. Littlejohn: *On the classification of differential equations having orthogonal polynomial solutions II.* Ann. Mat. Pura Appl. (4) **149** (1987) 77–102.

[42] H.L. Krall: *Certain differential equations for Tchebycheff polynomials.* Duke Math. J. **4** (1938) 705–718.

[43] H.L. Krall: *On orthogonal polynomials satisfying a certain fourth order differential equation.* The Pennsylvania State College Studies, No. 6, The Pennsylvania State College, State College, Pa., 1940.

[44] K.H. Kwon and L.L. Littlejohn: *Classification of classical orthogonal polynomials,* J. Korean Math. Soc. **34** (1997) 973–1008.

[45] K.H. Kwon,L.L. Littlejohn and G.J. Yoon: *Bochner-Krall orthogonal polynomials.* In: Proceedings of the International Workshop on Special Functions. Hong Kong 1999. World Scientific. Singapore, New Jersey, London, Hong Kong, pp. 181–193.

[46] K.H. Kwon,L.L. Littlejohn and G.J. Yoon: *Orthogonal polynomial solutions to spectral type differential equations: Magnus' conjecture.* J. Approx. Theory **112** (2001), 189–215.

[47] L.L. Littlejohn: *The Krall polynomials: A new class of orthogonal polynomials.* Quaestiones Math. 5 (1982) 255–265.

[48] L.L. Littlejohn: *An application of a new theorem on orthogonal polynomials and differential equations.* Quaestiones Math. **10** (1986) 49–61.

[49] L.L. Littlejohn: *Orthogonal polynomial solutions to ordinary and partial differential equations.* In: Proceedings of an International Symposium on Orthogonal Polynomials and their Applications. Segovia, Spain 1986. Lecture Notes in Mathematics, No. 1329, 1988, Springer-Verlag, New York, pp. 98–124.

[50] F. Marcellán, M. Alfaro and M.L. Rezola: *Orthogonal polynomials on Sobolev spaces: old and new directions.* J. Comp. Appl. Math. **48** (1993) 113–131.

[51] F. Marcellán and A. Ronveaux: *On a class of polynomials orthogonal with respect to a Sobolev inner product.* Indag. Math. N.S. **1** (1990) 451–464.

[52] H.G. Meijer and H. Bavinck: *An ultraspherical generalization of Krall polynomials.* In: Monografias de la Academia de Ciencias (Zaragoza) Second International Symposium (Segovia 1986) *On orthogonal polynomials and their applications,* pp. 103–108.

Department of Applied Mathematical Analysis
Delft University of Technology
Mekelweg 4
2628CD Delft
The Netherlands
`bavinck@its.tudelft.nl`

Exchange of Ideas in Applied Mathematics, Otmar Alt (2000)

Advanced Problems in Constructive Approximation
(Eds.) M.D. Buhmann and D.H. Mache
International Series of Numerical Mathematics Vol. 142, 21–26

(0, 1) Pál-type Interpolation: A General Method for Regularity

Marcel G. de Bruin & Detlef H. Mache

Abstract

Hermite-Birkhoff interpolation and Pál-type interpolation have been receiving much attention over the years. Also during the previous 15 year the subject of interpolation in non-uniformly distributed nodes has been looked into.

The methods of proof of regularity often were quite dependent on the problem at hand, and the purpose of this note is to treat a possible 'general' method of finding polynomial pairs that lead to a regular interpolation problem; for sake of simplicity so-called $(0, 1)$ Pál-type interpolation is looked into.

Keywords: Pál-type interpolation, regularity

AMS Subject classification: 41A05

1 Introduction

The study of Hermite-Birkhoff interpolation is a well-known subject (cf. the excellent book [1]). Recently the regularity of some interpolation problems on non-uniformly distributed nodes on the unit circle have been studied.

Along with the continuing interest in interpolation in general, a number of papers on Pál-type interpolation have appeared, cf. [2], [3], [4], [5], [6], [7].

In this paper the attention will be focused on interpolation problems of the following kind:

given two polynomials $p(z)$ resp. $q(z)$, with simple zeroes
$\{z_i\}_{i=1}^n \in \mathbf{C}$ resp. $\{w_j\}_{j=1}^m \in \mathbf{C}$ (*nodes*),
given *data* $\{c_i\}_{i=1}^n$, $\{d_j\}_{j=1}^m \in \mathbf{C}$,
find $P_k \in \Pi_k$, $k = m + n - 1$ with $P_k(z_i) = c_i$ $(1 \leq i \leq n)$
and $P_k'(w_j) = d_j$ $(1 \leq j \leq m)$.

Here Π_k is the set of polynomials of degree at most k with complex coefficients.

Although very often the method of proof of regularity depends on the problem at hand, one can, nevertheless, distinguish two main tools as indicated in [5]:

1. Prove that the square system of homogeneous linear equations for the unknown coefficients of the polynomial P_k has a non-vanishing determinant.
2. Find a differential equation for P_k (or for a factor of P_k) and show that if this equation has a polynomial solution, the solution must be the trivial one.

The aim of this paper is to study the second method in a 'general setting'. The layout of the paper is as follows: in Section 2 the results will be given with some examples, followed by the proofs in Section 3. Finally a (short) list of references is given.

2 Main results and examples

Consider the node-generating polynomial for the values

$$p(z) = \prod_{i=1}^{n} (z - z_i) \tag{1}$$

and that for the values of the first derivative

$$q(z) = \prod_{j=1}^{m} (z - w_j), \tag{2}$$

each having simple zeroes.

Remark. It is allowed that p and q have (a) common zero(es).

We then have the following result

Theorem 2.1 *If there exist polynomials $g(z)$, $r_1(z)$, $r_2(z)$ such that*

$$p(z) = (\alpha_0 + \alpha_1 z)g(z) + r_1(z)q(z), \tag{3}$$

$$p'(z) = \beta_0 g(z) + r_2(z)q(z), \ \beta_0 \neq 0, \tag{4}$$

satisfying the condition

$$g(w_j) \neq 0, \ 1 \leq j \leq m, \tag{5}$$

then $(0,1)$ Pál-type interpolation on the zeroes of $\{p(z), q(z)\}$ is regular

1. *for* $\alpha_1 = 0$,

2. *for* $\alpha_1 \neq 0$ *if and only if* $-\beta_0/\alpha_1 \notin \{1, 2, \ldots, m-1\}$.

Examples

1. Let $p_1(z)$, $p_2(z)$ be any two co-prime polynomials with simple zeroes, then the (0, 1) Pál-type interpolation problem on $\{p_1(z)p_2(z), p_2(z)\}$ is regular.

Put $\alpha_0 = \alpha_1 = 0$, $\beta_0 = 1$ and $r_1(z) = p_1(z)$, $q(z) = p_2(z)$, $p(z) = p_1(z)p_2(z)$. For any $r_2(z)$, the polynomial $g(z)$ follows from (4) and therefore satisfies $g(w_j) = p'(w_j)$: because the zeroes of q are also zeroes of p and moreover simple, then $p'(w_j) \neq 0$ and (5) is satisfied.

2. For $p(z) = z^n - \alpha^n$, $\alpha \neq 0$ and $q(z)$ any divisor of

$$z^n - n\eta z^{n-1} - \alpha^n, \; (n-1)^{n-1}\eta^n + \alpha^n \neq 0,$$

the (0, 1) Pál-type interpolation problem on $\{p(z), q(z)\}$ is regular.

Put $\alpha_1 = 0, \alpha_0 \neq 0$ and $\eta = \alpha_0/\beta_0$ in ((3),(4)). The condition (5) follows from (4) as $w_j \neq 0$. For the record: the choice

$$g(z) = \frac{n}{\beta_0} z^{n-1} - \frac{1}{\beta_0} r_2(z)(z^n - n\eta z^{n-1} - \alpha^n)$$

with arbitrary $r_2(z)$ and $r_1(z) = 1 + \eta r_2(z)$ leads to ((3),(4)) for q the full polynomial as indicated above; in case q is a divisor, the polynomials r_1, r_2 have to be multiplied by the complementary factor of q leading to the full polynomial. The condition on η implies that the zeroes of $q(z)$ are simple.

3. For $p(z) = z^n - \alpha^n$, $\alpha \neq 0$ and $q(z)$ any divisor of $z^{n-1} + \eta\alpha^n$, $\eta \neq 0$, the (0, 1) Pál-type interpolation problem on $\{p(z), q(z)\}$ is regular.

This is the case α_0, $\alpha_1 \neq 0$, $\beta_0 = n\alpha_1$: for the full polynomial $r_1(z) = \beta_0$, $r_2(z) = \alpha_1 z + \alpha_0 - 1/\beta_0$. As $-\beta_0/\alpha_1 = -n < 0$ the condition on the quotient is fulfilled; the calculation of g is left to the reader.

The next theorem is an example of what could be done in a very general setting: we use a simple connection between the coefficient of $g(z)$ from (6) and from (7).

Theorem 2.2 *If there exist polynomials* $g(z)$, $r_1(z)$, $r_2(z)$ *such that*

$$p(z) = (\alpha_0 + \alpha_1 z + \alpha_2 z^2)g(z) + r_1(z)q(z), \tag{6}$$

with $\alpha_0 + \alpha_1 z + \alpha_2 z^2$ *having two different (complex) roots* z_1, z_2, *and*

$$p'(z) = (\alpha_1 + 2\alpha_2 z)g(z) + r_2(z)q(z), \; \beta_0 \neq 0, \tag{7}$$

satisfying the conditions

$$g(w_j) \neq 0, \; 1 \leq j \leq m, \tag{8}$$

and

$$\int_{z_1}^{z_2} q(\zeta)d\zeta \neq 0, \tag{9}$$

then the $(0,1)$ *Pál-type interpolation problem on the zeroes of* $\{p(z), q(z)\}$ *is regular.*

To keep a long story short, two simple examples will be given only.

Examples

1. For n even, the $(0,1)$ Pál-type interpolation problem on

$$\{z^n - \alpha^n, (n-2)z^n - n\xi^2 z^{n-2} + 2\alpha^n\},$$

with α, $\xi \neq 0$ and $\alpha^n \neq \xi^n$ is regular.

For n odd, the conditions on α, ξ are: α, $\xi \neq 0$, $\alpha^n \neq \pm\xi^n$, $\xi^n \neq 2(n+1)\alpha^n$.

The proof uses $\alpha_0 + \alpha_1 z + \alpha_2 z^2 = z^2 - \xi^2$, $r_1(z) = -z/\xi^2$, $r_2(z) = -1/\xi^2$ and $g(z) = \{(n-2)z^n/2 + \alpha^n\}/\xi^2$. The conditions come in to ensure that q has simple zeroes and because of (8), (9).

2. Under the condition $\alpha^n \neq 1/(n+1)$, the $(0,1)$ Pál-type interpolation problem on the pair $p(z) = z(z-1)(z^{n+1}/(n+1) - \alpha^n z + \alpha^{n+1})$, $q(z) = z^n - \alpha^n$ is regular.

The proof uses $\alpha_0 + \alpha_1 z + \alpha_2 z^2 = z^2 - z$, $r_1(z) = -(z^2-z)^2(2z-1)$, $r_2(z) = -4(z^2-z)^2$, $g(z) = \{2z^{n+3} - 3z^{n+2} + (n+2)z^{n+1}/(n+1) - 2\alpha^n z^3 + 3\alpha^n z^2 - 2\alpha^n z + \alpha^{n+1}$.

The condition on α originates from (9). The zeroes of $p(z)$ are automatically simple ($p(z)$ satisfies the differential equation

$$(z^2-2)p'(z) - (2z-1)p(z) = q(z)$$

with solution

$$(z^2 - z)(z^{n+1}/(n+1) - \alpha^n z + C);$$

multiple zeroes of the second factor are zeroes of $z^n - \alpha^n$ – with absolute value α – and the choice $C = \alpha^{n+1}$ does the trick).

3 Proofs

The interpolation problem has been formulated in the introduction as:

- given polynomials $p(z)$ and $q(z)$ of degrees n and m respectively with simple zeroes
- find a polynomial $P(z)$ of degree at most $n+m-1$ with

$$P(z_i) = 0 \ (z_i \text{ the zeroes of } p(z), \ P'(w_j) = 0 \ (w_j \text{ the zeroes of } q(z). \tag{10}$$

Because of the first condition in (10), we can write

$$P(z) = p(z)Q(z), \text{ degree } Q(z) \leq m - 1. \tag{11}$$

The second condition of (10) then leads to

$$p(w_j)Q'(w_j) + p'(w_j)Q(w_j) = 0, \text{ with } w_j \text{ the } m \text{ zeroes of } q(z). \tag{12}$$

Proof of Theorem 2.1. Inserting ((3),(4)) into (12) and using (5) we find

$$(\alpha_0 + \alpha_1 w_j)Q'(w_j) + \beta_0 Q(w_j) = 0, \ 1 \leq j \leq m. \tag{13}$$

Because of the degree restriction on Q, at most $m-1$, this immediately implies

$$(\alpha_0 + \alpha_1 z)Q'(z) + \beta_0 Q(z) = 0. \tag{14}$$

Solving this linear first order ordinary differential equation for the cases $\alpha_1 = 0$ (distinguishing $\alpha_0 = 0$ or $\alpha_0 \neq 0$) and $\alpha_1 \neq 0$, we find that $Q(z)$ has to be identically zero under the condition stated in the theorem ($\alpha_1 \neq 0$ was the only case that (14) really had a non-trivial polynomial solution of degree at most $m-1$; that is where $-\beta_0/\alpha_1 \notin \{1, 2, \ldots, m-1\}$ comes in). □

Proof of Theorem 2.2. Proceeding as in the previous proof, but now the degree of the polynomial on the left-hand side of the equation could be equal to the degree of $q(z)$,we arrive at the differential equation

$$(\alpha_0 + \alpha_1 z + \alpha_2 z^2)Q'(z) + (\alpha_1 + 2\alpha_2 z)Q(z) = Cq(z) \tag{15}$$

for the polynomial Q of degree at most $m-1$. The equation (15) can be integrated at once and we find

$$(\alpha_0 + \alpha_1 z + \alpha_2 z^2)Q(z) = C \int_{z_1}^{z} q(\zeta)d\zeta + D. \tag{16}$$

Now the left-hand side has a zero for $z = z_1$ and $z = z_2$; the first gives $D = 0$ and the second, in view of the condition stated in (9), that $C = 0$. Thus $Q \equiv 0$, implying $P \equiv 0$. □

References

[1] G.G. Lorentz, S.D. Riemenschneider and K. Jetter, *Birkhoff Interpolation*, Addison Wesley Pub. Co., Mass. USA, 1983.

[2] Y. Bao, On simultaneous approximation to a differentiable function and its derivative by inverse Pál-type interpolation polynomials, Approx. Theory Appl. (N.S.) **11** (1995), 15–23.

[3] T.F. Xie and S.P. Zhou, On simultaneous approximation to a differentiable function and its derivative by Pál-type interpolation polynomials, Acta Math. Hungar. **69** (1995), 135–147.

[4] Z.F. Sebestyén, Pál-type interpolation on the roots of Hermite polynomials, Pure math. Appl. **9** (1998), 429–439.

[5] Marcel G. de Bruin, Interpolation on non-uniformly distributed nodes, Nonlinear Analysis **17** (2001), 1931–1940.

[6] Marcel G. de Bruin, Incomplete Pál-type interpolation, *submitted.*

[7] Marcel G. de Bruin and Detlef H. Mache, $(0,2)$ Pál-type interpolation: A General Method for regularity, *preprint.*

Marcel G. de Bruin
Department of Applied Mathematical Analysis
Delft University of Technology
P.O. Box 5031, 2600 GA Delft, The Netherlands
Email address: M.G.deBruin@ITS.TUDelft.nl

Detlef H. Mache
Institute for Applied Mathematics (LS VIII)
University of Dortmund
Vogelpothsweg 87, D-44221 Dortmund, Germany
Email address: mache@math.uni-dortmund.de

Advanced Problems in Constructive Approximation
(Eds.) M.D. Buhmann and D.H. Mache
International Series of Numerical Mathematics Vol. 142, 27–38

De la Vallée Poussin Means for the Hankel Transform

Wolfgang zu Castell & Frank Filbir

Abstract

We give a construction of a de la Vallée Poussin kernel for the Hankel transform based on the convolution structure on the space $L^1(\mathbf{R}_+, \mu_\nu)$. In contrast to the classical way to define such a kernel, our construction directly leads to an approximate identity for the underlying space.

1 Introduction

One of the most important problems in Fourier analysis deals with the difficulty that the Fourier transform

$$F(\xi) = \int_{-\infty}^{\infty} f(x)e^{-i\xi x}\, dx, \qquad \xi \in \mathbf{R},$$

of a function $f \in L^1(\mathbf{R})$ need not belong to $L^1(\mathbf{R})$ itself and therefore the inverse Fourier integral may not exist. Nevertheless, in summability theory one tries to attack the problem by introducing summability kernels into the inverse Fourier integral leading to approximate solutions. Let us briefly recall this fundamental concept.

In the classical setting we usually start with a function $h \in L^1(\mathbf{R})$ with $\int_{-\infty}^{\infty} h(x)\, dx = 1$ which is given as the inverse Fourier transform of a function $H \in L^1(\mathbf{R})$, i.e.,

$$h(x) = \frac{1}{2\pi} \int_{-\infty}^{\infty} H(\xi)e^{ix\xi}\, d\xi, \qquad x \in \mathbf{R}.$$

To obtain a summability kernel from h we set $h_\lambda(x) = \lambda h(\lambda x)$, $\lambda > 0$. This naturally gives rise to an approximation process in $L^1(\mathbf{R})$ of the following form:

$$\lim_{\lambda\to\infty} \|f - h_\lambda * f\|_{L^1(\mathbf{R})} = 0, \qquad f \in L^1(\mathbf{R}).$$

Note that the convolution operator can be written as an inverse Fourier integral. The Fourier transform of the summability kernel thereby acts as a mollifier:

$$f * h_\lambda(x) = \frac{1}{2\pi}\int_{-\infty}^{\infty} H\left(\frac{\xi}{\lambda}\right) F(\xi) e^{ix\xi}\, d\xi, \qquad x \in \mathbf{R}.$$

The most prominent examples in classical Fourier analysis are the Fejér kernel, the de la Vallée Poussin kernel and the Bochner-Riesz kernel. In all of these cases the defining function H is compactly supported; to be precise the Fejér kernel σ_1 is given by its Fourier transform $C_1(\xi) = \max(1 - |\xi|, 0)$ and the Bochner-Riesz kernel analogously by the function $S_\delta(\xi) = \max(1 - |\xi|^2, 0)^\delta$, $\delta > 0$. Note that the Fejér kernel is the special case $\delta = 1$ of the Cesàro kernel $C_\delta(\xi) = \max(1 - |\xi|, 0)^\delta$, $\delta > 0$.

The de la Vallée Poussin kernel usually is defined as $v(x) = 2\sigma_1(2x) - \sigma_1(x)$, $x \in \mathbf{R}$. It is a well-known fact that the Fejér kernel and the de la Vallée Poussin kernel can be expressed as a convolution of two characteristic functions, i.e., the function $\chi_{[0,1]}$ convolved with itself for the Fejér kernel and the function $\chi_{[0,1]}$ convolved with $\chi_{[0,2]}$ for the de la Vallée Poussin kernel. Nevertheless, this observation is crucial for our construction. For further information on the classical theory we refer the reader to the monographs [3] and [9].

In the present paper we deal with the construction of summability methods for the *Hankel transform*

$$\widehat{f}(\xi) = \int_0^\infty f(x)\mathcal{J}_\nu(\xi x)d\mu_\nu(x), \qquad \xi \in \mathbf{R}_+,$$

the kernel of which is given by a Bessel function of the first kind and of order $\nu > -\frac{1}{2}$ (cf. (2.1) below for the definition of the function J_ν)

$$\mathcal{J}_\nu(x) = \Gamma(\nu+1)\left(\frac{x}{2}\right)^{-\nu} J_\nu(x), \qquad x \in \mathbf{R}_+, \tag{1.1}$$

with $d\mu_\nu(x)$ denoting the measure $(2^\nu\Gamma(\nu+1))^{-1}x^{2\nu+1}dx$. The transform is well defined for all functions $f \in L^1(\mathbf{R}_+, \mu_\nu)$.

We want to construct a de la Vallée Poussin kernel as well as a Fejér kernel. Taking the obvious definition for these kernels as in the classical case does not lead to approximate identities since the inverse Hankel transform of the function $\max(1-\xi, 0)$, $\xi \in \mathbf{R}_+$, does not belong to $L^1(\mathbf{R}_+, \mu_\nu)$ for all $\nu > -\frac{1}{2}$. We therefore have to choose another way to define analogue summability kernels.

This problem is known from orthogonal polynomials in the algebraic case. In [6] Themistoclakis and the second named author have defined a de la Vallée Poussin kernel for expansions in terms of orthogonal polynomials using the convolution structure for the underlying space of functions. We want to follow their idea in the continuous case, i.e., we will use the convolution structure on $L^1(\mathbf{R}_+, \mu_\nu)$ to define approximate identities for the Hankel transform. In a forthcoming paper the authors will investigate the approximation properties of this new kernel.

To make the paper self-contained we recall some facts from special functions in the second section which will be needed in the sequel. We then introduce classical summability kernels for the Hankel transform. In this section we will briefly sketch the basic material about the convolution structure related to the Hankel transform. In the last section we finally give the construction of the de la Vallée Poussin kernel.

2 Facts from special functions

The *Bessel function* of the first kind and of index ν can be defined by its series representation

$$J_\nu(x) = \left(\frac{x}{2}\right)^\nu \sum_{k=0}^{\infty} \frac{(-1)^k}{k!\,\Gamma(\nu+k+1)} \left(\frac{x}{2}\right)^{2k}, \qquad x \in \mathbf{R}_+. \tag{2.1}$$

Here and throughout the paper we assume $\nu > -\frac{1}{2}$.

Working with the Hankel transform it is more convenient to use the following modified definition

$$\mathcal{J}_\nu(x) = \Gamma(\nu+1) \sum_{k=0}^{\infty} \frac{(-1)^k}{k!\,\Gamma(\nu+k+1)} \left(\frac{x}{2}\right)^{2k} = {}_0F_1\left[\begin{matrix} \\ \nu+1 \end{matrix} \middle| -\frac{x^2}{4}\right], \; x \in \mathbf{R}_+, \tag{2.2}$$

giving the relation (1.1).

Let us briefly introduce the hypergeometric notation. The function ${}_pF_q$ with p numerator parameters $a_1, \ldots, a_p$ and q denominator parameters $b_1, \ldots, b_q$ in $\mathbf{C}$, $p, q \in \mathbf{N}_0$, is defined by the following formal power series.

$${}_pF_q\left[\begin{matrix} a_1, \ldots, a_p \\ b_1, \ldots, b_q \end{matrix} \middle| z\right] = \sum_{k=0}^{\infty} \frac{(a_1)_k \cdots (a_p)_k}{(b_1)_k \cdots (b_q)_k} \cdot \frac{z^k}{k!}, \qquad z \in \mathbf{C},$$

where we used the *Pochhammer symbol* $(a)_\nu = a \cdot (a+1) \cdots (a+\nu-1), \nu \in \mathbf{N}, (a)_0 = 1$ for $a \in \mathbf{C}$.

The series converges for all $z \in \mathbf{C}$, if $p \leq q$ and for $|z| < 1$, if $p = q + 1$. For $p > q+1$ the only point of convergence is $z = 0$. Further information especially concerning the calculus for hypergeometric functions can be found in the monograph [7].

The most prominent example is *Gauss' hypergeometric function* ${}_2F_1$. We will need the following two well-known properties of this function.

The first is known as *Euler's integral representation.* For $\Re c > \Re b > 0$ and $|x| < 1$, it states:

$$ {}_2F_1\left[\begin{matrix} a,b \\ c \end{matrix} \,\middle|\, x\right] = \frac{\Gamma(c)}{\Gamma(b)\Gamma(c-b)} \int_0^1 \tau^{b-1}(1-\tau)^{c-b-1}(1-x\tau)^{-a}\,d\tau. \tag{2.3} $$

There are several relations between hypergeometric functions of squared argument and functions of a single argument known as *quadratic transformations.* One of these reads (cf. [1] (3.1.11))

$$ {}_2F_1\left[\begin{matrix} a,b \\ 2b \end{matrix} \,\middle|\, \frac{4x}{(1+x)^2}\right] = (1+x)^{2a}\,{}_2F_1\left[\begin{matrix} a, a-b+\frac{1}{2} \\ b+\frac{1}{2} \end{matrix} \,\middle|\, x^2\right], \qquad |x| < 1. \tag{2.4} $$

Using both of these properties we can show a modified version of an integral representation which can be found in the tables (cf. for example [8], p. 55).

Lemma 2.1 *Let $a, b \in \mathbf{C}$ and $\Re b > 0$. Then the following representation holds true.*

$$ {}_2F_1\left[\begin{matrix} a, a-b+\frac{1}{2} \\ b+\frac{1}{2} \end{matrix} \,\middle|\, x^2\right] = \tag{2.5} $$
$$ = \frac{\Gamma\left(b+\frac{1}{2}\right)}{\sqrt{\pi}\,\Gamma(b)} \int_0^\pi \sin^{2b-1}\phi \cdot [1 - 2x\cos\phi + x^2]^{-a}\,d\phi, \qquad |x| < 1. $$

Proof. From (2.3) we have using (2.4)

$$ \begin{aligned} {}_2F_1\left[\begin{matrix} a, a-b+\frac{1}{2} \\ b+\frac{1}{2} \end{matrix} \,\middle|\, x^2\right] &= (1+x)^{-2a}\,{}_2F_1\left[\begin{matrix} a,b \\ 2b \end{matrix} \,\middle|\, \frac{4x}{(1+x)^2}\right] = \\ &= \frac{\Gamma(2b)}{\Gamma^2(b)}(1+x)^{-2a} \int_0^1 [\tau(1-\tau)]^{b-1}\left(1 - \frac{4x\tau}{(1+x)^2}\right)^{-a} d\tau \\ &= \frac{\Gamma(2b)}{\Gamma^2(b)} \int_0^1 [\tau(1-\tau)]^{b-1}[1 - 2(2\tau-1)x + x^2]^{-a}\,d\tau. \end{aligned} $$

Setting $\cos\phi = 2\tau - 1$ and using the *duplication formula* for the gamma function

$$ \Gamma(2z) = \frac{2^{2z-1}}{\sqrt{\pi}}\Gamma(z)\Gamma\left(z+\frac{1}{2}\right), \qquad z \in \mathbf{C}\setminus -\mathbf{N}, \tag{2.6} $$

to simplify the constant in front of the integral completes the proof. □

3 Classical summability kernels for the Hankel transform

In this section we present some classical summability kernels related to the Hankel transform. To have a good reference as well as for the sake of completeness we will first of all present some basic facts on convolution structures generated by the Bessel functions and the related Hankel transform.

Let $d\mu_\nu(x)$ denote the measure $\frac{1}{c_\nu}x^{2\nu+1}\ dx$ on the positive real axis $\mathbf{R}_+$, where $c_\nu = 2^\nu\Gamma(\nu+1)$. The following *product formula* for Bessel functions plays a key role in the present paper.

$$\mathcal{J}_\nu(ax)\mathcal{J}_\nu(bx) = \frac{\Gamma(\nu+1)}{\sqrt{\pi}\Gamma(\nu+\frac{1}{2})}\int_0^\pi \mathcal{J}_\nu(cx)\sin^{2\nu}\phi\ d\phi, \qquad x\in\mathbf{R}_+, \tag{3.1}$$

where $a,b\in\mathbf{R}_+$ and $c=\sqrt{a^2+b^2-2ab\cos\phi}$.

If we interpret the parameters a and b as the lengths of two sides of a triangle adjacent to the angle ϕ, c is the length of the third side of this triangle.

To introduce the convolution structure it is more convenient to rewrite the product formula (3.1) in the following way

$$\mathcal{J}_\nu(ax)\mathcal{J}_\nu(bx) = \int_0^\infty \mathcal{J}_\nu(cx)K_\nu(a,b,c)\ d\mu_\nu(c), \qquad x\in\mathbf{R}_+, \tag{3.2}$$

where the kernel is defined by

$$K_\nu(a,b,c) = \frac{\Gamma^2(\nu+1)}{2^\nu\sqrt{\pi}\Gamma\left(\nu+\frac{1}{2}\right)}\frac{[(a+b)^2-c^2]_+^{\nu-\frac{1}{2}}[c^2-(a-b)^2]_+^{\nu-\frac{1}{2}}}{(abc)^{2\nu}}, \qquad a,b,c\in\mathbf{R}_+,$$

where $(t)_+=\max(t,0)$. Note that from (3.2) we obtain $\int_0^\infty K_\nu(a,b,c)\ d\mu_\nu(c)=1$, i.e., $K_\nu(a,b,c)\ d\mu_\nu(c)$ is a probability measure on $\mathbf{R}_+$.

We now define a *generalized translation operator* on the space of all continuous functions with compact support $C_c(\mathbf{R}_+)$ by

$$T_yf(x) = \int_0^\infty f(z)K_\nu(x,y,z)\ d\mu_\nu(z), \qquad x,y\in\mathbf{R}_+.$$

It can be shown that this operator can be extended to all the spaces $L^p(\mathbf{R}_+,\mu_\nu)$,$1\le p\le\infty$, and to the space of bounded continuous functions $C_b(\mathbf{R}_+)$. Moreover, this operator is a bounded linear operator on all of these spaces with norm not greater than one. Furthermore, the measure $d\mu_\nu$ is invariant with respect to this operator, i.e.,

$$\int_0^\infty f(z)\ d\mu_\nu(z) = \int_0^\infty T_yf(z)\ d\mu_\nu(z),$$

for all $f \in C_c(\mathbf{R}_+)$ and $y \in \mathbf{R}_+$. In a more abstract context the measure is interpreted as Haar measure for the underlying algebraic structure.

Using this translation we define the *convolution* of two functions $f, g \in L^1(\mathbf{R}_+, \mu_\nu)$ by

$$f * g(y) = \int_0^\infty f(x) T_y g(x)\, d\mu_\nu(x), \qquad y \in \mathbf{R}_+. \tag{3.3}$$

It can easily be checked that the convolution is commutative and that $\|f * g\|_1 \leq \|f\|_1\, \|g\|_1$ for all $f, g \in L^1(\mathbf{R}_+, \mu_\nu)$. Thus $L^1(\mathbf{R}_+, \mu_\nu)$ becomes a commutative Banach algebra. The Gelfand transform with respect to this Banach algebra is given by the Hankel transform, i.e.,

$$\begin{aligned} \widehat{f}(\xi) &= \int_0^\infty f(x) \mathcal{J}_\nu(\xi x)\, d\mu_\nu(x) \\ &= \xi^{-\nu} \int_0^\infty f(x) J_\nu(\xi x) x^{\nu+1}\, dx, \quad \xi \in \mathbf{R}_+. \end{aligned} \tag{3.4}$$

Note that the transform is self-inverse, i.e.,

$$f(x) = \int_0^\infty \widehat{f}(\xi) \mathcal{J}_\nu(x\xi)\, d\mu_\nu(\xi), \qquad x \in \mathbf{R}_+.$$

From this facts we immediately get the important *convolution theorem*

$$(f * g)\widehat{\ }(\xi) = \widehat{f}(\xi) \cdot \widehat{g}(\xi), \qquad \xi \in \mathbf{R}_+, \quad f, g \in L^1(\mathbf{R}_+, \mu_\nu).$$

Let us now introduce summability kernels for the Hankel transform. As mentioned above the general concept to construct a summability kernel h_λ is to start with a function $h \in L^1(\mathbf{R}_+, \mu_\nu)$ with $\int_0^\infty h(x)\, d\mu_\nu(x) = 1$ and then to define $h_\lambda(x) = \lambda h(\lambda x)$. We will now state the underlying functions corresponding to the analogues of the classical summability kernels of approximation theory.

- The *Cesàro* kernel is defined by its Hankel transform

  $$\widehat{\mathcal{C}}_\delta(\xi) = (1 - \xi)^\delta \chi_{[0,1]}(\xi), \qquad \xi \in \mathbf{R}_+,\ \delta > 0,$$

  where $\chi_{[0,1]}$ denotes the characteristic function of the interval $[0, 1]$. Let us mention two special cases. For $\delta = 0$ we get the *Dirichlet kernel* while $\delta = 1$ leads to the *Fejér kernel.*

- The Hankel transform of the *Bochner-Riesz* kernel is given by

  $$\widehat{\mathcal{R}}_\delta(\xi) = (1 - \xi^2)^\delta \chi_{[0,1]}(\xi), \qquad \xi \in \mathbf{R}_+,\ \delta > 0.$$

We now give explicit representations of the above defined kernels in terms of special functions. An important tool for getting this representations is the *beta function integral*

$$B(a,b)=\frac{\Gamma(a)\Gamma(b)}{\Gamma(a+b)}=\int_0^1(1-t)^{a-1}t^{b-1}\,dt,\qquad a,b\in\mathbf{C}\setminus-\mathbf{N}.$$

For the inverse Hankel transform of the Cesàro kernel we obtain the following series

$$\begin{aligned}\mathcal{C}_\delta(x) &= \int_0^\infty \widehat{\mathcal{C}_\delta}(\xi)\mathcal{J}_\nu(x\xi)\,d\mu_\nu(\xi)\\ &= \frac{1}{2^\nu}\sum_{k=0}^{\infty}\frac{(-1)^k}{k!\,\Gamma(\nu+k+1)}\cdot\left(\frac{x}{2}\right)^{2k}\int_0^1(1-\xi)^\delta\xi^{2k+2\nu+1}\,d\xi\\ &= \frac{1}{2^\nu}\sum_{k=0}^{\infty}\frac{(-1)^k}{k!\,\Gamma(\nu+k+1)}\cdot\left(\frac{x}{2}\right)^{2k}\frac{\Gamma(2k+2\nu+2)\Gamma(\delta+1)}{\Gamma(2k+2\nu+\delta+3)}.\end{aligned}$$

Note that the function $\mathcal{J}_\nu$ is analytic in the whole complex plane. Interchanging summation and integration is therefore satisfied. Using the duplication formula for the gamma function (2.6) and the relation $(a)_\nu=\frac{\Gamma(a+\nu)}{\Gamma(a)}$ we can further conclude that for $x\in\mathbf{R}_+$,

$$\begin{aligned}\mathcal{C}_\delta(x) &= \frac{\Gamma(\delta+1)}{2^{\nu+\delta+1}}\sum_{k=0}^{\infty}\frac{(-1)^k\Gamma\left(k+\nu+\frac{3}{2}\right)}{k!\,\Gamma\left(k+\nu+\frac{\delta+3}{2}\right)\Gamma\left(k+\nu+\frac{\delta+4}{2}\right)}\left(\frac{x}{2}\right)^{2k}\\ &= \frac{\Gamma(\delta+1)\Gamma\left(\nu+\frac{3}{2}\right)}{2^{\nu+\delta+1}\Gamma\left(\nu+\frac{\delta+3}{2}\right)\Gamma\left(\nu+\frac{\delta+4}{2}\right)}\sum_{k=0}^{\infty}\frac{(-1)^k}{k!}\frac{\left(\nu+\frac{3}{2}\right)_k}{\left(\nu+\frac{\delta+3}{2}\right)_k\left(\nu+\frac{\delta+4}{2}\right)_k}\left(\frac{x^2}{4}\right)^k\\ &= \frac{\Gamma(\delta+1)\Gamma\left(\nu+\frac{3}{2}\right)}{2^{\nu+\delta+1}\Gamma\left(\nu+\frac{\delta+3}{2}\right)\Gamma\left(\nu+\frac{\delta+4}{2}\right)}\,{}_1F_2\left[\begin{matrix}\nu+\frac{3}{2}\\ \nu+\frac{\delta+3}{2},\nu+\frac{\delta+4}{2}\end{matrix}\,\middle|\,-\frac{x^2}{4}\right].\end{aligned}\tag{3.5}$$

In the introduction we mentioned that the Fejér kernel does in general not define an approximate identity in $L^1(\mathbf{R}_+,\mu_\nu)$. To show why this is the case, let us look at the asymptotic behavior of the kernel. From ([7], 5.11.2(4) and 5.11.1(19)) we have

$$\begin{aligned}&{}_1F_2\left[\begin{matrix}\nu+\frac{3}{2}\\ \nu+\frac{\delta+3}{2},\nu+\frac{\delta+4}{2}\end{matrix}\,\middle|\,-\frac{x^2}{4}\right]\sim\\ &\quad\sim\frac{\Gamma\left(\nu+\frac{\delta+3}{2}\right)\Gamma\left(\nu+\frac{\delta+4}{2}\right)}{\Gamma\left(\nu+\frac{3}{2}\right)}\frac{(-1)^\gamma}{2^{2\gamma+1}\sqrt{\pi}}e^{ix}x^{2\gamma}+\mathcal{O}\left(x^{2\gamma-\frac{1}{2}}\right),\end{aligned}$$

for large $x\in\mathbf{R}_+$, where $\gamma=-\frac{\nu}{2}-\frac{\delta}{2}-\frac{3}{4}$. It follows that $\mathcal{C}_\delta\in L^1(\mathbf{R}_+,\mu_\nu)$ for $\delta>\nu+\frac{1}{2}$.

For the Bochner-Riesz kernel, we have

$$\begin{aligned}
\mathcal{R}_\delta(x) &= \int_0^\infty \widehat{\mathcal{R}}_\delta(\xi)\mathcal{J}_\nu(x\xi)\,d\mu_\nu(\xi) \\
&= \frac{1}{2^\nu}\sum_{k=0}^\infty \frac{(-1)^k}{k!\,\Gamma(\nu+k+1)}\left(\frac{x}{2}\right)^{2k}\int_0^1 (1-\xi^2)^\delta \xi^{2k+2\nu+1}\,d\xi \\
&= \frac{\Gamma(\delta+1)}{2^{\nu+1}}\sum_{k=0}^\infty \frac{(-1)^k}{k!\,\Gamma(\nu+\delta+k+2)}\left(\frac{x}{2}\right)^{2k} \\
&= \frac{\Gamma(\delta+1)}{2^{\nu+1}\Gamma(\nu+\delta+1)}\mathcal{J}_{\nu+\delta+1}(x), \qquad x\in\mathbf{R}_+. \qquad (3.6)
\end{aligned}$$

From ([12], 7.21(1)) we have for large $x \in \mathbf{R}_+$ the asymptotic expansion for the Bessel functions

$$J_\nu(x) \sim (2\pi x)^{-\frac{1}{2}}\left(\cos\left(x-\frac{\nu}{2}\pi-\frac{\pi}{4}\right)+\mathcal{O}\left(\frac{1}{x}\right)\right). \qquad (3.7)$$

Taking into account the order of magnitude of the measure μ_ν for large $x \in \mathbf{R}_+$ it follows that $\mathcal{R}_\delta(x) \sim \mathcal{O}\left(x^{\nu-\delta-\frac{1}{2}}\right)$. Thus the kernel belongs to $L^1(\mathbf{R}_+,\mu_\nu)$ and therefore defines an approximate identity, if $\delta > \nu + \frac{1}{2}$.

The Hankel transform of order $\nu = \frac{d-2}{2}$, $d \in \mathbf{N}$, is also called *Fourier-Bessel transform*. It is the Fourier transform of radial functions on $\mathbf{R}^d$ which is a well-studied topic in classical Fourier analysis. The bound $\delta > \nu + \frac{1}{2} = \frac{d-1}{2}$ is called *critical index* (cf. [9], Cor. 4.16). Chanillo & Muckenhoupt [4] studied weak type estimates for Bochner-Riesz means of radial functions in $L^p(\mathbf{R}^d)$. Further extensions have been given by Colzani, Travaglini & Vignati (cf. [5] and the references therein).

Other summability kernels for the Hankel transform like the Gaussian kernel and the Poisson kernel and their behavior with respect to pointwise convergence have been studied in some detail by Stempak (cf. [10] and [11]). We further mention the work of Betancor and Rodríguez-Mesa. They studied questions of norm convergence for the Hankel transform (cf. for example [2] and the references cited there).

4 Construction of the de la Vallée Poussin kernel

We are now able to define the *de la Vallée Poussin kernel* for the Hankel transform:

$$\widehat{\mathcal{V}_s^r}(\xi) = \chi_{[0,r]} * \chi_{[0,s]}(\xi), \qquad \xi \in \mathbf{R}_+, \quad r,s>0. \qquad (4.1)$$

Recall that $*$ denotes the convolution associated with the Hankel transform as given by (3.3).

To derive an explicit representation of the de la Vallée Poussin kernel we will use the opposite direction. We will first calculate the Hankel transform of the characteristic function $\chi_{[0,r]}$ and then use the convolution theorem to get the desired result.

Proposition 4.1 *The Hankel transform of order $\nu > -\frac{1}{2}$ of the characteristic function $\chi_{[0,r]}$, $r > 0$, is the Bessel function*

$$\widehat{\chi}_{[0,r]}(\xi) = \frac{r^{2\nu+2}}{2^{\nu+1}\Gamma(\nu+2)}\, \mathcal{J}_{\nu+1}(r\xi), \qquad \xi \in \mathbf{R}_+. \tag{4.2}$$

Proof. From the representation (2.1) it directly follows that

$$\frac{d}{dz}\left[z^{\nu} J_{\nu}(z)\right] = z^{\nu} J_{\nu-1}(z).$$

Since $z^{\nu} J_{\nu-1}(z)|_{z=0} = 0$ we have that

$$\int_0^1 \tau^{\nu+1} J_{\nu}(t\tau)\, d\tau = J_{\nu+1}(t), \qquad t \in \mathbf{R}_+.$$

Using this formula gives

$$\begin{aligned}\widehat{\chi}_{[0,r]}(\xi) &= \xi^{-\nu}\int_0^r x^{\nu+1} J_{\nu}(\xi x)\, dx = \frac{r^{\nu+2}}{\xi^{\nu}}\int_0^1 x^{\nu+1} J_{\nu}(\xi r x)\, dx \\ &= r^{2\nu+2}(r\xi)^{-(\nu+1)} J_{\nu+1}(r\xi) = \frac{r^{2\nu+2}}{2^{\nu+1}\Gamma(\nu+2)}\mathcal{J}_{\nu+1}(r\xi), \qquad \xi \in \mathbf{R}_+.\end{aligned}$$

□

Let us just remark that equation (4.2) also follows from (3.6) by letting $\delta \to 0$. From the proposition we immediately have

Corollary 4.2 *For $r, s > 0$ and $\nu > -\frac{1}{2}$ we have*

$$\mathcal{V}_s^r(x) = \frac{(rs)^{2\nu+2}}{2^{2\nu+2}\Gamma^2(\nu+2)}\, \mathcal{J}_{\nu+1}(rx) \cdot \mathcal{J}_{\nu+1}(sx), \qquad x \in \mathbf{R}_+. \tag{4.3}$$

Again the asymptotic relation (3.7) leads to an estimate for the order of magnitude for large arguments. The kernel $\mathcal{V}_s^r(x)$ thus satisfies $\mathcal{V}_s^r(x) \sim \mathcal{O}\left(x^{-2}\right)$ for $x \to \infty$. Since $\mathcal{J}_{\nu+1}(0) = 1$ we can conclude that the integral

$$\int_0^{\infty} \left|\mathcal{J}_{\nu+1}(rx)\mathcal{J}_{\nu+1}(sx)\right|\, d\mu_{\nu}(x), \qquad r, s > 0,$$

exists and is finite, i.e., the kernel belongs to $L^1(\mathbf{R}_+, \mu_\nu)$. Therefore the de la Vallée Poussin kernel as defined by equation (4.1) generates an approximate identity in the sense described above.

We are now able to prove the explicit representation for the Hankel transform of the de la Vallée Poussin kernels for $\xi < r - s$.

Theorem 4.3 *For $r > s > 0$ and $\nu > -\frac{1}{2}$ the de la Vallée Poussin kernel vanishes for all $\xi > r + s$. The kernel decreases for $r - s < \xi \leq r + s$ while for $\xi < r - s$ we have*

$$\widehat{\mathcal{V}_s^r}(\xi) = \frac{1}{2^{\nu+1}\Gamma(\nu+2)} s^{2\nu+2} {}_2F_1\left[\begin{array}{c} \nu+1, 0 \\ \nu+2 \end{array} \middle| \frac{s^2}{r^2}\right], \qquad \xi < r - s. \tag{4.4}$$

Proof. Since the inverse of the Hankel transform is the transform itself, we have

$$\begin{aligned}\widehat{\mathcal{V}_s^r}(\xi) &= \int_0^\infty \mathcal{V}_s^r(x)\mathcal{J}_\nu(\xi x)\, d\mu_\nu(x) \\ &= \frac{(rs)^{2\nu+2}}{2^{3\nu+2}\Gamma(\nu+1)\Gamma^2(\nu+2)} \int_0^\infty \mathcal{J}_{\nu+1}(rx)\mathcal{J}_{\nu+1}(sx)\mathcal{J}_\nu(\xi x)x^{2\nu+1}\, dx, \ \xi \in \mathbf{R}_+.\end{aligned}$$

Setting $\omega = \sqrt{r^2 + s^2 - 2rs\cos\phi}$ and applying the product formula (3.1) we have

$$\begin{aligned}\widehat{\mathcal{V}_s^r}(\xi) &= \frac{(rs)^{2\nu+2}}{2^{3\nu+2}\sqrt{\pi}\Gamma(\nu+1)\Gamma(\nu+2)\Gamma\left(\nu+\frac{3}{2}\right)} \int_0^\pi \sin^{2\nu+2}\phi \times \\ &\quad \times \int_0^\infty \mathcal{J}_\nu(\xi x)\mathcal{J}_{\nu+1}(\omega x)x^{2\nu+1}\, dx\, d\phi.\end{aligned}$$

For the inner integral we can use the following formula (cf. [12], p. 406)

$$\int_0^\infty J_\nu(a\tau)J_{\nu+1}(b\tau)\, d\tau = \begin{cases} a^\nu b^{-(\nu+1)}, & \text{if } 0 < a < b, \\ (2a)^{-1}, & \text{if } a = b, \\ 0, & \text{if } a > b, \end{cases} \tag{4.5}$$

to get

$$\begin{aligned}\int_0^\infty \mathcal{J}_\nu(\xi x)\mathcal{J}_{\nu+1}(\omega x)x^{2\nu+1}\, dx &= \Gamma(\nu+1)\Gamma(\nu+2) \times \\ &\times \left(\frac{\xi}{2}\right)^{-\nu}\left(\frac{\omega}{2}\right)^{-\nu-1} \int_0^\infty J_\nu(\xi x)J_{\nu+1}(\omega x)\, dx \\ &= 2^{2\nu+1}\Gamma(\nu+1)\Gamma(\nu+2)\begin{cases} \omega^{-2\nu-2}, & \text{if } 0 < \xi \leq \omega, \\ \frac{1}{2}\omega^{-2\nu-2}, & \text{if } \xi = \omega, \\ 0, & \text{if } \xi > \omega. \end{cases}\end{aligned}$$

We modify the value of this function at $\xi = \omega$ to $\omega^{-2\nu-2}$. We therefore have

$$\begin{aligned}\widehat{\mathcal{V}_s^r}(\xi) &= \frac{1}{2^{\nu+1}\sqrt{\pi}\Gamma\left(\nu+\frac{3}{2}\right)}(rs)^{2\nu+2}\int_0^\pi \omega^{-2\nu-2}\sin^{2\nu+2}\phi\, d\phi \\ &= \frac{s^{2\nu+2}}{2^{\nu+1}\sqrt{\pi}\Gamma\left(\nu+\frac{3}{2}\right)}\int_0^\pi \frac{\sin^{2\nu+2}\phi}{(1-2\alpha\cos\phi+\alpha^2)^{\nu+1}}\, d\phi, \qquad (4.6)\end{aligned}$$

where $\alpha = \frac{s}{r}$. Since $r > s$ we have $0 < \alpha < 1$. We can therefore apply (2.5) to conclude

$$\int_0^\pi \omega^{-2\nu-2}\sin^{2\nu+2} d\phi = \frac{\sqrt{\pi}\Gamma\left(\nu+\frac{3}{2}\right)}{\Gamma(\nu+2)}\, {}_2F_1\left[\begin{array}{c} \nu+1, 0 \\ \nu+2 \end{array}\,\middle|\, \frac{s^2}{r^2}\right].$$

Incorporating the result into (4.6) completes the proof. □

Let $V_s^r = \left(\int_0^\infty \mathcal{V}_s^r(x)d\mu_\nu(x)\right)^{-1}\mathcal{V}_s^r$ and $V_{s,\lambda}^r = \lambda V_s^r(\lambda x)$. A standard argument from approximation theory then gives

Corollary 4.4 *For $f \in L^1(\mathbf{R}_+, \mu_\nu)$, $\nu > -\frac{1}{2}$, and $r, s > 0$, we have*

$$\lim_{\lambda\to\infty} \left\| f - V_{s,\lambda}^r * f \right\|_{L^1(\mathbf{R}_+,\mu_\nu)} = 0. \qquad (4.7)$$

References

[1] Andrews, G., R. Askey & R. Roy, *Special functions*, Cambridge University Press, Cambridge, 1999.

[2] Betancor, J.J. & L. Rodríguez-Mesa, *L_1-convergence and strong summability of Hankel transforms*, Publ. Math. Debrecen **55** No. 3-4 (1999), 437–451.

[3] Butzer, P.L. & R.J. Nessel, *Fourier analysis and approximation*, Vol. I, Birkhäuser, Basel, 1971.

[4] Chanillo, S. & B. Muckenhoupt, *Weak type estimates for Bochner-Riesz spherical summation multipliers*, Trans. Amer. Math. Soc. **294** No. 2 (1986), 693–703.

[5] Colzani, L., G. Travaglini & M. Vignati, *Bochner-Riesz means of functions in weak-L^p*, Mh. Math. **115** (1993), 35–45.

[6] Filbir, F. & W. Themistoclakis, *On the construction of de la Vallée Poussin means for orthogonal polynomials using convolution structures*, to appear in J. Comput. Anal. Appl.

[7] Luke, Y.L., *The special functions and their approximations*, Vol. I, Academic Press, New York, 1969.

[8] Magnus, W., F. Oberhettinger & R.P. Soni, *Formulas and theorems for the special functions of mathematical physics*, Springer, Berlin, 1966.

[9] Stein, E.M. & G. Weiss, *Introduction to Fourier analysis on Euclidean spaces*, Princeton University Press, Princeton, 1971.

[10] Stempak, K., *The Littlewood-Paley theory for the Fourier-Bessel transform*, University of Wrocław, Preprint no. 45, 1985.

[11] Stempak, K., *La théorie de Littlewood-Paley pour la transformation de Fourier-Bessel*, C.R. Acad. Sc. Paris **303** Série I (1986), 15–18.

[12] Watson, G.N., *A treatise on the theory of Bessel functions*, reprint of the 2nd edition, Cambridge University press, Cambridge, 1996.

Institute of Biomathematics and Biometry
GSF – National Research Center for Environment and Health
Ingolstaedter Landstrasse 1
D-85764 Neuherberg, Germany
Email address: castell@gsf.de, filbir@gsf.de

Advanced Problems in Constructive Approximation
(Eds.) M.D. Buhmann and D.H. Mache
International Series of Numerical Mathematics Vol. 142, 39–52

Polynomial Bases on the Sphere

Noemí Laín Fernández

Abstract

Considering that the well-known basis of spherical harmonics of degree at most n is not localized on the sphere, we construct better localized polynomial bases by means of *reproducing kernels*. Such a construction leads to the problem of finding sets of $(n+1)^2$ points on the sphere that admit unique polynomial interpolation. Finally, we present a possible construction of polynomial wavelets on the sphere.

1 Introduction

Let $\Omega := \{x \in \mathbf{R}^3 : \ \|x\|_2 = 1\}$ denote the unit sphere embedded in the Euclidean space $\mathbf{R}^3$ and let $\Psi : [0,\pi] \times [0, 2\pi) \longrightarrow \mathbf{R}^3$, $(\rho,\theta) \longmapsto (\sin\rho\cos\theta, \sin\rho\sin\theta, \cos\rho)$ be its parameterization in spherical coordinates (ρ,θ). Corresponding to the surface element $dw(\xi)$, we have the inner product and $L^2(\Omega)$-norm

$$\begin{aligned}\langle F, G\rangle &:= \int_\Omega F(\xi)\,\overline{G(\xi)}dw(\xi) = \int_0^{2\pi}\int_0^{\pi} \sin\rho\; F(\Psi(\rho,\theta))\,\overline{G(\Psi(\rho,\theta))}\,d\rho\,d\theta,\\ \|F\|^2 &:= \langle F, F\rangle.\end{aligned}$$

Furthermore, let $\mathrm{Harm}_n(\mathbf{R}^3)$ denote the space of harmonic homogeneous polynomials of degree n in three variables. Restricting these functions to Ω, we obtain the so-called *spherical harmonics* of order n. Throughout this paper, we will concentrate on the space $V_n := \Pi_n|_\Omega$. It can be shown that

$$V_n = \bigoplus_{k=0}^{n} \mathrm{Harm}_k(\Omega), \tag{1}$$

where this direct sum decomposition has to be understood in the sense that any spherical polynomial of degree $\leq n$ is the restriction of a harmonic polynomial of degree less or equal to n to the sphere. Since dim $\mathrm{Harm}_k(\Omega) = 2k+1$, it follows that $N := \dim V_n = \sum_{k=0}^{n}(2k+1) = (n+1)^2$. An $L^2(\Omega)$-orthonormal basis of V_n that is not localized on the sphere is given by

$$\left\{ Y_k^j(\rho,\theta) = \sqrt{\frac{2k+1}{4\pi}} P_k^{|j|}(\cos\rho)\, e^{ij\theta},\ k = 0,\ldots,n,\ j = -k,\ldots,k \right\}, \tag{2}$$

where

$$P_k^j(t) = \left(\frac{(k-j)!}{(k+j)!}\right)^{1/2} (1-t^2)^{j/2} \frac{d^j}{dt^j} P_k(t), \quad j = 0,\ldots,k,\ t \in [-1,1],$$

denote the associated Legendre functions and P_k stands for the Legendre polynomial of degree k normalized according to the condition $P_k(1) = 1$. From now on, this basis will be referred to as the *basis of spherical harmonics.*

A way of constructing better localized bases is by means of *reproducing kernels.* Let $\{Y_k^j : j = -k,\ldots,k,\ k = 0,\ldots,n\}$ be an arbitrary $L^2(\Omega)$-orthonormal basis of V_n. It is straightforward to check that the reproducing kernel of $\mathrm{Harm}_k(\Omega)$ is given by

$$G_k(\xi,\eta) := \sum_{j=-k}^{k} Y_k^{-j}(\xi) Y_k^j(\eta), \quad \xi,\eta \in \Omega.$$

Using now the addition theorem (see [4]) for $\mathrm{Harm}_k(\Omega)$, one comes up with the following theorem

Theorem 1.1 *The unique reproducing kernel of V_n is given by*

$$K_n(\xi,\eta) := \sum_{k=0}^{n} G_k(\xi,\eta) = \sum_{k=0}^{n} \frac{2k+1}{4\pi} P_k(\xi\cdot\eta) =: k_n(\xi\cdot\eta), \qquad \xi,\eta \in \Omega. \tag{3}$$

In particular, $K_n(\xi,\xi) = \left(\frac{n+1}{4\pi}\right)^2$ for all $\xi \in \Omega$. It should be observed that $K_n(\xi,\eta) = k_n(\xi\cdot\eta)$ as a zonal function, only depends on the Euclidean product of the vectors ξ and η. Therefore, it is invariant with respect to rotations, i.e., transformations of the group $SO(3)$.

2 The space V_n

2.1 Scaling functions

In contrast to the spherical harmonics Y_k^j introduced in (2), the functions $K_n(\cdot,\cdot) : \Omega\times\Omega \longrightarrow \mathbf{R}$ defined in (3) have the property of being the spherical polynomials with minimal $L^2(\Omega)$-norm among all spherical polynomials of degree $\leq n$ that attain the value 1 when evaluated at a prescribed point. The following theorem establishes this localization property.

Theorem 2.1 *Let $\xi \in \Omega$. Then*

$$\left\| \frac{K_n(\xi,\cdot)}{K_n(\xi,\xi)} \right\| = \min\left\{ \|P\| \, : P \in V_n, \, P(\xi) = 1 \right\}. \tag{4}$$

Proof. Let $\{Y_k^j, \, k = 0, \ldots, n, \, j = -k, \ldots, k\}$ be an arbitrary $L^2(\Omega)$-orthonormal basis of V_n and let $P \in V_n$ with $P(\xi) = 1$. The polynomial P can be expressed in terms of its Fourier sum

$$P(\xi) = \sum_{k=0}^{n} \sum_{j=-k}^{k} \langle P, Y_k^j \rangle Y_k^j(\xi), \qquad \xi \in \Omega.$$

As a consequence of the Cauchy-Schwarz inequality and the addition theorem, we obtain

$$\begin{aligned} 1 &= (P(\xi))^2 = \left(\sum_{k=0}^{n} \sum_{j=-k}^{k} \langle P, Y_k^j \rangle Y_k^j(\xi) \right)^2 \\ &\leq \left(\sum_{k=0}^{n} \sum_{j=-k}^{k} |\langle P, Y_k^j \rangle|^2 \right) \left(\sum_{k=0}^{n} \sum_{j=-k}^{k} |Y_k^j(\xi)|^2 \right) \\ &= \|P\|^2 \, K_n(\xi,\xi) = \|P\|^2 \left(\frac{K_n(\xi,\xi)}{\|K_n(\xi,\cdot)\|} \right)^2, \end{aligned}$$

where the last equality follows from the fact that

$$\|K_n(\xi,\cdot)\| = \left(\langle K_n(\xi,\cdot), K_n(\xi,\cdot) \rangle \right)^{1/2} = \left(K_n(\xi,\xi) \right)^{1/2}. \qquad \square$$

Our aim is to study the problem of characterizing sets of points $\{\eta_i, \, i=1,\ldots,N\}$ such that the functions $\{\varphi_i^n := K_n(\eta_i,\cdot), \, i = 1, \ldots, N\}$ constitute a basis of the space V_n. The functions $\{\varphi_i^n, \, i = 1, \ldots, N\}$ will be called *scaling functions.* As the following observation shows, the linear independence of the scaling functions is reflected in the regularity of an $N \times N$ matrix. Given $\{\eta_i, \, i = 1, \ldots, N\} \subset \Omega$, we can construct the interpolation matrix

$$\mathbf{A}_n := \begin{pmatrix} Y_0^0(\eta_1) & Y_0^0(\eta_2) & Y_0^0(\eta_3) & \ldots & Y_0^0(\eta_N) \\ Y_1^{-1}(\eta_1) & Y_1^{-1}(\eta_2) & Y_1^{-1}(\eta_3) & \ldots & Y_1^{-1}(\eta_N) \\ Y_1^0(\eta_1) & Y_1^0(\eta_2) & Y_1^0(\eta_3) & \ldots & Y_1^0(\eta_N) \\ Y_1^1(\eta_1) & Y_1^1(\eta_2) & Y_1^1(\eta_3) & \ldots & Y_1^1(\eta_N) \\ \vdots & \vdots & \vdots & \ddots & \vdots \\ Y_n^{-n}(\eta_1) & Y_n^{-n}(\eta_2) & Y_n^{-n}(\eta_3) & \ldots & Y_n^{-n}(\eta_N) \\ \vdots & \vdots & \vdots & \ddots & \vdots \\ Y_n^n(\eta_1) & Y_n^n(\eta_2) & Y_n^n(\eta_3) & \ldots & Y_n^n(\eta_N) \end{pmatrix}. \tag{5}$$

By virtue of the addition theorem, one can directly see that the symmetric positive semidefinite matrix $\mathbf{\Phi}_n := \mathbf{A}_n^* \mathbf{A}_n$ has the entries

$$\mathbf{\Phi}_n(r,s) = \sum_{k=0}^{n} \sum_{j=-k}^{k} Y_k^{-j}(\eta_r)\, Y_k^{j}(\eta_s) = K_n(\eta_r, \eta_s) = \langle K_n(\eta_r, \cdot),\, K_n(\eta_s, \cdot)\rangle .$$

Therefore, the matrix $\mathbf{\Phi}_n$ is a Gram matrix and will be positive definite, in particular regular, if and only if the functions involved, i.e., the scaling functions, are linearly independent. As det $\mathbf{\Phi}_n = |\det \mathbf{A}_n|^2$, we can study the regularity of either $\mathbf{A}_n$ or $\mathbf{\Phi}_n$ to determine whether the scaling functions constitute a basis of V_n or not.

Definition 2.2 *A set of points $\{\eta_i,\ i = 1, \ldots, N\} \subset \Omega$ for which the associated scaling functions constitute a basis of V_n is called a* fundamental system *of V_n.*

2.2 Examples

2.2.1 Linear polynomials

As we deal with low-dimensional matrices (dim $V_1 = 4$), we can give a complete characterization of the F.S. of V_n and state conditions under which the scaling functions constitute an orthogonal basis of V_n.

Theorem 2.3 *Four points $\{\eta_1, \eta_2, \eta_3, \eta_4\} \subset \Omega$ form a fundamental system of V_1 if and only if they do not lie on a circle.*

Proof. Let $\{\eta_k = (\eta_k^1, \eta_k^2, \eta_k^3),\ k = 1, \ldots, 4\}$ be four points on the sphere. The Gram matrix $\mathbf{\Phi}_1 = (K_1(\eta_i, \eta_j))_{i,j=1,\ldots,4}$ attains the form

$$\mathbf{\Phi}_1 = \frac{1}{4\pi}\begin{pmatrix} 1+3\eta_1\cdot\eta_1 & 1+3\eta_1\cdot\eta_2 & 1+3\eta_1\cdot\eta_3 & 1+3\eta_1\cdot\eta_4 \\ 1+3\eta_2\cdot\eta_1 & 1+3\eta_2\cdot\eta_2 & 1+3\eta_2\cdot\eta_3 & 1+3\eta_2\cdot\eta_4 \\ 1+3\eta_3\cdot\eta_1 & 1+3\eta_3\cdot\eta_2 & 1+3\eta_3\cdot\eta_3 & 1+3\eta_3\cdot\eta_4 \\ 1+3\eta_4\cdot\eta_1 & 1+3\eta_4\cdot\eta_2 & 1+3\eta_4\cdot\eta_3 & 1+3\eta_4\cdot\eta_4 \end{pmatrix}.$$

It is immediate to see that we can decompose $\mathbf{\Phi}_1$ into the product $\mathbf{\Phi}_1 = \frac{1}{4\pi}\mathbf{A}^T\mathbf{A}$ with

$$\mathbf{A} := \begin{pmatrix} 1 & 1 & 1 & 1 \\ \sqrt{3}\eta_1^1 & \sqrt{3}\eta_2^1 & \sqrt{3}\eta_3^1 & \sqrt{3}\eta_4^1 \\ \sqrt{3}\eta_1^2 & \sqrt{3}\eta_2^2 & \sqrt{3}\eta_3^2 & \sqrt{3}\eta_4^2 \\ \sqrt{3}\eta_1^3 & \sqrt{3}\eta_2^3 & \sqrt{3}\eta_3^3 & \sqrt{3}\eta_4^3 \end{pmatrix}.$$

As det $\mathbf{\Phi}_1 = \left(\frac{1}{4\pi}\right)^4 (\det \mathbf{A})^2$, the matrix $\mathbf{\Phi}_1$ is regular if and only if $\mathbf{A}$ is regular. But $\mathbf{A}$ is equivalent to the matrix

$$\begin{pmatrix} 1 & 0 & 0 & 0 \\ \sqrt{3}\eta_1^1 & \sqrt{3}(\eta_2^1-\eta_1^1) & \sqrt{3}(\eta_3^1-\eta_1^1) & \sqrt{3}(\eta_4^1-\eta_1^1) \\ \sqrt{3}\eta_1^2 & \sqrt{3}(\eta_2^2-\eta_1^2) & \sqrt{3}(\eta_3^2-\eta_1^2) & \sqrt{3}(\eta_4^2-\eta_1^2) \\ \sqrt{3}\eta_1^3 & \sqrt{3}(\eta_2^3-\eta_1^3) & \sqrt{3}(\eta_3^3-\eta_1^3) & \sqrt{3}(\eta_4^3-\eta_1^3) \end{pmatrix},$$

which will be regular if and only if the vectors $\overrightarrow{\eta_1\eta_2}, \overrightarrow{\eta_1\eta_3}, \overrightarrow{\eta_1\eta_4}$ are not coplanar. □

Theorem 2.4 *The scaling functions* $\{K_1(\eta_i,\cdot),\ i=1,\ldots,4\}$ *are orthogonal if and only if* $\{\eta_i,\ i=1,\ldots,4\}$ *are the vertices of a regular tetrahedron inscribed in* Ω. *In this case, the matrix* $\mathbf{\Phi}_1$ *assumes the diagonal form* $\frac{1}{\pi}\,\mathbf{I}_4$.

Proof. "⇒": Orthogonality of the scaling functions implies that

$$\delta_{ij} = \langle \varphi_i^1, \varphi_j^1 \rangle = K_1(\eta_i,\eta_j) = \mathbf{\Phi}_1(i,j) = \frac{1}{4\pi}(1+3\,\eta_i\cdot\eta_j), \quad i,j=1,\ldots,4. \quad (6)$$

Hence, we get the following system of six linear equations

$$\eta_i\cdot\eta_j = -\frac{1}{3} \quad \text{for} \quad 1\le i<j\le 4. \quad (7)$$

As K_1 is invariant with respect to rotations, we can assume without loss of generality that $\eta_1=(0,0,1)$. The first three equations (for $i=1$) in (7) enforce the points η_2,η_3 and η_4 to lie on a circle parallel to the equator at latitude $\theta=\arccos\left(-\frac{1}{3}\right)$, i.e.,

$$\eta_k = \left(\frac{2\sqrt{2}}{3}\cos\theta_k, \frac{2\sqrt{2}}{3}\sin\theta_k, -\frac{1}{3}\right) \quad \text{for } k=2,3,4. \quad (8)$$

Accordingly, our system (7) becomes

$$\begin{aligned}\cos(\theta_2-\theta_3) &= -\frac{1}{2},\\ \cos(\theta_2-\theta_4) &= -\frac{1}{2},\\ \cos(\theta_3-\theta_4) &= -\frac{1}{2},\end{aligned}$$

which yields

$$\rho_2=\alpha,\ \rho_3=\alpha+\frac{2\pi}{3},\ \rho_4=\alpha+\frac{4\pi}{3} \quad \text{with } \alpha\in[0,2\pi).$$

Combining this fact with (8), we conclude that the points $\{\eta_i,\ i=1,\ldots,4\}$ are the vertices of a regular tetrahedron inscribed in the sphere.
"⇐": As $K_1(\cdot,\cdot)$ is invariant with respect to rotations, we can assume without loss of generality that

$$\eta_1=(0,0,1)\quad,\qquad \eta_2=\left(\frac{2\sqrt{2}}{3},0,-\frac{1}{3}\right),$$
$$\eta_3=\left(-\frac{\sqrt{2}}{3},\frac{\sqrt{6}}{3},-\frac{1}{3}\right)\quad,\qquad \eta_4=\left(-\frac{\sqrt{2}}{3},-\frac{\sqrt{6}}{3},-\frac{1}{3}\right),$$

are the vertices of a regular tetrahedron. A straightforward calculation shows that the matrix $\mathbf{\Phi}_1$ attains the desired form. □

2.2.2 General degree n

With dim $V_n = (n+1)^2$, the structure of the Gram matrices becomes very complicated and it is not possible to give a complete characterization of the fundamental systems. Nevertheless, one can prove the following result

Theorem 2.5 *If* $\{\eta_i,\ i = 1, \ldots, N\} \subset \Omega$ *lie on a circle, then they do not form a fundamental system.*

Proof. Basically, the proof is a direct consequence of Theorem 2.3. We have that

$$\mathbf{A}_n = \begin{pmatrix} Y_0^0(\eta_1) & \cdots & Y_0^0(\eta_4) & Y_0^0(\eta_5) & \cdots & Y_0^0(\eta_N) \\ Y_1^{-1}(\eta_1) & \cdots & Y_1^{-1}(\eta_4) & Y_1^{-1}(\eta_5) & \cdots & Y_1^{-1}(\eta_N) \\ Y_1^0(\eta_1) & \cdots & Y_1^0(\eta_4) & Y_1^0(\eta_5) & \cdots & Y_1^0(\eta_N) \\ Y_1^1(\eta_1) & \cdots & Y_1^1(\eta_4) & Y_1^1(\eta_5) & \cdots & Y_1^1(\eta_N) \\ \hline Y_2^{-2}(\eta_1) & \cdots & Y_2^{-2}(\eta_4) & Y_2^{-2}(\eta_5) & \cdots & Y_2^{-2}(\eta_N) \\ \vdots & \ddots & \vdots & \vdots & \ddots & \vdots \\ Y_n^n(\eta_1) & \cdots & Y_n^n(\eta_4) & Y_n^n(\eta_5) & \cdots & Y_n^n(\eta_N) \end{pmatrix}.$$

If the N points lie on a circle, then any four of them also lie on a circle. Therefore using Theorem 2.3, the submatrix constituted by the first four rows of $\mathbf{A}_n$ will have rank less or equal to three. Consequently, the entire matrix $\mathbf{A}_n$ will have rank less or equal to $N-1$ and hence be singular. □

With n growing, the analysis of the regularity of the matrices $\mathbf{A}_n$ becomes inaccessible, so we have to restrict our analysis to specific choices of point constellations. A possible way of constructing fundamental systems is due to B. Sünderman [6]. A similar but more general result is found in M. v. Golitschek and W.A. Light [3]. Another description of specific sets of points which admit unique polynomial interpolation is also given in Y. Xu [7].

3 Wavelets

For the particular case of equidistant nodes on symmetric latitudes, the matrices involved attain an accessible form and it is possible to carry out the following construction of polynomial wavelets on the sphere.
Given $l \in \mathbf{N}$, we can define the spaces

$$W_n := V_{n+l} \ominus V_n = \operatorname{span}\left\{Y_k^j,\ k = n+1, \ldots, n+l,\ j = -k, \ldots, k\right\}.$$

Note that $d :=$ dim $W_n =$ dim $V_{n+l} -$ dim $V_n = l(l + 2n + 2)$. Again, the goal is to identify functions, then called *wavelets*, which form a localized basis of the space W_n. In accordance with the definition of the scaling functions, we define the wavelets in terms of reproducing kernels.

Definition 3.1 *Let $S := \{\xi_i,\ i = 1, \ldots, d\} \subset \Omega$. We call*

$$\psi_i^n(\xi) := \varphi_i^{n+l}(\xi) - \varphi_i^n(\xi) = \sum_{s=n+1}^{n+l} \frac{2s+1}{4\pi} P_s(\xi_i \cdot \xi), \quad i = 1, \ldots, d, \quad \xi \in \Omega, \quad (9)$$

the wavelet functions *corresponding to the set S.*

The wavelet functions have the following properties, which do not depend on the choice of the set S of points.

Theorem 3.2 *Let $S := \{\xi_i,\ i = 1, \ldots, d\} \subset \Omega$ and let $\{\psi_i^n,\ i = 1, \ldots, d\}$ be the corresponding wavelet functions.*

(i) *The inner product of wavelets may be calculated as follows*

$$\langle \psi_i^n, \psi_j^n \rangle = \psi_j^n(\xi_i), \quad i, j = 1, \ldots, d.$$

(ii) *Let $\{\varphi_j^n,\ j = 1, \ldots, N\}$ denote the scaling functions with respect to $\{\eta_j,\ j = 1, \ldots, N\} \subset \Omega$. The wavelets and the scaling functions are orthogonal to each other:*

$$\langle \psi_i^n, \varphi_j^n \rangle = 0, \quad i = 1, \ldots, d,\ j = 1, \ldots, N.$$

(iii) *The wavelet ψ_i^n is localized around ξ_i :*

$$\left\| \frac{\psi_i^n}{\psi_i^n(\xi_i)} \right\| = \min \left\{ \|P\| \ : P \in W_n,\ P(\xi_i) = 1 \right\}. \quad (10)$$

Proof. To verify (i), we show that the wavelets satisfy a reproduction property in W_n. Let $Q \in W_n$. Then

$$\langle \psi_i^n, Q \rangle = \langle K_{n+l}(\xi_i, \cdot), Q \rangle - \langle K_n(\xi_i, \cdot), Q \rangle = Q(\xi_i).$$

Consequently, the reproducing kernel of W_n is represented by the wavelets $\{\psi_i^n,\ i = 1, \ldots, d\}$. Assertion (ii) follows directly from the definition of the participating functions. The proof of (iii) is along the same lines of Theorem 2.1. □

Analogously, we can now ask ourselves for which sets S, the wavelets constitute a basis of the space W_n. As the next subsection shows, it is possible to give an answer for the case $l=2$.

3.1 A possible construction for the case $l = 2$

By definition $W_n := V_{n+2} - V_n$ and $d = \dim W_n = 4n+8$. A possible multiresolution construction consists in adding at each level n the totality of $(2n+3)+(2n+5) = 4n+8$ points distributed equidistantly on two symmetric latitudes. As the following theorem shows, the totality of these points forms a fundamental system of W_n.

Theorem 3.3 *Let $\rho \in (0,\pi)$ such that*

$$P_l^m(\pm\cos\rho) \neq 0 \quad \textit{for all } l = n+1, n+2,\ m = 0,\ldots,l. \tag{11}$$

Then $S := \{\eta_k := \Psi(\rho, \frac{2\pi k}{2n+3}),\ k = 1,\ldots,2n+3\} \cup \{\xi_j := \Psi(\pi-\rho, \frac{2\pi j}{2n+5}),\ j = 1,\ldots,2n+5\}$ constitutes a fundamental system of W_n.

Proof. In order to prove the linear independence of the wavelet functions corresponding to the set S, we have to study the regularity of the matrix

$$\mathbf{B}_n := [\mathbf{Y}_{n+1}^{-(n+1)},\ldots,\mathbf{Y}_{n+1}^{n+1},\mathbf{Y}_{n+2}^{-(n+2)},\ldots,\ \mathbf{Y}_{n+2}^{n+2}]^T, \tag{12}$$

where $\mathbf{Y}_k^j$ denotes the column vector given by

$$\mathbf{Y}_k^j := (Y_k^j(\eta_1),\ldots,Y_k^j(\eta_{2n+3}),Y_k^j(\xi_1),\ldots,Y_k^j(\xi_{2n+5}))^T.$$

Making use of the fact that the spherical harmonics are functions with separated variables, we can transform $\mathbf{B}_n$ into an equivalent block matrix by multiplication with regular matrices. First we construct the diagonal block matrix

$$\mathbf{F} := \begin{pmatrix} F_{2n+3} & 0 \\ 0 & F_{2n+5} \end{pmatrix}, \tag{13}$$

where $F_n = \frac{1}{\sqrt{n}}\left(e^{\frac{2\pi i}{n}(1-j)(k-1)i}\right)_{k,j=1,\ldots,n} \in \mathbf{C}^{n\times n}$ is the $n\times n$-dimensional Fourier matrix. Second we employ the permutation matrix $\mathbf{P}_1$ such that $\mathbf{P}_1\,\mathbf{B}_n$ assumes the form

$$\begin{aligned} \mathbf{P}_1\mathbf{B}_n \;=\; & [\,\mathbf{Y}_{n+1}^{n+1},\mathbf{Y}_{n+2}^{n+1},\mathbf{Y}_{n+2}^{-(n+2)},\mathbf{Y}_{n+1}^{-(n+1)},\mathbf{Y}_{n+2}^{-(n+1)},\mathbf{Y}_{n+2}^{n+2}, \\ & \mathbf{Y}_{n+1}^{-n},\mathbf{Y}_{n+2}^{-n},\ldots,\mathbf{Y}_{n+1}^{0},\mathbf{Y}_{n+2}^{0},\ldots,\mathbf{Y}_{n+1}^{n},\mathbf{Y}_{n+2}^{n}\,]^T. \end{aligned} \tag{14}$$

The matrix $\mathbf{B}_n$ is regular if and only if the product $\mathbf{P}_1\,\mathbf{B}_n\,\mathbf{F}$ is regular. Let $x := \cos\rho$ and let $\mathbf{v}_k^j := \left(\mathbf{Y}_k^j\right)^T$ denote the transpose of the column vector $\mathbf{Y}_k^j$. In the way we have chosen the nodes on the two latitudes $\cos\rho$ and $\cos(\pi-\rho) = -\cos\rho$, we have that $\mathbf{v}_k^j := \left((\mathbf{v}_k^j)^1,(\mathbf{v}_k^j)^2\right)$, where

$$(\mathbf{v}_k^j)^1 = \left(\sqrt{\frac{2k+1}{4\pi}}\,P_k^{|j|}(x)\,e^{\frac{2\pi isj}{2n+3}}\right)_{s=1,\ldots,2n+3} \in \mathbf{C}^{2n+3},$$

and

$$(\mathbf{v}_k^j)^2 = \left(\sqrt{\frac{2k+1}{4\pi}}\,P_k^{|j|}(-x)\,e^{\frac{2\pi isj}{2n+5}}\right)_{s=1,\ldots,2n+5} \in \mathbf{C}^{2n+5},$$

contain the first $2n+3$ and second $2n+5$ components of $\mathbf{v}_k^j$ respectively. Furthermore, we will denote by $\mathbf{u}(r)$ the r-th entry of a vector $\mathbf{u}$. Let us now compute the entries of the matrix $\mathbf{P}_1\,\mathbf{B}_n\,\mathbf{F}$. Given a row $\mathbf{v}_k^j$ of $\mathbf{P}_1\mathbf{B}_n$, the r-th entry of $\mathbf{v}_k^j\,\mathbf{F}$ is given by

(i) if $1 \le r \le 2n+3$

$$\begin{aligned}
\mathbf{v}_k^j\ \mathbf{F}(r) &= \frac{1}{\sqrt{2n+3}} \sum_{s=1}^{2n+3} Y_k^j(\rho,\theta_s)\, e^{\frac{2\pi i}{2n+3}(1-r)(s-1)} \\
&= \sqrt{\frac{(2k+1)}{4\pi(2n+3)}} P_k^{|j|}(\cos\rho) \sum_{s=1}^{2n+3} e^{\frac{2\pi i}{2n+3}(js-rs+r+s-1)} \\
&= \sqrt{\frac{(2k+1)}{4\pi(2n+3)}} P_k^{|j|}(\cos\rho)\, e^{\frac{2\pi i}{2n+3}(r-1)} \sum_{s=1}^{2n+3} e^{\frac{2\pi i}{2n+3}(j-r+1)s},
\end{aligned}$$

(ii) if $2n+4 \le r \le 4n+8$, i.e., $r = 2n+3+l$ ($l = 1,\dots,2n+5$)

$$\begin{aligned}
\mathbf{v}_k^j\ \mathbf{F}(r) &= \frac{1}{\sqrt{2n+5}} \sum_{s=1}^{2n+5} Y_k^j(\pi-\rho,\theta_s)\, e^{\frac{2\pi i}{2n+5}(1-l)(s-1)} \\
&= \sqrt{\frac{(2k+1)}{4\pi(2n+5)}} P_k^{|j|}(\cos(\pi-\rho)) \sum_{s=1}^{2n+5} e^{\frac{2\pi i}{2n+5}(js-ls+l+s-1)} \\
&= \sqrt{\frac{(2k+1)}{4\pi(2n+5)}} P_k^{|j|}(\cos(\pi-\rho))\, e^{\frac{2\pi i}{2n+5}(l-1)} \sum_{s=1}^{2n+5} e^{\frac{2\pi i}{2n+5}(j-l+1)s}.
\end{aligned}$$

Observe that

$$\sum_{s=1}^{2n+3} e^{\frac{2\pi i}{2n+3}(j-r+1)s} = \begin{cases} 2n+3 & \text{if } j-r+1 \equiv 0 \mod (2n+3), \\ 0 & \text{otherwise,} \end{cases} \tag{15}$$

and

$$\sum_{s=1}^{2n+5} e^{\frac{2\pi i}{2n+5}(j-l+1)s} = \begin{cases} 2n+5 & \text{if } j-l+1 \equiv 0 \mod (2n+5), \\ 0 & \text{otherwise.} \end{cases} \tag{16}$$

Let

$$a_k := \left(\frac{(2n+3)(2k+1)}{4\pi}\right)^{1/2} e^{\frac{2\pi i}{2n+3}(r-1)}, \tag{17}$$

and

$$b_k := \left(\frac{(2n+5)(2k+1)}{4\pi}\right)^{1/2} e^{\frac{2\pi i}{2n+5}(l-1)} \tag{18}$$

for $j=-k,\dots,k$ and $k=n+1,n+2$.

In view of (15) and (16), we can conclude that $\mathbf{v}_k^j\,\mathbf{F}$ has only two nonzero entries. To be precise, we obtain for $j \geq 0$

$$\mathbf{v}_k^j\,\mathbf{F}(r) = \begin{cases} a_k\,P_k^j(x) & \text{in } r = j+1, \\ b_k\,P_k^j(-x) & \text{in } r = 2n+j+4, \\ 0 & \text{otherwise,} \end{cases}$$

and for $j < 0$

$$\mathbf{v}_k^j\,\mathbf{F}(t) = \begin{cases} a_k\,P_k^{-j}(x) & \text{in } r = 2n+4+j, \\ b_k\,P_k^{-j}(-x) & \text{in } r = 4n+9+j, \\ 0 & \text{otherwise.} \end{cases}$$

For $0 \leq j \leq n$, we observe that the nonzero entries of $\mathbf{v}_k^j\,\mathbf{F}$ are located between

$$1 \leq r \leq n+1 \quad \text{and} \quad 2n+4 \leq r \leq 3n+4. \tag{19}$$

For $-n \leq j \leq -1$, nonzero entries occur at positions

$$n+4 \leq r \leq 2n+3 \quad \text{and} \quad 3n+9 \leq r \leq 4n+8. \tag{20}$$

Therefore, the only rows of $\mathbf{P}_1\mathbf{B}_n\mathbf{F}$ (for fixed j with $-n \leq j \leq n$) with nonzero entries at the same positions are the ones corresponding to the multi-indices $(n+1, j)$ and $(n+2, j)$.

For $j = \pm(n+1), \pm(n+2)$, it can be seen that

$$\mathbf{v}_{n+1}^{n+1}\mathbf{F}(r) = \begin{cases} a_{n+1}\,P_{n+1}^{n+1}(x) & \text{if } r = n+2, \\ b_{n+1}\,P_{n+1}^{n+1}(-x) & \text{if } r = 3n+5, \\ 0 & \text{otherwise,} \end{cases}$$

$$\mathbf{v}_{n+2}^{n+1}\mathbf{F}(r) = \begin{cases} a_{n+2}\,P_{n+1}^{n+1}(x) & \text{if } r = n+2, \\ b_{n+2}\,P_{n+1}^{n+1}(-x) & \text{if } r = 3n+5, \\ 0 & \text{otherwise,} \end{cases}$$

$$\mathbf{v}_{n+2}^{-(n+2)}\mathbf{F}(r) = \begin{cases} a_{n+2}\,P_{n+2}^{n+2}(x) & \text{if } r = n+2, \\ b_{n+2}\,P_{n+2}^{n+2}(-x) & \text{if } r = 3n+7, \\ 0 & \text{otherwise,} \end{cases}$$

$$\mathbf{v}_{n+1}^{-(n+1)}\mathbf{F}(r) = \begin{cases} a_{n+1}\,P_{n+1}^{n+1}(x) & \text{if } r = n+3, \\ b_{n+1}\,P_{n+1}^{n+1}(-x) & \text{if } r = 3n+8, \\ 0 & \text{otherwise,} \end{cases}$$

$$\mathbf{v}_{n+2}^{-(n+1)}\mathbf{F}(r) = \begin{cases} a_{n+2}\,P_{n+1}^{n+1}(x) & \text{if } r = n+3, \\ b_{n+2}\,P_{n+1}^{n+1}(-x) & \text{if } r = 3n+8, \\ 0 & \text{otherwise,} \end{cases}$$

$$\mathbf{v}_{n+2}^{n+2}\mathbf{F}(r) = \begin{cases} a_{n+2}\,P_{n+2}^{n+2}(x) & \text{if } r = n+3, \\ b_{n+2}\,P_{n+2}^{n+2}(-x) & \text{if } r = 3n+6, \\ 0 & \text{otherwise.} \end{cases} \tag{21}$$

Hence, the rows $\mathbf{v}_k^j\,\mathbf{F}$ corresponding to the first three and the last three expressions of (21) have nonzero entries at the same positions. In view of (19), (20) and (21), let $\mathbf{P}_2$ be the permutation matrix

$$\begin{aligned}\mathbf{P}_2 := [&e_{n+2}, e_{3n+5}, e_{3n+7}, e_{n+3}, e_{3n+6}, e_{3n+8}, e_{n+4}, e_{3n+9}, \dots,\\ &\dots, e_{2n+3}, e_{4n+8}, e_1, e_{2n+4}, e_2, e_{2n+5}, \dots, e_{n+1}, e_{3n+4}].\end{aligned} \tag{22}$$

Then the product $\mathbf{P}_1\mathbf{B}_n\mathbf{F}\,\mathbf{P}_2$ is the diagonal block matrix

$$\mathbf{P}_1\mathbf{B}_n\mathbf{F}\,\mathbf{P}_2 = \operatorname{diag}(A_1, A_2, B_n, B_{n-1}, \dots, B_0, \dots, B_{n-1}, B_n),$$

where $A_1, A_2 \in \mathbf{R}^{3\times 3}$, and $B_m \in \mathbf{R}^{2\times 2}$ with $m = 1, \dots, n$, are given by

$$A_1 = \begin{pmatrix} a_{n+1}\,P_{n+1}^{n+1}(x) & b_{n+1}\,P_{n+1}^{n+1}(-x) & 0 \\ a_{n+2}\,P_{n+2}^{n+1}(x) & b_{n+2}\,P_{n+2}^{n+1}(-x) & 0 \\ a_{n+2}\,P_{n+2}^{n+2}(x) & 0 & b_{n+2}\,P_{n+2}^{n+2}(-x) \end{pmatrix},$$

$$A_2 = \begin{pmatrix} a_{n+1}\,P_{n+1}^{n+1}(x) & 0 & b_{n+1}\,P_{n+1}^{n+1}(-x) \\ a_{n+2}\,P_{n+2}^{n+1}(x) & 0 & b_{n+2}\,P_{n+2}^{n+1}(-x) \\ a_{n+2}\,P_{n+2}^{n+2}(x) & b_{n+2}\,P_{n+2}^{n+2}(-x) & 0 \end{pmatrix},$$

and

$$B_m = \begin{pmatrix} a_{n+1}\,P_{n+1}^{m}(x) & b_{n+1}\,P_{n+1}^{m}(-x) \\ a_{n+2}\,P_{n+2}^{m}(x) & b_{n+2}\,P_{n+2}^{m}(-x) \end{pmatrix}.$$

In order to prove the regularity of $\mathbf{B}_n$, we now simply have to guarantee the regularity of the matrices A_1, A_2 and B_m $(m = 0, \dots, n)$. Making use of the different parity of the functions P_{n+1}^m and P_{n+2}^m $(m = 0, \dots, n)$ and bearing in mind the definition of $a_{k,j}$ and $b_{k,j}$ in (17) and (18), respectively, we are in a position to establish the equivalence of the matrices

$$B_m \quad (m = 0, \dots, n) \qquad \text{and} \qquad \begin{pmatrix} 2P_{n+1}^m(x) & 0 \\ 0 & 2P_{n+2}^m(x) \end{pmatrix}.$$

Hence, B_m will be regular if and only if $P_k^m(x) \neq 0$ for all $k = n+1, n+2$, $m = 0, \dots, n$. Expanding the determinant of A_1 by the last column and respectively the determinant of A_2 by the second column, we obtain analogously that the matrices A_1 and A_2 are regular if and only if $P_{n+2}^{n+2}(x) \neq 0$ and $P_{n+1}^{n+1}(x), P_{n+2}^{n+1}(x) \neq 0$. □

3.2 Matrix notation

Let

$$\begin{aligned}\mathbf{v}_n^T &= \left(Y_0^0(\xi), Y_1^{-1}(\xi), Y_1^0(\xi), Y_1^1(\xi), \dots, Y_n^{-n}(\xi), \dots, Y_n^n(\xi)\right),\\ \mathbf{z}_n^T &= \left(Y_{n+1}^{-(n+1)}(\xi), \dots, Y_{n+1}^{n+1}(\xi), \dots, Y_{n+l}^{-(n+l)}(\xi), \dots, Y_{n+l}^{n+l}(\xi)\right).\end{aligned}$$

Furthermore, let $\{\eta_i,\ i = 1, \dots, N\}$ and $\{\xi_j,\ j = 1, \dots, d\}$ be fundamental systems of V_n and W_n. Since the corresponding scaling and wavelet functions constitute bases of the spaces V_n and W_n, for any $F_n \in V_n$ and $G_n \in W_n$ there exist coefficient vectors $\mathbf{a}^n = (a_1^n, \dots, a_N^n) \in \mathbf{C}^N$ and $\mathbf{b}^n = (b_1^n, \dots, b_d^n) \in \mathbf{C}^d$, such that F_n and G_n admit a representation in terms of the scaling and wavelet functions as

$$\begin{aligned} F_n(\xi) &= \sum_{s=1}^{N} a_s^n \varphi_s^n(\xi) = \sum_{s=1}^{N} a_s^n \sum_{k=0}^{n} \frac{2k+1}{4\pi} P_k(\eta_s \cdot \xi) \\ &= \sum_{s=1}^{N} \sum_{k=0}^{n} \sum_{j=-k}^{k} a_s^n Y_k^{-j}(\eta_s) Y_k^j(\xi), \end{aligned} \tag{23}$$

$$\begin{aligned} G_n(\xi) &= \sum_{s=1}^{d} b_s^n \psi_s^n(\xi) = \sum_{s=1}^{d} b_s^n \sum_{k=n+1}^{n+l} \frac{2k+1}{4\pi} P_k(\xi_s \cdot \xi) \\ &= \sum_{s=1}^{d} \sum_{k=n+1}^{n+l} \sum_{j=-k}^{k} a_s^n Y_k^{-j}(\xi_s) Y_k^j(\xi). \end{aligned} \tag{24}$$

Introducing matrix notation, we obtain the following lemma

Lemma 3.4 *Let* $\mathbf{A}_n$ *and* $\mathbf{B}_n$ *be the matrices introduced in* (5) *and* (12) *and let* $F_n \in V_n$ *and* $G_n \in W_n$ *be functions with an expansion as in* (23) *and* (24). *Then*

(i) $F_n(\xi) = \sum_{s=1}^{N} a_s^n \varphi_s^n(\xi) = \bar{\mathbf{y}}_n^T \mathbf{A}_n \mathbf{a}^n$ and $\mathrm{G}(\xi) = \sum_{r=1}^{d} \mathrm{b}_r^n \psi_r(\xi) = \bar{\mathbf{z}}_n^T \mathbf{B}_n \mathbf{b}^n$.

(ii) $(\varphi_1^n, \dots, \varphi_N^n) = \mathbf{A}_n^* \mathbf{y}_n^T$ and $(\psi_1^n, \dots, \psi_d^n) = \mathbf{B}_n^* \mathbf{z}_n^T$.

3.3 Two-scale relations and decomposition

In this section, we will only study the case $l = 2$. Let $F_{n+2} \in V_{n+2}$. Furthermore, consider $(n+3)^2$ points $\{\eta_i,\ i = 1, \dots, (n+1)^2\} \cup \{\xi_j,\ j = 1, \dots, 4n+8\}$ distributed on $n+3$ latitudes $z_k = \cos \rho_k$ $(k = 1, \dots, n+3)$, where the latitude at height z_k contains $(2k+1)$ equidistantly distributed points and the last two latitudes are chosen symmetric to the equator, i.e., $\cos \rho_{n+2} = -\cos \rho_{n+3}$. On account of Theorem 2.1 in [3], the totality of these points constitutes a F.S. for $V_{n+2} = V_n \oplus W_n$. Moreover, due to Theorem 3.3 of the previous section, we are in a position to affirm that the points $\{\xi_j,\ j = 1, \dots, 4n+8\}$ constitute a F.S. of W_n. In this section, we work out the relationship between the coefficient vectors $\mathbf{a}^{n+2}$, $\mathbf{a}^n$ and $\mathbf{b}^n$ in the so-called two-scale relation

$$\begin{aligned} F_{n+2}(\xi) &= \sum_{k=1}^{(n+3)^2} a_k^{n+2} \varphi_k^{n+2}(\xi) = \sum_{k=1}^{(n+1)^2} a_k^n \varphi_k^n(\xi) + \sum_{k=1}^{4n+8} b_k^n \psi_k^n(\xi) \\ &= F_n(\xi) + G_n(\xi), \end{aligned} \tag{25}$$

where $F_n \in V_n$, $G_n \in W_n$ and the functions $\{\varphi_i^n,\ i = 1,\ldots,N\}$ and $\{\psi_j^n,\ j = 1,\ldots,4n+8\}$ are the scaling and wavelet functions corresponding to the fundamental systems $\{\eta_i,\ i=1,\ldots,N\}$ and $\{\xi_j,\ j=1,\ldots,4n+8\}$.
Using the matrix notation introduced above, equation (25) can be rewritten as

$$\bar{\mathbf{y}}_{n+2}^T \mathbf{A}_{n+2}\mathbf{a}^{n+2} = \bar{\mathbf{y}}_n^T \mathbf{A}_n \mathbf{a}^n + \bar{\mathbf{z}}_n^T \mathbf{B}_n \mathbf{b}^n.$$

That is

$$\mathbf{A}_{n+2}\ \mathbf{a}^{n+2} = \begin{pmatrix} \mathbf{A}_n & \\ & \mathbf{B}_n \end{pmatrix} \begin{pmatrix} \mathbf{a}^n \\ \mathbf{b}^n \end{pmatrix}. \tag{26}$$

The following lemma establishes how in view of (25) we can decompose a function of V_{n+2} into wavelets of W_n and scaling functions of V_n.

Lemma 3.5 *Let the scaling functions $\{\varphi_i^n,\ i = 1,\ldots,N\}$, the wavelets $\{\psi_j,\ j = 1,\ldots,4n+8\}$ and the corresponding matrices $\mathbf{A}_{n+2}, \mathbf{A}_n$ and $\mathbf{B}_n$ be based on a fundamental system $\{\eta_i,\ i=1,\ldots,N\}\cup\{\xi_j,\ j=1,\ldots,4n+8\}$ of the form presented in Theorem 3.3.*

1. *(Reconstruction) Let the coefficient vectors $\mathbf{a}^n$ and $\mathbf{b}^n$ in (25) be given. Then*

$$\mathbf{a}^{n+2} = \mathbf{A}_{n+2}^{-1} \begin{pmatrix} \mathbf{A}_n & \\ & \mathbf{B}_n \end{pmatrix} \begin{pmatrix} \mathbf{a}^n \\ \mathbf{b}^n \end{pmatrix}. \tag{27}$$

2. *(Decomposition) Let the coefficient vector $\mathbf{a}^{n+2}$ in (25) be given. Then*

$$\begin{pmatrix} \mathbf{a}^n \\ \mathbf{b}^n \end{pmatrix} = \begin{pmatrix} \mathbf{A}_n^{-1} & \\ & \mathbf{B}_n^{-1} \end{pmatrix} \mathbf{A}_{n+2}\ \mathbf{a}^{n+2}.$$

The proof follows directly from (26). Given a fundamental system defined as in Theorem 3.3, we can give the explicit expression of $\mathbf{B}_n^{-1}$.

Theorem 3.6 *Let $\mathbf{B}_n$ be the matrix in (12) corresponding to the fundamental system defined in Theorem 3.3 with $x = \cos\rho$. Its inverse $\mathbf{B}_n^{-1}$ is given by $\mathbf{F}\,\mathbf{P}_2\,\mathbf{C}\,\mathbf{P}_1$, where $\mathbf{F}$, $\mathbf{P}_1$ and $\mathbf{P}_2$ are as in (13), (14) and (22) and $\mathbf{C}$ is the diagonal block matrix*

$$\mathit{diag}\ (D_1, D_2, C_n, C_{n-1}, \ldots, C_0, \ldots, C_{n-1}, C_n)$$

with

$$C_m = \begin{pmatrix} \dfrac{a_{n+1}^{-1}}{2\,P_{n+1}^m(x)} & \dfrac{a_{n+2}^{-1}}{2\,P_{n+2}^m(x)} \\ \dfrac{b_{n+1}^{-1}}{2\,P_{n+1}^m(-x)} & \dfrac{b_{n+2}^{-1}}{2\,P_{n+2}^m(-x)} \end{pmatrix} \in \mathbf{R}^{2\times 2},$$

$$D1 = \begin{pmatrix} \dfrac{a_{n+1}^{-1}}{2\,P_{n+1}^{n+1}(x)} & \dfrac{-a_{n+2}^{-1}}{2\,P_{n+2}^{n+1}(-x)} & 0 \\ \dfrac{b_{n+1}^{-1}}{2P_{n+1}^{n+1}(x)} & \dfrac{b_{n+2}^{-1}}{2\,P_{n+2}^{n+1}(-x)} & 0 \\ \dfrac{-a_{n+2}\ a_{n+1}^{-1}\ b_{n+2}^{-1}}{2\ P_{n+1}^{n+1}(x)} & \dfrac{b_{n+2}^{-1}}{2\ P_{n+2}^{n+1}(-x)} & \dfrac{b_{n+2}^{-1}}{P_{n+2}^{n+2}(-x)} \end{pmatrix} \in \mathbf{R}^{3\times 3},$$

and

$$D_2 = \begin{pmatrix} \frac{a_{n+1}^{-1}}{2\ P_{n+1}^{n+1}(x)} & \frac{-a_{n+2}^{-1}}{2\ P_{n+2}^{n+1}(-x)} & 0 \\ \frac{-a_{n+2}\ a_{n+1}^{-1}\ b_{n+2}^{-1}}{2\ P_{n+2}^{n+1}(-x)} & \frac{b_{n+2}^{-1}}{2\ P_{n+2}^{n+1}(-x)} & \frac{b_{n+2}^{-1}}{P_{n+2}^{n+2}(-x)} \\ \frac{b_{n+1}^{-1}}{2\ P_{n+1}^{n+1}(x)} & \frac{b_{n+2}^{-1}}{2\ P_{n+2}^{n+1}(-x)} & 0 \end{pmatrix} \in \mathbf{R}^{3\times 3}.$$

The constants a_k and b_k for $k = n+1, n+2$ and $j = -k, \ldots, k$ are defined as in (17) *and* (18).

The analysis of other values of l is still a question of ongoing research. Also the question of which latitudes to choose to obtain well-conditioned Riesz bases remains open and will be our main focus of interest in the future.

References

[1] Fischer, B., Prestin, J.: *Wavelets based on orthogonal polynomials,* Math. Comp, **66**, (1997), 1593–1618.

[2] Freeden, W., Gervens, T. and Schneider, M.: *Constructive Approximation on the Sphere: with Applications to Geomathematics,* Clarendon Press: Oxford, (1998).

[3] Golitschek M. v. and Light. W.A.: *Interpolation by Polynomials and Radial Basis Functions on Spheres,* Constr. Approx., **17**,(2001), 1–18.

[4] Müller, C.: *Spherical Harmonics,* Springer: Berlin, Heidelberg, New York, (1966).

[5] Reimer, M.: *Constructive Theory of Multivariate Functions: with an application to tomography,* Mannheim; Wien; Zürich: BI-Wiss.-Verl., (1990).

[6] Sündermann, B.: Projektionen auf Polynomräumen in mehreren Veränderlichen, Diss. Dortmund (1983).

[7] Xu, Y.: *Polynomial Interpolation on the unit sphere,* submitted.

Department of Mathematics
University of Lübeck
Wallstraße 40
D–23560 Lübeck, Germany
Email address: fernande@math.uni-luebeck.de

Advanced Problems in Constructive Approximation
(Eds.) M.D. Buhmann and D.H. Mache
International Series of Numerical Mathematics Vol. 142, 53–62

A Shooting Method for Symbolic Computation of Splines

Christoph Fredebeul

Abstract

Similar to the case of linear ODEs, where boundary value problems are turned into initial value problems by application of the shooting method, an algorithm is presented allowing recursive calculation of the polynomials that piecewise determine a spline. Tests indicating the usefulness of the new algorithm are given.

1 Introduction

When attempting to calculate a spline of given degree that should interpolate a given set of data, two approaches are under consideration. From the numerical point of view, having to deal with roundoff errors and stability, B-splines are preferred [1]. In case of symbolic computation, e.g. with MAPLE, one is interested in calculating the polynomials defining the spline piecewise. Their coefficients result as the solution of a large system of linear equations. Now, if the amount of data is large (more than 100, say), this approach becomes inefficient, although special structures of the data (equidistant grid, rational numbers) are exploited.

In this paper an algorithm based on a recursion is presented that avoids solving a linear system of equations. Hence, the storage requirement is drastically reduced. This is of particular importance in the case of symbolic computation, since the length of a number may vary during the calculation due to exact arithmetic! Similar to the case of linear ordinary differential equations, where boundary value problems are turned into initial value problems by application of the shooting method, the presented algorithm first calculates basic solutions all satisfying the left boundary conditions. Now, the desired spline results from a linear combination of these basic solutions. The free parameters are determined by the boundary conditions of the right end of the interval. The efficiency of the new algorithm compared to the one given in MAPLE is documented. The source code is listed in the appendix.

2 Symbolic shooting for splines

Let $d, N \in \mathbf{N}$, $d << N$, and a set of data $D := \{(x_k, y_k) \in \mathbf{R}^2 | k = 0, \ldots, N\}$ be given. Now, a function $s \in C^{d-1}[x_0, x_N]$ is called interpolating spline (of degree d), if $s(x_k) = y_k$, $k = 0, \ldots, N$, and $s|_{[x_{k-1}, x_k]} \in \Pi_d$, $k = 1, \ldots, N$ holds, i.e.,

$$s|_{[x_{k-1},x_k]}(x) \;=\; p_{[k]}(x) \;=\; \sum_{i=0}^{d} a_{[k],i} x^i, \quad 1 \le k \le N.$$

However, to be uniquely determined, $d-1$ additional conditions have to be posed to s. Here, instead of discussing all possibilities, we restrict ourselves to the cases given in MAPLE. For odd d, the resulting s are known as natural splines, i.e., determined by the boundary conditions

$$s^{(l)}(x_k) = 0, \quad l = \frac{d-1}{2} + 1, \ldots, d-1, \quad k = 0, N.$$

For even d, a similar (but nonsymmetric) set of conditions is taken.

For simplicity, we present the algorithm for the case of d odd, $d \ge 3$. The handling of the exceptions $d = 1, 2$ as well as the case of d even may be obtained directly from the source code.

The basic idea

As we have seen above, in each interval $I_k := [x_{k-1}, x_k]$ a spline of degree d is represented by a polynomial $p_{[k]}$ of degree d. Hence, due to the smoothness conditions in x_k and the interpolation conditions in x_k and x_{k+1}, the polynomial $p_{[k+1]}$ is given by

$$p_{[k+1]}(x) := p_{[k]}(x) + c_{[k]}(x - x_k)_+^d, \quad c_{[k]} := \frac{y_{k+1} - p_{[k]}(x_{k+1})}{(x_{k+1} - x_k)^d}. \tag{1}$$

However, there is one obstacle preventing us from using (1) directly as a recursion. The first polynomial $p_{[1]}$ is not uniquely determined by the boundary conditions in x_0 and the interpolation conditions in x_0 and x_1. It still possesses $\delta := (d-1)/2$ degrees of freedom, $\alpha := (\alpha_1, \ldots, \alpha_\delta)$ say. Since the coefficients $a_{[1]i}$, $i = 0, \ldots, d$, of $p_{[1]}$ depend linearly on α we have

$$a_{[1]i}(\alpha) = a_{[1]i,0} + \sum_{l=1}^{\delta} \alpha_l a_{[1]i,l} \tag{2}$$

and hence

$$p_{[1]} = p_{[1]0} + \sum_{l=1}^{\delta} \alpha_l p_{[1]l} \tag{3}$$

with uniquely determined polynomials $p_{[1]l}$, $l = 0, \dots, \delta$. Now, supposing that $p_{[k]}$ depends linearly on α, we see from (1) that the same is true for $c_{[k]}$, i.e.

$$c_{[k]} = c_{[k]0} + \sum_{l=1}^{\delta} \alpha_l c_{[k]l}, \tag{4}$$

with

$$c_{[k]0} := \frac{y_{k+1} - p_{[k]0}(x_{k+1})}{(x_{k+1} - x_k)^d}, \quad c_{[k]l} := \frac{-p_{[k]l}(x_{k+1})}{(x_{k+1} - x_k)^d}, \quad l = 1, \dots, \delta. \tag{5}$$

Hence, by induction it follows that

$$p_{[k]} = p_{[k]0} + \sum_{l=1}^{\delta} \alpha_l p_{[k]l}, \quad k = 1, \dots, N. \tag{6}$$

with uniquely determined polynomials $p_{[k]l}$. Now we are able to exploit the recursive structure indicated by (1).

The algorithm

For a more compact presentation, we introduce some abbreviations. For given data $\{(x_k, y_k),\ k = 0, \dots, N\}$, let us denote

$$m(x) := (1, x, \dots, x^d)^T \tag{7}$$

$$\text{and} \quad A_k := \left(a_{[k],i,l}\right), \quad i = 0, \dots, d, \quad l = 0, \dots, \delta, \tag{8}$$

$$\bar{\alpha} := (1, \alpha_1, \dots, \alpha_\delta)^T \tag{9}$$

$$a_k := A_k \bar{\alpha}, \tag{10}$$

$$t_k := \left(\binom{d}{i}(-x_k)^{d-i}\right), \quad i = 0, \dots, d, \tag{11}$$

$$\omega_k := \frac{1}{(x_{k+1} - x_k)^d}, \tag{12}$$

for $k = 1, \dots, N$. Observe that $p_{[k]}(x) = (m(x))^T A_k \bar{\alpha}$ and $(x - x_k)^d = (m(x))^T t_k$. Now, the algorithm consists of the following steps.

1. Let $\alpha := (a_{[1]2}, \dots, a_{[1]\delta}, a_{[1]d})$, $d > 3$ odd, resp. $\alpha = (a_{[1]3})$, $d = 3$ the free coefficients of $p_{[1]}$. Determine A_1 from the initial conditions $p_{[1]}^{(l)}(x_0) = 0$, $l = \delta + 1, \dots, d - 1$, and the interpolatory conditions $p_{[1]}(x_k) = y_k$, $k = 0, 1$.

2. For $k = 1, \ldots, N-1$, define A_{k+1} recursively by

$$A_{k+1} := A_k + \omega_k t_k \left(y_{k+1} e_1^T - (m(x_{k+1}))^T A_k \right) \tag{13}$$

with $e_1 := (1, 0, \ldots, 0)^T \in \mathbf{R}^{\delta+1}$. It is not necessary to store the A_k, $k = 2, \ldots, N-1$. Only A_1 and A_N are needed.

3. Determine α from the boundary conditions $p_N^{(l)}(x_N) = 0$, $l = \delta+1, \ldots, d-1$, where $p_{[N]}(x) = (m(x))^T A_N \bar{\alpha}$.

4. Finally, starting with $a_1 := A_1 \bar{\alpha}$, we recursively gain

$$a_{k+1} := a_k + \omega_k \left(y_{k+1} - (m(x_{k+1}))^T a_k \right) t_k \ , \tag{14}$$

$k = 1, \ldots, N-1$, and hence the desired coefficients of the polynomials $p_{[k]} = (m(x))^T a_k$, $k = 1, \ldots, N$.

All these calculations can be done efficiently using standard MAPLE commands, e.g. from the Linalg-package. Symbolic computation is useful especially in steps 1 and 3, whereas it does not seem to be required in steps 2 and 4. But the opposite is true. Rewriting (14) yields

$$a_{k+1} = \left(I - \omega_k t_k (m(x_{k+1}))^T \right) a_k + \omega_k y_{k+1} t_k$$

which is stable iff arbitrary products $\prod_{k=i}^{j} M_k$, $1 \le i \le j \le N$, of the matrices

$$M_k := \left(I - \omega_k t_k (m(x_{k+1}))^T \right)$$

are bounded. Computations of the eigenvalues of some products in the case of an equidistant grid indicate that this is not true. With growing length of the product one eigenvalue also grows rapidly, giving rise to growing errors in numerical computation. Indeed, this approach is highly unstable when floating point arithmetic is used. But, of course, no instability arises in case of symbolic computation. Hence, as demonstrated in the next section, this algorithm is useful in the case that exact arithmetic is demanded.

3 Comparison with MAPLE

From a careful examination of the new algorithm `fastspline` (see Appendix A) with respect to the expressions that have to be calculated we see that the asymptotic computatitional cost is $\mathcal{O}(\frac{3}{2}Nd^2)$. Since the structure of the linear system is exploited in Maple, this cost is about $\mathcal{O}(4Nd^2)$. Roughly, we may expect an average factor of about 3. In what follows we compare the run times (given in seconds) of two test examples calculated for different d and N. All calculations were done on a Pentium III 650 MHz PC with 128 MB RAM, using MAPLE V, R4.

As a first example, the function $f(x) = x^5$ was evaluated on the grid $x_k := k/N$, $k = 0, \ldots, N$, all done in exact arithmetic. In Table 1, the run times of **fastspline** versus **spline** are given. Here, the average of the new algorithm is better than expected; it even grows with increasing N, probably as a result of the overhead in storing numbers of varying length (due to the use of exact arithmetic). Due to their direct calculation, especially the values of the new algorithm with respect to degrees one and two are drastically better than those from **spline**.

Table 1: Run times of **spline** (upper) and **fastspline** (lower), $f(x) = x^5$

$N \ddots d$	1	2	3	4	5	6	7
25	0.055	0.170	0.330	0.610	0.960	1.505	2.180
	0.005	0.105	0.235	0.250	0.345	0.395	0.525
50	0.230	0.555	1.250	2.270	3.520	6.500	10.165
	0.010	0.180	0.500	0.570	0.790	1.115	1.725
100	0.880	2.200	4.815	8.750	15.205	28.650	50.480
	0.020	0.360	1.100	1.355	2.720	4.900	8.225
200	3.485	9.265	20.035	36.080	70.685	135.170	251.710
	0.045	0.725	2.750	4.370	12.470	26.985	52.425
400	14.405	39.615	87.360	165.000	375.245	766.715	1498.210
	0.140	1.520	8.865	16.235	78.965	187.720	377.675
800	60.910	176.740	414.050	827.820	2226.729		
	0.315	3.360	30.910	81.110	559.484		

As a second example we took the function $g(x) = \sin(x)$. Hence, the f-values are no longer rational numbers and continuing the calculation with symbolic expressions produces ugly high run times in MAPLE (see Table 2).

Table 2: Run times of **spline** (upper[1]) and **fastspline** (lower row), $f(x) = sin(x)$, symbolic calculation

$N \ddots d$	1	2	3	4	5	6	7
25	0.875	10.265	43.235	106.352	260.384	421.240	743.773
	0.199	.395	1.355	1.980	3.135	5.020	6.185
50	3.592	35.395	179.833	711.383	2140.325	—[1]	—[1]
	0.444	1.935	6.855	17.180	23.495	79.170	213.665
100	18.317	217.533	—[1]	—[1]	—[1]		
	0.514	5.230	63.839	211.090	1133.008		
200	—[1]	—[1]	—[1]				
	0.950	21.704	1748.510				

Hence, we evaluate the f-values as real numbers to a given precision and convert them into rational numbers without loss of accuracy. In Table 3, the run

[1] In the case $N = 100$, $d = 3$, and $N = 200$, $d = 1$, the following MAPLE error message occurs: *Error, (in collect/series) too many levels of recursion*

times are given for different d and N after approximating the f-values within 10 digits (MAPLE-standard) of accuracy using MAPLE's `evalf` and `convert` routines. Again, there is an obvious advantage of `fastspline` (see `proc cover-fastspline`) compared to the rational variant of `spline` (see `proc cover-spline`), this time concerning both d and N. Even in comparison to `spline` (using floating point arithmetic) the new algorithm yields comparable run times.

Table 3: Run times of `spline` (top): floating point arithmetic[2], `cover-spline` (middle) and `cover-fastspline` (lower row): floats converted to rational numbers, $f(x) = \sin(x)$

$N \cdot\cdot d$	1	2	3	4	5	6	7
25	0.095	0.060	0.120	0.180	0.460	0.585	0.805
	0.540	0.245	0.480	0.815	1.240	2.265	2.860
	0.010	0.145	0.285	0.375	0.430	0.480	0.660
50	0.080	0.260	0.520	0.840	1.320	1.880	2.720
	0.280	0.780	1.551	2.855	4.640	7.935	11.136
	0.015	0.225	0.635	0.730	0.995	1.385	2.069
100	0.330	0.770	1.785	3.110	4.785	6.905	9.760
	1.060	2.840	5.914	10.785	18.840	31.694	50.105
	0.030	0.450	1.383	1.905	3.500	5.695	9.190
200	1.240	3.195	6.995	12.265	19.206	28.300	40.259
	4.170	11.073	24.445	45.955	86.130	150.986	273.315
	0.065	0.995	3.915	7.325	15.620	30.085	57.680
400	5.936	15.535	32.985	58.020	92.060	138.435	199.929
	19.715	52.931	118.770	230.945	473.682	883.808	1703.076
	0.270	2.360	17.050	35.721	100.960	201.721	410.415
800	17.805	52.068	120.862	—[2]	365.878		
	61.461	182.189	472.326	1053.501	2380.871		
	0.450	3.720	67.220	197.924	655.977		

Conclusions

As been indicated by the tests above, the new algorithm is fast, especially faster than the one implemented in MAPLE and robust in the sense that it is able to solve some problems whereas Maple does not. The restriction to non-numeric computation (i.e. no floating point arithmetic should be used!) is not critical since the run times of `cover - fastspline` and `spline` are still comparable. Hence, in the case of many points, the new algorithm is very recommendable.

Acknowledgement: The author would like to thank Ralf Tenberg for his patience and his advise in many fruitful discussions.

References

[1] C. de Boor: *A Practical Guide to Splines* Springer, New York, 1978

[2] In the case $N = 800$, $d = 4$, the following MAPLE error message occurs: *Error, (in spline) unable to compute solutions*

4 Source code

```
fastspline := proc(X,Y,z,d)
  local  dm1h,dh,N,i,k,l,a,p,b,b1,m,monome,bincof,koef,loesvec,hilfmat,mip2,
         hilfvect,koef2,hilfvec2,hilf;

 monome := proc(x, d)
 local   l,mon;
    mon := linalg[vector](d+1);
    mon[1] := 1;
    for l from 1 to d do
        mon[l+1] := x*mon[l]; od;
    RETURN(evalm(mon));
 end:

# begin of the usual testings, taken from the original
Maple-"spline" ....
 if nargs = 3 then RETURN(fastspline(X,Y,z,3)) fi;
 if d = ('linear') then RETURN(fastspline(X,Y,z,1)) fi;
 if d = ('quadratic') then RETURN(fastspline(X,Y,z,2)) fi;
 if d = ('cubic') then RETURN(fastspline(X,Y,z,3)) fi;
 if d = ('quartic') then RETURN(fastspline(X,Y,z,4)) fi;
 if type([X, Y],[vector, vector])
   then N := linalg['vectdim'](X);
        if linalg['vectdim'](Y) <> N then ERROR(`incompatible dimensions`) fi
   elif type([X, Y],[list, list])
      then N := nops(X);
           if nops(Y) <> N then ERROR(`incompatible dimensions`) fi
      else ERROR(`1st and 2nd arguments must be two lists or two vectors`) fi;
 if not type(z,name)
    then ERROR(`3rd argument (the variable) must be a name`) fi;
 if not type(d,posint)
    then ERROR(`4th argument (the degree) must be a positive integer`) fi;
 N := N-1;
 for i to N do
    if type(X[i],numeric) and type(X[i+1],numeric) and X[i+1] <= X[i]
        then ERROR(`X values (knots) must be in strictly ascending`) fi
 od;
 # .... end of the usual testings, taken from the original
 Maple-"spline"

 # begin of special case d=1 ....
  if d = 1 then
     p:=[];
     for i from 1 to N-1 do
         hilf := (Y[i+1]-Y[i])/(X[i+1]-X[i]);
         p := [op(p), [z < X[i+1], simplify(hilf*(z-X[i])+Y[i])]];
     od;
     hilf := (Y[N+1]-Y[N])/(X[N+1]-X[N]);
     p := [op(p), [simplify(hilf*(z-X[N])+Y[N])]];
     if nops(p) = 1 then RETURN(op(p[1]))
                    else RETURN('piecewise'(seq(op(i), i=p))) fi:
  fi;
 # .... end of special case d=1
```

```
# begin of special case d=2 .....
 if d = 2 then
    hilf := (Y[2]-Y[1])/(X[2]-X[1])^2;
    b := evalm([Y[1]+hilf*X[1]^2, -2*X[1]*hilf, hilf]);
    mip2:=monome(X[2],d);
    bincof:=([seq((-1)^(d-k)*binomial(d,k), k=0..d)]);
    m := linalg[vector]([seq(z^k, k=0..d)]);
    p := [];
    for i from 1 to N-1 do
         p := [op(p), [z < X[i+1], linalg[multiply](m,b)]];
         for l from 1 to d+1 do
              hilfvec2[l] := mip2[d+2-l]*bincof[l]; od;
         mip2:=monome(X[i+2],d);
         hilf := linalg[multiply](mip2,b);
         hilf := (Y[i+2]-hilf)/(X[i+2]-X[i+1])^d;
         for l from 1 to d+1 do
              b[l] := b[l]+hilfvec2[l]*hilf; od:
    od;
    p := [op(p), [linalg[multiply](m,b)]];
    if nops(p) = 1 then RETURN(op(p[1]))
                  else RETURN('piecewise'(seq(op(i), i=p))) fi:
 fi;
# .... end of special case d = 2

# begin of general case d >= 3 .... #     begin of step 1
(left boundary conditions) ....
 p := convert([seq(a[i]*z^i, i=0..d)], `+`);
 dm1h := floor((d-1)/2);
 dh := floor(d/2);
 koef := seq(a[k], k=2..dm1h), a[d];
 loesvec := solve({seq(simplify(subs(z=X[1], diff(p,z$k))), k=dm1h+1..d-1)},
                  {seq(a[k],k=dm1h+1..d-1)});
 p := subs(op(loesvec), p);
 loesvec := solve({seq(subs(z=X[k], p)-Y[k],k=1..2)}, {a[0], a[1]});
 p := subs(op(loesvec), p);
 m := linalg[genmatrix]({p}, [koef], hilfvect);
 m := linalg[augment](-hilfvect, m);
 hilfmat := linalg[matrix](d+1, dm1h+1);
 for k from 1 to dm1h+1 do
      for l from 0 to d do
          hilfmat[l+1,k] := coeff(m[1,k], z, l); od:
 od:
 b1 := evalm(hilfmat);
#       .... end of step 1    (left boundary conditions)

#       begin of step 2  (first main loop) ....
 b := evalm(b1);
 bincof := ([seq((-1)^(d-k)*binomial(d,k), k=0..d)]);
 mip2 := monome(X[2], d);
 for i from 1 to N-1 do
    for l from 1 to d+1 do
        hilfvec2[l] := mip2[d+2-l]*bincof[l]; od;
    mip2 := monome(X[i+2],d);
    hilfvect := linalg[multiply](mip2,b);
```

```
    hilfvect[1] := hilfvect[1]-Y[i+2];
    for l from 1 to d+1 do
        for k from 1 to dm1h+1 do
            hilfmat[l,k] := hilfvec2[l]*hilfvect[k]; od:
    od:
    b := linalg[matadd](b, hilfmat, 1, -1/(X[i+2]-X[i+1])^d);
od;
#       .... end of step 2    (first main loop)

#       begin of step 3  (right boundary conditions) ....
koef2 := linalg[vector]([1,koef]);
m := linalg[vector]([seq(z^k, k=0..d)]);
p := linalg[multiply](m,linalg[multiply](b,koef2));
loesvec := solve({seq(simplify(subs(z=X[N+1], diff(p,z$k))), k=dh+1..d-1)},
                 {koef});
hilfvect := subs(op(loesvec), evalm(koef2));
#       .... end of step 3    (right boundary conditions)

#       begin of step 4  (second main loop) ....
b := linalg[multiply](b1,hilfvect);
mip2 := monome(X[2],d);
p := [];
for i from 1 to N-1 do
    p := [op(p),[z < X[i+1],linalg[multiply](m,b)]];
    for l from 1 to d+1 do
        hilfvec2[l] := mip2[d+2-l]*bincof[l]; od;
    mip2 := monome(X[i+2],d);
    hilf := linalg[multiply](mip2,b);
    hilf := (Y[i+2]-hilf)/(X[i+2]-X[i+1])^d;
    for l from 1 to d+1 do
        b[l] := b[l]+hilfvec2[l]*hilf; od:
od;
p := [op(p), [linalg[multiply](m,b)]];
if nops(p) = 1 then op(p[1]) else 'piecewise'(seq(op(i), i=p)) fi:
end:
#       .... end of step 4    (second main loop) # .... end
of general case d >= 3 # .... end of "fastspline"

# Here are the additional test routines:

#          convert floating point numbers into rational
numbers #          and start "fastspline" afterwards
cover_fastspline := proc(X,Y,z,d)
    local Xr, Yr;
    Xr:=convert(X,rational,exact);
    Yr:=convert(Y,rational,exact);
    evalf(fastspline(Xr,Yr,z,d));
end:

#          the same as above for "spline" cover_spline :=
proc(X,Y,z,d)
    .....

#          generate run times by applying the following
procedure #          in the case of non-rational data ....
```

```
zeiten := proc(knoten,maxgrad)
    local   grad, xdat, ydat, k, n, ab, s3, sc3, fs3, zeit;
    zeit:=linalg[matrix](3,maxgrad):
    n:=knoten:
    for grad from 1 to maxgrad do
        xdat:=evalf([seq((k-1)/n,k=1..n+1)]):
        ydat:=evalf([seq(sin(xdat[k]),k=1..n+1)]):
        ab:=time():
        s3:=spline(xdat,ydat,x,grad):
        zeit[1,grad]:=time()-ab;
        ab:=time():
        sc3:=cover_spline(xdat,ydat,x,grad):
        zeit[2,grad]:=time()-ab;
        ab:=time():
        fs3:=cover_fastspline(xdat,ydat,x,grad):
        zeit[3,grad]:=time()-ab;
    od:
    RETURN(evalm(zeit));
end:

# .... and by a similar procedure in the case of rational
arithmetic: # drop the evalfs, evaluate x^5 instead of
sin(x), # drop cover_spline and the following two lines #
and call fastspline instead of cover_fastspline.
```

Department of Applied Mathematics
University of Dortmund
D-44221 Dortmund, Germany
Email address: christoph.fredebeul@math.uni-dortmund.de

Advanced Problems in Constructive Approximation
(Eds.) M.D. Buhmann and D.H. Mache
International Series of Numerical Mathematics Vol. 142, 63–78

Error Estimates for the Carathéodory-Fejér Method in Polynomial Approximation

Manfred Hollenhorst

Abstract

In the Carathéodory-Fejér method one computes – starting from a complex power series absolutely convergent on the unit disk – a polynomial which is (hopefully) a better approximation to the function given by the power series than the truncated power series (with respect to the supremum norm). In this article we show that – under fairly restrictive conditions on the coefficients of the power series – the Carathéodory-Fejér method gives an asymptotically optimal approximation and in some cases is really a better approximation than the truncated power series.

1 Introduction

The Carathéodory-Fejér method in polynomial approximation can be regarded as a nonlinear approximation operator. We start from the development of a function f into a power series:

$$f(z) = c_0 + c_1 z + c_2 z^2 + \cdots \quad , \tag{1}$$

which we assume to be absolutely convergent on S^1, the boundary of the unit disc in the complex plane $\mathbf{C}$, and from the polynomial p_n of degree n which forms the beginning of this series:

$$p_n(z) = c_0 + c_1 z + c_2 z^2 + \cdots + c_n z^n \quad .$$

After choosing n and m the coefficients $c_{n+1}, \ldots, c_{n+m+1}$ are utilized to modify the coefficients of p_n, and f is then approximated by the modified polynomial

$C_{n,m}$ (the Carathéodory-Fejér approximating polynomial). We consider the error of this approximation with respect to the supremum norm on S^1, namely

$$||F|| = \sup\{|F(z)| : |z| = 1\} \; .$$

We define the minimum deviation with respect to this norm by

$$E_n(f) = \min\{||f - P|| : \; P \in \Pi_n\} \quad ,$$

where Π_n denotes the space of polynomials of degree $\leq n$.

In this article an estimate is given which shows that the error norm in the approximation by $C_{n,m}$ is asymptotically optimal, i.e.,

$$||f - C_{n,m}|| \leq E_n(f)(1 + o(1))$$

if $m \to \infty$ as $n \to \infty$, under the condition that $|c_{n+j}| \leq g^{j-1}|c_{n+1}|$ for all positive integers j, where

$$g \leq (\sqrt{13} - 1)/6 \approx 0.43426 \quad .$$

We also prove that if $c_k \geq 0$ for all $k > n$, if $0 < c_{n+j+1} \leq \gamma c_{n+j}$ for $j = 1, 2, \cdots, m$ with some $\gamma < 1$, and if n sufficiently large, but $m(n) = o(n/\log n)$, then

$$||f - C_{n,m}|| < ||f - p_n|| \; .$$

This means that under the fairly restrictive assumptions made above we can guarantee that $C_{n,m}$ is a better approximation to f than p_n, i.e., the truncated power series of f.

2 Carathéodory-Fejér approximation

The starting point of this method is a theorem about the approximation of complex polynomials by meromorphic functions of a certain class, namely the famous

Theorem 2.1 (Carathéodory and Fejér [1])
Let $f_{n,m}$ be a polynomial with complex coefficients of the form

$$f_{n,m}(z) = c_{n+1}z^{n+1} + c_{n+2}z^{n+2} + \cdots + c_{n+m+1}z^{n+m+1}$$

where n and m are positive integers. Let G_n be the set of functions which are defined and continuous on $A = \{z : z \in \mathbf{C}$ and $|z| \geq 1\}$ and which have a development of the form

$$g(z) = \sum_{k=-\infty}^{n} g_k z^k$$

uniformly convergent on A.

(i) *Then a function g_* exists in G_n with*

$$||f_{n,m} - g_*|| = \inf\{||f_{n,m} - g|| : g \in G_n\} \quad .$$

(ii) *A function g_* is a best approximation to $f_{n,m}$ from G_n in the sense of (i) if and only if*

$$f_{n,m}(z) - g_*(z) = \epsilon_{n,m} z^{n+m+1} \frac{\overline{q}_0 z^m + \overline{q}_1 z^{m-1} + \cdots + \overline{q}_m}{q_0 + q_1 z + \cdots + q_m z^m} \tag{2}$$

holds for all $z \in S^1$, where $\epsilon_{n,m}$ is one of the numbers with largest possible modulus fulfilling (2).

Remark 2.2 *From* (2) *we see that $f_{n,m} - g_*$ is of constant modulus $|\epsilon_{n,m}|$ on S^1 and can be written as $\epsilon_{n,m} z^{n+m+1}$ times the reciprocal of a Blaschke product. It has poles only in the interior of the unit circle.*

Let us return to the case of a function f with a development (1) absolutely convergent on S^1. Then the Carathéodory-Fejér approximation method proceeds as follows:

1.) The part $c_{n+m+2} z^{n+m+2} + c_{n+m+3} z^{n+m+3} + \cdots$ is neglected.

2.) From $f_{n,m}(z) = c_{n+1} z^{n+1} + c_{n+2} z^{n+2} + \cdots + c_{n+m+1} z^{n+m+1}$ we compute g_*, the best approximation from G_n, by solving a singular value problem. But for polynomial approximation of course only the terms with nonnegative exponents of z can be used. So if

$$g_*(z) = \sum_{k=-\infty}^{n} a_k z^k$$

is a solution of the above Carathéodory-Fejér problem, we utilize

$$\tilde{g}_{n,m}(z) = \sum_{k=0}^{n} a_k z^k$$

and discard

$$H_{n,m}(z) = \sum_{k=-\infty}^{-1} a_k z^k$$

3.) The $n+1$ terms from the beginning of the series (1) are then added to $\tilde{g}_{n,m}(z)$ in order to form the Carathéodory-Fejér approximation polynomial to f:

$$C_{n,m} := p_n + \tilde{g}_{n,m}$$

The Blaschke product form of the error in the Carathéodory-Fejér problem was – to the knowledge of the author – first exploited for polynomial approximation by S. Darlington [2].

3 Error estimate for finite series

In order to show that $C_{n.m}$ is a "good" polynomial approximation to f, we have to ensure that $H_{n,m}$ and $c_{n+m+2}z^{n+m+2}+c_{n+m+3}z^{n+m+3}+\cdots$ are negligible relative to $E_n[f)$. We first consider only the approximation of polynomials of higher degree (i.e., the $f_{n,m}$ in the Carathéodory-Fejér theorem), so we only need to estimate $||H_{n,m}||$.

We now derive three systems of equations by comparing coefficients in (2) in order to determine the coefficients of $\tilde{g}_{n,m}$ and $H_{n,m}$ explicitly. First we multiply (2) with the denominator of the right-hand side:

$$\left(\sum_{k=n+1}^{n+m+1} c_k z^k - \sum_{k=-\infty}^{n} a_k z^k\right)\sum_{j=0}^{m} q_j z^j = \epsilon_{n,m} z^{n+m+1}\sum_{j=0}^{m} \overline{q}_j z^{m-j}$$

Equating the coefficients of $z^{n+m+k+1}$ we get for $k = 0, 1, \ldots, m$:

$$c_{n+m+1}q_k + c_{n+m}q_{k+1} + \cdots + c_{n+k+1}q_m = \epsilon_{n,m}\overline{q}_{m-k} \tag{3}$$

Comparing the coefficients of z^{n+l+1} yields for $l = 0, 1, \ldots, m-1$

$$a_n q_{l+1} + a_{n-1}q_{l+2} + \cdots + a_{n+l-m+1}q_m = c_{n+1}q_l + c_{n+2}q_{l-1} + \cdots + c_{n+l+1}q_0 =: r_l \tag{4}$$

The coefficients of $z^n, z^{n-1}, \ldots$ give rise to an infinite set of recursion formulae for the coefficients a_{k-m} $(k = n, n-1, \ldots)$:

$$a_k q_0 + a_{k-1}q_1 + \cdots + a_{k-m}q_m = 0 \tag{5}$$

From (3) follows that the column vector

$$u := (q_0, q_1, \cdots, q_m, \overline{q}_0, \overline{q}_1, \cdots, \overline{q}_m)'$$

is a solution of the eigenvalue problem

$$\Gamma u = \epsilon_{n,m} u \tag{6}$$

with

$$\Gamma := \begin{pmatrix} 0 & \cdots & \cdots & 0 & 0 & \cdots & 0 & \overline{c}_{n+m+1} \\ \vdots & \vdots & & \vdots & \vdots & 0 & & \vdots \\ \vdots & \vdots & & \vdots & 0 & \overline{c}_{n+m+1} & \cdots & \overline{c}_{n+2} \\ 0 & \cdots & \cdots & 0 & \overline{c}_{n+m+1} & \overline{c}_{n+m} & \cdots & \overline{c}_{n+1} \\ 0 & \cdots & 0 & c_{n+m+1} & 0 & \cdots & \cdots & 0 \\ \vdots & 0 & & \vdots & \vdots & \vdots & & \vdots \\ 0 & c_{n+m+1} & \cdots & c_{n+2} & \vdots & \vdots & & \vdots \\ c_{n+m+1} & c_{n+m} & \cdots & c_{n+1} & 0 & \cdots & \cdots & 0 \end{pmatrix}.$$

Remark 3.1 *Equation* (6) *can also be formulated as a singular value problem with half the number of rows, but in the sequel we will exploit some well-known properties of eigenvalue problems.*

We now come to the central estimate of the neglected part of the series solution of the Carathéodory-Fejér problem.

Proposition 3.2 *Assume that*

$$f_{n,m}(z) = c_{n+1}z^{n+1} + c_{n+2}z^{n+2} + \cdots + c_{n+m+1}z^{n+m+1}$$

has complex coefficients c_k satisfying for $j = 1, \ldots, m$

$$|c_{n+1+j}| \leq g^j |c_{n+1}| \quad , \tag{7}$$

where $g \leq (\sqrt{13} - 1)/6 \approx 0.43426$, and let

$$h := g(2 - 3g^2)/(1 - 2g^2)$$

Let furthermore

$$g_*(z) = \sum_{k=-\infty}^{n} a_k z^k$$

be a solution of the Carathéodory Fejér problem for $f_{n,m}$, i.e.

$$||f_{n,m} - g_*|| = \inf\{||f_{n,m} - g|| : g \in G_n\} \quad ,$$

and let

$$H_{n,m}(z) = \sum_{k=-\infty}^{-1} a_k z^k \quad .$$

Then

$$||H_{n,m}|| \leq \frac{g\, h^{n+1}\, |c_{n+1}|}{1 - 2g - 2g^2 + 3g^3} \quad .$$

Proof. We choose

$$w := (0, \cdots, 0, 1/\sqrt{2}, 0, \cdots, 0, \frac{c_{n+1}}{\sqrt{2}\,|c_{n+1}|})'$$

where the nonzero elements are the $(m+1)$ st and the $(2m+2)$ nd components of the vector. Then we have

$$|\epsilon_{n,m}| \geq w^*\Gamma w = \frac{\overline{c}_{n+1}c_{n+1} + c_{n+1}\overline{c}_{n+1}}{2|c_{n+1}|} = |c_{n+1}| \quad . \tag{8}$$

The asterisk denotes the Hermitian transpose of a vector or matrix.

$\epsilon_{n,m}$ can be chosen positive because any complex factor of modulus 1 in a solution of (2) can be incorporated in the coefficients $q_0, q_1, \cdots, q_m$.

If we choose on the other hand

$$v := (v_0, v_1, \cdots, v_{m-1}, 0, \overline{v}_0, \overline{v}_1, \cdots, \overline{v}_{m-1}, 0)' \qquad \text{with} \qquad 2\sum_{j=0}^{m-1} v_j \overline{v}_j = 1$$

we have

$$\begin{aligned} |v^*\Gamma v| &\leq 2|c_{n+m+1}|\,|v_1\overline{v}_{m-1} + v_2\overline{v}_{m-2} + \cdots + v_{m-1}\overline{v}_1| \\ &\quad + 2|c_{n+m}|\,|v_2\overline{v}_{m-1} + v_3\overline{v}_{m-2} + \cdots + v_{m-1}\overline{v}_2| + \cdots + 2|c_{n+3}|\,|v_{m-1}\overline{v}_{m-1}| \\ &\leq |c_{n+1}|((m-1)g^m + (m-2)g^{m-1} + \cdots + 2g^3 + g^2) \\ &< |c_{n+1}|g^2\frac{d}{dg}\left(\frac{1}{1-g}\right) = |c_{n+1}|\frac{g^2}{(1-g)^2} < |c_{n+1}| \end{aligned}$$

where the last inequality holds because of $g < 1/2$.

So the eigenvector u consisting of the coefficients q_k and their complex conjugates cannot have zeroes in its $(m+1)$ st and $(2m+2)$ nd components, and we can assume $q_m = 1$. From the first equation in (3) we obtain

$$q_0 = \overline{c}_{n+m+1}/\epsilon_{n,m}$$

and consequently

$$|q_0| \leq g^m \quad .$$

$2m-2$ of the remaining equations in (5) yield the following system of linear equations:

$$(I - T)\hat{q} = \gamma$$

with

$$\hat{q} := (q_1, q_2, \cdots, q_{m-1}, \overline{q}_1, \overline{q}_2, \cdots, \overline{q}_{m-1})'$$

$$\gamma := 1/\epsilon_{n,m} \cdot (\overline{c}_{n+m}, \cdots, \overline{c}_{n+2}, c_{n+m}, \ldots, c_{n+2})'$$

$$T := \frac{1}{\epsilon_{n,m}} \begin{pmatrix} 0 & \cdots & \cdots 0 & 0 & \cdots 0 & \overline{c}_{n+m+1} \\ \vdots & \vdots & \vdots & \vdots & & \vdots \\ \vdots & \vdots & \vdots & 0 & \overline{c}_{n+m+1} \cdots & \overline{c}_{n+4} \\ 0 & \cdots & 0 & \overline{c}_{n+m+1} & \overline{c}_{n+m} \cdots & \overline{c}_{n+3} \\ 0 & \cdots 0 & c_{n+m+1} & 0 & \cdots & 0 \\ \vdots & & \vdots & \vdots & \vdots & \vdots \\ 0 & c_{n+m+1} \cdots & c_{n+4} & \vdots & \vdots & \vdots \\ c_{n+m+1} & c_{n+m} \cdots & c_{n+3} & 0 & \cdots & 0 \end{pmatrix}$$

In the following we shall denote by a vector or a matrix enclosed in vertical bars the vector or matrix of the corresponding absolute values. Comparisons between vectors and matrices, which will be denoted by $\leq$, will be carried out component by component.

From (7) and (8) we have the estimates

$$|T| \le \hat{T}$$

with

$$\hat{T} := \begin{pmatrix} 0 & \dots & \dots & 0 & 0 & \dots & 0 & g^m \\ \vdots & \vdots & & \vdots & \vdots & 0 & & \vdots \\ \vdots & \vdots & & \vdots & 0 & g^m & \dots & g^3 \\ 0 & \dots & \dots & 0 & g^m & g^{m-1} & \dots & g^2 \\ 0 & \dots & 0 & g^m & 0 & \dots & \dots & 0 \\ \vdots & 0 & & \vdots & \vdots & \vdots & & \vdots \\ 0 & g^m & \dots & g^3 & \vdots & \vdots & & \vdots \\ g^m & g^{m-1} & \dots & g^2 & 0 & \dots & \dots & 0 \end{pmatrix}$$

and

$$|\gamma| \le (g^{m-1}, g^{m-2}, \dots, g^2, g, g^{m-1}, g^{m-2}, \dots, g^2, g)' =: \hat{\gamma} \quad .$$

If we define $\hat{T} \cdot \hat{\gamma} =: \delta$, we have for $j = 1, \dots, m-1$

$$\frac{\delta_j}{\hat{\gamma}_j} = (g^m g^j + g^{m-1} g^{j-1} + \dots + g^{m-j+1} g) g^{-m+j} = g^{2j} + g^{2j-2} + \dots + g^2$$

and correspondingly for $j = m, \dots, 2m-2$

$$\frac{\delta_j}{\hat{\gamma}_j} = (g^m g^{j-m+1} + g^{m-1} g^{j-m} + \dots + g^{2m-j} g) g^{j-2m+1} = g^{2j-2m+2} + \dots + g^2 \quad .$$

For

$$\tau := \max\{\delta_j / \hat{\gamma}_j : j = 1, \dots, 2m-2\}$$

we have therefore

$$\tau \le \frac{g^2}{1-g^2} \quad .$$

So we can estimate $\hat{q}$ by the product of the inverse of the coefficient matrix and the right-hand side of the system of equations defining it:

$$|\hat{q}| = |(I-T)^{-1} \gamma| \le |\sum_{i=0}^{\infty} T^i | \hat{\gamma} \le \sum_{i=0}^{\infty} \hat{T}^i \hat{\gamma} \le \frac{1}{1-\tau} \hat{\gamma} \tag{9}$$

Explicitly we have for $j = 1, \dots, m-1$

$$|q_j| \le \frac{1-g^2}{1-2g^2} g^{m-j} \quad .$$

Now we can give an estimate for the right-hand sides in (4) :

$$|r_l| \le \sum_{j=0}^{l} |c_{n+1+j}|\,|q_{l-j}| \le |c_{n+1}|\frac{1-g^2}{1-2g^2}\sum_{j=0}^{l} g^j g^{m+j-l} < |c_{n+1}|\frac{g^{m-l}}{1-2g^2}$$

Especially we have

$$|a_n| = |r_{m-1}| \le \frac{|c_{n+1}|\,g}{1-2g^2} \quad .$$

We then determine estimates for the a_k by induction.
Let us now assume that

$$|a_{n-j}| \le \frac{|c_{n+1}|\,g\,h^j}{1-2g^2}$$

for $j = 1, 2, \ldots, l-1$ with $l \le m-1$. Then we have from (4)

$$\begin{aligned}
|a_{n-l}| &= |r_{m-1-l} - \sum_{j=0}^{l-1} a_{n-j} q_{m-l+j}| \\
&\le \frac{|c_{n+1}|\,g}{1-2g^2}\left(g^l + \sum_{j=0}^{l-1} g^j \left(1+\frac{1-g^2}{1-2g^2}\right)^j g^{l-j}\frac{1-g^2}{1-2g^2}\right) \\
&= \frac{|c_{n+1}|\,g^{l+1}}{1-2g^2}\left(1+\frac{1-g^2}{1-2g^2}\frac{\left(\frac{2-3g^2}{1-2g^2}\right)^l - 1}{1+\frac{1-g^2}{1-2g^2}-1}\right) \\
&= \frac{|c_{n+1}|\,g}{1-2g^2}g^l\left(1+\frac{1-g^2}{1-2g^2}\right)^l = \frac{|c_{n+1}|g}{1-2g^2}h^l \quad .
\end{aligned}$$

For $l = m, m+1, \ldots$ the induction proceeds as above and yields

$$\begin{aligned}
|a_{n-l}| &= |-\sum_{j=1}^{m} a_{n-l+j}\, q_{m-j}| \\
&\le \frac{|c_{n+1}|\,g}{1-2g^2}\sum_{j=1}^{m}\left(g^{l-j}\left(1+\frac{1-g^2}{1-2g^2}\right)^{l-j} g^j\frac{1-g^2}{1-2g^2}\right) \\
&= \frac{|c_{n+1}|\,g^{l+1}}{1-2g^2}\,\frac{1-g^2}{1-2g^2}\left(\frac{2-3g^2}{1-2g^2}\right)^{l-m}\frac{\left(\frac{2-3g^2}{1-2g^2}\right)^m - 1}{1+\frac{1-g^2}{1-2g^2}-1} \\
&< \frac{|c_{n+1}|\,g^{l+1}}{1-2g^2}\left(1+\frac{1-g^2}{1-2g^2}\right)^l = \frac{|c_{n+1}|\,g}{1-2g^2}h^l \quad .
\end{aligned}$$

Summing over all a_l with negative l we have

$$||H_{n.m}|| \leq \sum_{l=-\infty}^{-1} |a_l| \leq \frac{|c_{n+1}|\,g}{1-2g^2}\frac{h^{n+1}}{1-h} = \frac{|c_{n+1}|\,g\,h^{n+1}}{1-2g-2g^2+3g^3} \quad . \qquad \square$$

4 Error estimates for infinite power series

With the aid of the above estimate of the "neglected" part in the Carathéodory-Fejér problem we can prove the following result about polynomial approximations, which is a modification of theorems from the author's dissertation [5], which was written under the auspices of Prof. Dr. G. Meinardus:

Theorem 4.1 *Assume that the series* $f(z) = c_0 + c_1 z + c_2 z^2 + \cdots$ *is absolutely convergent on the unit disc* $\{z : |z| \leq 1\}$ *, let* p_n, $C_{n,m}$ *, and* $E_n(f)$ *be defined as above.*

(i) *Assume* $|c_{n+j}| \leq g^{j-1}|c_{n+1}|$ *for all positive integers* j*, where*

$$g \leq (\sqrt{13}-1)/6 \approx 0.43426 \quad ,$$

and let

$$h := g(2-3g^2)/(1-2g^2) \quad .$$

Then

$$||f - C_{n,m}|| \leq E_n(f)\left(1 + \frac{2\,g\,h^{n+1}}{1-2g-2g^2+3g^3}\right)\left(1+\frac{2g^m}{1-g}\right) \quad .$$

(ii) *If, in addition to the assumptions of (i),* $m \to \infty$ *as* $n \to \infty$*, then*

$$||f - C_{n,m}|| \leq E_n(f)(1+o(1)) \quad .$$

(iii) *If for all* $k > n$ *we have* $c_k \geq 0$ *, if* $0 < c_{n+j+1} \leq \gamma\, c_{n+j}$ *for* $j = 1, 2, \cdots, m$ *with some* $\gamma < 1$*, and if* n *sufficiently large, but* $m(n) = o(n/\log n)$*, then*

$$||f - C_{n,m}|| < ||f - p_n|| \quad .$$

Remark 4.2 *Figure 1 shows the estimate of* $||f - C_{n,m}||$ *given in (i) divided by* $E_n(f)$ *depending on* g *for degrees* $n = 5$ *(solid line),* $n = 10$ *(dot-dashed), and* $n = 15$ *(dashed) with* $m = n$.

Proof. ad (i): Let

$$f_{n,m}(z) = c_{n+1}z^{n+1} + c_{n+2}z^{n+2} + \cdots + c_{n+m+1}z^{n+m+1}$$

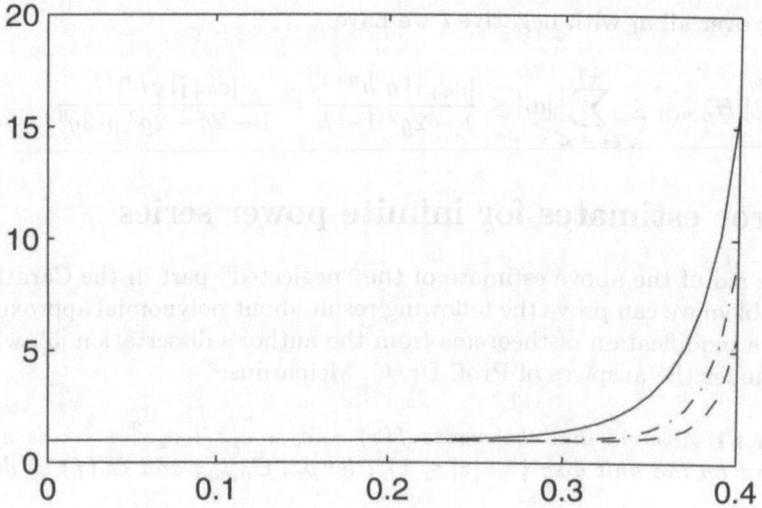

Figure 1: Estimation factor in (i) of Theorem 1

as above; then due to the constant modulus of the error curve (see, e.g., Klotz [8]) $\tilde{g}_{n,m}(z) = \sum_{k=0}^{n} a_k z^k$ is the polynomial of best approximation to $f_{n,m} - H_{n,m}$ of degree n, and the error norm is

$$||f_{n,m} - H_{n,m} - \tilde{g}_{n,m}|| = |\epsilon_{n,m}|$$

Applying the general inequality

$$|E_n(f) - E_n(g)| \leq ||f - g||$$

we get

$$|\, ||f_{n,m} - H_{n,m} - \tilde{g}_{n,m}|| - E_n(f_{n,m})| \leq ||H_{n,m}|| \quad ,$$

and consequently

$$|\, ||f_{n,m} - \tilde{g}_{n,m}|| - E_n(f_{n,m})| \leq 2||H_{n,m}|| \quad .$$

The linear functional L defined for all functions ϕ continuous on S^1 by

$$L(\phi) := \frac{1}{2\pi i} \oint_{S^1} \phi(z) z^{-n-2} dz$$

supplies a lower estimate of the approximation error, so that

$$|c_{n+1}| \leq E_n(f_{n,m}) \quad .$$

So under the assumptions of the theorem we have from Proposition 3.2

$$\frac{||f_{n,m} - \tilde{g}_{n,m}|| - E_n(f_{n,m})}{E_n(f_{n,m})} \le 2\frac{||H_{n,m}||}{|c_{n+1}|} \le \frac{2\,g\,h^{n+1}}{1-2g-2g^2+3g^3} \quad .$$

Another consequence of the general inequality cited above is

$$\begin{aligned} |E_n(f) - E_n[f_{n,m})| &= |E_n(f) - E_n(f_{n,m}+p_n)| \le ||f - (f_{n,m}+p_n)|| \\ &= ||\sum_{k=n+m+2}^{\infty} c_k z^k|| \quad . \end{aligned}$$

We also employ a simple estimate of the remainder of the series development of f

$$||\sum_{k=n+m+2}^{\infty} c_k z^k|| \le \frac{g^{m+1}}{1-g}|c_{n+1}| \le \frac{g^{m+1}}{1-g}E_n(f) \quad .$$

Combining the last three estimates we have

$$\begin{aligned} ||f - C_{n,m}|| &\le ||f_{n,m} + p_n - (\tilde{g}_{n,m}+p_n)|| + ||\sum_{k=n+m+2}^{\infty} c_k z^k|| \\ &\le E_n(f_{n,m})\left(1 + 2\frac{||H_{n,m}||}{|c_{n+1}|}\right) + ||\sum_{k=n+m+2}^{\infty} c_k z^k|| \\ &\le E_n(f)\left(1 + 2\frac{||H_{n,m}||}{|c_{n+1}|}\right) + ||\sum_{k=n+m+2}^{\infty} c_k z^k||\left(2 + 2\frac{||H_{n,m}||}{|c_{n+1}|}\right) \\ &\le E_n(f)\left(\left(1 + 2\frac{||H_{n,m}||}{|c_{n+1}|}\right) + \frac{g^{m+1}}{1-g}\left(2 + 2\frac{||H_{n,m}||}{|c_{n+1}|}\right)\right) \\ &\le E_n(f)\left(1 + 2\frac{g^{m+1}}{1-g}\right)\left(1 + \frac{2\,g\,h^{n+1}}{1-2g-2g^2+3g^3}\right) \quad . \end{aligned}$$

From this (i) and the asymptotic estimate in (ii) follow under the conditions listed in the theorem.

ad (iii): We first show that under the condition of monotonically decreasing coefficients as given in (iii) we have

$$q_{j-1} < \gamma\, q_j \tag{10}$$

for $j = 1, 2, \cdots, m$:

Because the c_k are real, we only need to consider a $m+1$ by $m+1$ eigenvalue problem to determine $\epsilon_{n,m}$ and $q_0, q_1, \ldots, q_m$. According to the Perron-Frobenius theorem (see e.g. Varga [12], p. 30) $q_0, q_1, \ldots, q_m$ are positive, and we can again choose $q_m = 1$. The condition of irreducibility in this theorem can be easily verified by drawing the graph of the corresponding matrix, which is strongly connected (see e.g. Varga [12], p. 20) because the last row and the last column contain only nonzero elements.

Now according to (3)

$$\begin{aligned} q_{m-k} &= (c_{n+m+1}q_k + c_{n+m}q_{k+1} + \cdots + c_{n+k+1}q_m)/\epsilon_{n,m} \\ &> (c_{n+m}q_{k+1} + \cdots + c_{n+k+1}q_m)/\epsilon_{n,m} \\ &\geq (c_{n+m+1}q_{k+1} + \cdots + c_{n+k+2}q_m)/(\epsilon_{n,m}\gamma) = q_{m-k-1}/\gamma \quad . \end{aligned}$$

From (3) with $k = 0$ we have the general estimate

$$\begin{aligned} \epsilon_{n,m} &\leq c_{n+1} + \sum_{k=2}^{m+1} c_{n+k}\,|q_{m+1-k}| \\ &\leq c_{n+1} + \max\{|q_l| : l = 0, 1, \ldots, m-1\} \sum_{k=2}^{m+1} c_{n+k} \quad . \end{aligned}$$

So from (10) we conclude that

$$\epsilon_{n,m} \leq c_{n+1} + \gamma \sum_{k=2}^{m+1} c_{n+k} \quad .$$

Again, from (3) we deduce

$$\begin{aligned} \epsilon_{n,m}q_{m-k} &= c_{n+m+1}q_k + c_{n+m}q_{k+1} + \cdots + c_{n+k+1}q_m \\ &\leq c_{n+m-k+2}\gamma^{k-1}q_k + c_{n+m-k+1}\gamma^{k-1}q_{k+1} + \cdots + c_{n+2}\gamma^{k-1}q_m \\ &\leq \gamma^{k-1} \sum_{j=2}^{m+1} c_{n+j} \quad . \end{aligned}$$

Now let $\alpha_1, \ldots, \alpha_m$ be the zeroes of

$$Q(z) = q_0 + q_1 z + \cdots + q_m z^m \quad .$$

Then from (10) and the Eneström-Kakeya theorem (see e.g. Specht [11], p. 31) or the Hurwitz [6] theorem results

$$\max\{|\alpha_j| : j = 1, \ldots, m\} \leq \gamma \quad .$$

From (2) follows further

$$g_*(z) = z^{n+1} \sum_{k=0}^{m-1} r_k z^k / Q(z) \quad . \tag{11}$$

We now estimate (4)

$$r_k = \sum_{j=0}^{k} c_{n+j+1} q_{k-j} \leq \sum_{j=0}^{k} c_{n+1} \gamma^j \gamma^{m-k+j-1} \frac{\sum_{l=2}^{m+1} c_{n+l}}{\epsilon_{n,m}}$$
$$\leq \frac{\gamma^{m-k-1}}{1-\gamma^2} \sum_{l=2}^{m+1} c_{n+l} \quad .$$

We utilize the development of $1/Q(z)$ by Wronskian functions (see e.g. Specht [11], p. 12):

$$1/Q(z) = \sum_{k=0}^{\infty} \sigma_k z^{-m-k}$$

with

$$\sigma_k = \sum \alpha_1^{\nu_1} \cdots \alpha_m^{\nu_m}$$

where we sum over those nonnegative integers $\nu_1, \ldots, \nu_m$ for which $\nu_1 + \cdots + \nu_m = k$. We can interpret these m-tuples $(\nu_1, \ldots, \nu_m)$ as distributions of k indistinguishible objects into m cells (namely k times adding 1 to the exponents of $\alpha_1, \ldots, \alpha_m$). There are

$$\binom{k+m-1}{m-1}$$

different ones among these distributions (see e.g. Riordan [9] p. 92), so we have

$$|\sigma_k| \leq \gamma^k \binom{k+m-1}{m-1} \quad .$$

From (11) we have

$$g_*(z) = z^{n+1} \sum_{k=0}^{m-1} r_k z^k / Q(z) = \frac{\sum_{k=0}^{m-1} r_k \, z^{n+1+k}}{\prod_{j=1}^{m} (z-\alpha_j)} = \sum_{k=0}^{m-1} r_k \sum_{l=0}^{\infty} \sigma_l z^{n+1+k-l-m}$$
$$= \sum_{j=-\infty}^{n} z^j \sum_{k=\max\{0,m-1-n+j\}}^{m-1} r_k \, \sigma_{n+1+k-m-j} \quad .$$

From the uniqueness of the power series development of the function $z^{-n} g_*(z)$ in $z = \infty$ we conclude

$$a_j = \sum_{k=\max\{0,m-1-n+j\}}^{m-1} r_k \, \sigma_{n+1+k-m-j} \quad .$$

So under the assumption $m < n$ we have the estimate

$$\begin{aligned}
||H_{n,m}|| &\leq \sum_{j=1}^{\infty} |a_{-j}| \leq \sum_{j=1}^{\infty}\sum_{k=0}^{m-1} \frac{\gamma^{m-1-k}\sum_{l=2}^{m+1} c_{n+l}}{1-\gamma^2}\binom{n+j+k}{m-1}\gamma^{n+1+k+j-m} \\
&\leq \frac{m\sum_{l=2}^{m+1} c_{n+l}}{(m-1)!(1-\gamma^2)}\sum_{j=1}^{\infty}\gamma^{n+j}\prod_{k=1}^{m-1}(n+k+j) \\
&= \frac{m\sum_{l=2}^{m+1} c_{n+l}}{(m-1)!(1-\gamma^2)}\sum_{j=1}^{\infty}\frac{d^{m-1}}{d\gamma^{m-1}}\gamma^{n+j+m-1} \\
&= \frac{m\sum_{l=2}^{m+1} c_{n+l}}{(m-1)!(1-\gamma^2)}\frac{d^{m-1}}{d\gamma^{m-1}}\left(\frac{\gamma^{n+m}}{1-\gamma}\right) \\
&= \frac{m\sum_{l=2}^{m+1} c_{n+l}}{(m-1)!(1-\gamma^2)}\sum_{j=0}^{m-1}\binom{m-1}{j}\frac{\gamma^{n+m-j}(m-j-1)!\prod_{l=0}^{m-1}(n+m-l)}{(1-\gamma)^{m-j}\prod_{k=j}^{m-1}(n+m-k)} \\
&\leq \frac{m\gamma^n\sum_{l=2}^{m+1} c_{n+l}}{(m-1)!(1-\gamma^2)}\frac{\gamma}{1-\gamma}\left(1+\frac{\gamma}{1-\gamma}\right)^{m-1}(n+m)^m \quad .
\end{aligned}$$

The last (very crude) estimate holds for $n > m$. By taking logarithms in the last term we see that it converges to zero, if divided by $\sum_{l=2}^{m+1} c_{n+l}$, with $n \to \infty$ under the condition $m(n) = o(n/\log n)$:

$$\log\left(\frac{\gamma^n m}{(m-1)!}\left(\frac{1}{1-\gamma}\right)^{m-1}(n+m)^m\right)$$

$$\leq n\log\gamma + \log m - \log(m-1)! - (m-1)\log(1-\gamma) + m\log(2n) \ .$$

This shows finally that under the conditions of (iii) for n large enough

$$\begin{aligned}
||f - C_{n,m}|| &\leq \epsilon_{n,m} + ||H_{n,m}|| + \sum_{k=n+m+2}^{\infty} c_k \\
&\leq c_{n+1} + \sum_{k=2}^{m+1} c_{n+k}\left(\gamma + \frac{||H_{n,m}||}{\sum_{l=2}^{m+1} c_{n+l}}\right) + \sum_{k=n+m+2}^{\infty} c_k \\
&< \sum_{k=n+1}^{\infty} c_k = ||f - p_n|| \quad . \qquad \square
\end{aligned}$$

5 Remarks

1.) One can transfer the above results to polynomial approximation on the real interval $[-1, +1]$ starting from a development of the function to be approximated into Chebyshev polynomials, as the author did in his dissertation.

Gutknecht and Trefethen showed that in this transfer one can – due to the reality of the coefficients – improve the estimates by truncating the series

$$g_*(z) = \sum_{k=-\infty}^{n} a_k z^k$$

at $-n$ instead of 0.

2.) In the end of the 1970s the Carathéodory-Fejér method was applied to polynomial and rational approximation independently by Gutknecht and Trefethen; this was treated in a series of articles including estimates as the above but requiring $g < 1/72$, see, e.g., [3], [4] , and the literature cited there.

3.) The conditions of the above theorem are fairly restrictive. For (i) and (ii) the power series of f must have a convergence radius $\geq 1/g$, and moreover one must choose as c_{n+1} a coefficient in this series which is "not too small". The conditions of (iii) are fulfilled if e.g. f with real coefficients in (1) has one "dominating" algebraic-logarithmic singularity on the boundary of its circle of convergence, which must be of radius greater than 1, see e.g. Jungen [7]. These restrictions were made precise in [10] by Saff and Totik who showed that the set of functions for which the Carathéodory-Fejér method "works" in the sense of giving a better approximation than p_n for infinitely many n, is of second category, i.e., it does not contain any ball with respect to the norm we used.

4.) The author wishes to thank the organizers of the IDoMAT conference for the opportunity to give a lecture and to publish his results. This encouraged him to resume his study of the Carathéodory-Fejér method and to extend it to rational approximation. He also thanks the referee for his suggestions to improve the paper.

References

[1] Carathéodory, C., and Fejér, L., *Über den Zusammenhang der Extremen von harmonischen Funktionen mit ihren Koeffizienten und über den Picard-Landauschen Satz, Rendiconti del circolo matematico di Palermo*, Vol. 32, (1911), 218–239.

[2] Darlington, S., *Analytical Approximations to Approximations in the Chebyshev Sense, The Bell System Technical Journal*, Vol. 49, (1970, 1–32.

[3] Gutknecht, M.H., *Rational Carathéodory-Fejér approximation on a disk, a circle, and an interval, Journal of Approximation Theory*, Vol. 41, (1984), 257–278.

[4] Gutknecht, M.H. and Trefethen, L.N., *Real polynomial Chebyshev approximation by the Carathéodory-Fejér method, SIAM Journal of Numerical Analysis*, Vol. 19, (1982), 358–371.

[5] Hollenhorst, M., *Nichtlineare Verfahren bei der Polynomapproximation, Dissertation, University of Erlangen-Nürnberg*, 1976.

[6] Hurwitz, A., *Über einen Satz des Herrn Kakeya, Tohoku Mathematical Journal*, VOL. 4, (1913–1914), 89–93.

[7] Jungen, R., *Sur les séries de Taylor n'ayant que des singularités algébro-logarithmiques sur leur cercle de convergence, Comm. Math. Helvet.*, VOL. 3, (1931), 266–306.

[8] Klotz, V., *Polynomiale und rationale Tschebyscheff-Approximation in der komplexen Ebene, Dissertation, University of Erlangen-Nürnberg*, (1974).

[9] Riordan, J., *An Introduction to Combinatorial Analysis, New York, London, Sydney: John Wiley*, (1958).

[10] Saff, E.B. and Totik, V., *Limitations of the Carathéodory-Fejér Method for Polynomial Approximation, Journal of Approximation Theory*, VOL. 58, (1989), 284–296.

[11] Specht, W. *Enzyklopädie der mathematischen Wissenschaften. Band I, 1. Teil, Heft 8: Algebraische Gleichungen mit reellen oder komplexen Koeffizienten, Stuttgart: Teubner*, (1958).

[12] Varga, R.S., *Matrix Iterative Analysis, Englewood Cliffs, N.J.: Prentice Hall*, (1962).

Computer Center
Justus Liebig University
Heinrich-Buff-Ring 44
D-35392 Giessen, Germany
Email address: hollenhorst@hrz.uni-giessen.de

Advanced Problems in Constructive Approximation
(Eds.) M.D. Buhmann and D.H. Mache
International Series of Numerical Mathematics Vol. 142, 79–93

Shape Preserving Widths of Weighted Sobolev-Type Classes

V.N. Konovalov and D. Leviatan

1 Introduction

Let X be a real linear space of vectors x with norm $\|x\|_X$, $W \subset X, W \neq \emptyset$, and $V \subset X$, $V \neq \emptyset$. Let L^n be a subspace in X of dimension $\dim L^n \leq n$, $n \geq 0$.and $M^n = M^n(x^0) := x^0 + L^n$ be a shift of the subspace L^n by an arbitrary vector $x^0 \in X$. Denote

$$E(x, M^n)_X := \inf_{y \in M^n} \|x - y\|_X,$$

and let

$$E(W, M^n)_X := \sup_{x \in W} E(x, M^n)_X,$$

denote the deviation of the set W from M^n.

The Kolmogorov n-width of W is defined by

$$d_n(W)_X := \inf_{M^n} E(W, M^n)_X, \quad n \geq 0.$$

If $M^n \cap V \neq \emptyset$, then we denote by

$$E(x, M^n \cap V)_X := \inf_{y \in M^n \cap V} \|x - y\|_X,$$

the best approximation of the vector $x \in X$ by $M^n \cap V$, and by

$$E(W, M^n \cap V)_X := \sup_{x \in W} E(x, M^n \cap V)_X,$$

the deviation of the set W from $M^n \cap V$.

The quantity

$$d_n(W, V)_X := \inf_{M^n} E(W, M^n \cap V)_X, \quad n \geq 0$$

is called the relative n-width of W with the constraint V in X. These widths were introduced by the first author in [1].

Evidently, if $V = X$, then the relative n-width $d_n(W,V)_X$ coincides with the Kolmogorov n-width $d_n(W)_X$. Clearly, $d_n(W,V)_X \geq d_n(W)_X$.

We also let $\Lambda(X, L^n)$ be the set of all linear maps $\Lambda : X \to L^n$. Then

$$E(W, L^n)_X^{lin} := \inf_{\Lambda \in \Lambda(X,L^n)} \sup_{x \in W} \|x - \Lambda x\|_X$$

denotes the best linear approximation of the set W by L^n. The linear n-width of W is defined by

$$d_n(W)_X^{lin} := \inf_{L^n \subset X} E(W, L^n)_X^{lin}, \quad n \geq 0.$$

Let $I = [a, b]$ be a finite interval in $\mathbb{R}$, and let $r \in \mathbb{N}$ and $0 \leq \alpha < \infty$. For $1 \leq p \leq \infty$, and $\rho(t) := \operatorname{dist}\{t, \partial I\}$, $t \in I$, we denote

$$W_{p,\alpha}^r := W_{p,\alpha}^r(I) := \{x : I \to \mathbb{R} \\ x^{(r-1)} \in AC_{loc}(I), \|x^{(r)}\rho^\alpha\|_{L_p(I)} \leq 1\}.$$

If $\alpha = 0$, then we write $W_p^r := W_p^r(I) := W_{p,0}^r(I)$. We also write L_q for $L_q(I)$.

Let

$$\Delta_\tau^s x(t) := \sum_{k=0}^{s} (-1)^{s-k} \binom{s}{k} x(t + k\tau), \quad \{t, t + s\tau\} \subset I, \\ s = 0, 1, \ldots,$$

and denote by $\Delta_+^s W_{p,\alpha}^r = \Delta_+^s W_{p,\alpha}^r(I)$, $s = 0, 1, \ldots$ the subclasses of functions $x \in W_{p,\alpha}^r$ for which $\Delta_\tau^s x(t) \geq 0$, for all $\tau > 0$ such that $[t, t + s\tau] \subseteq I$. Analogously we will use the notation $\Delta_+^s L_q$. In recent years shape preserving approximation has become a central subject especially in application. This is due to the fact that in CAGD and especially in questions of design, shape preservation is one of the main considerations. Our results below show what one may expect to achieve and what is beyond reach of any approximation process which involves approximation from linear n-dimensional manifolds, when we preserve the shape of the approximants.

2 Unconstrained Kolmogorov and linear widths

We begin with some asymptotic relations for unconstrained Kolmogorov and linear widths. First we have [2]

Theorem 2.1 *Let I be a finite interval and let $r \in \mathbb{N}$, $1 \leq p, q \leq \infty$ and $0 < \alpha < \infty$, be such that $r - \alpha - \frac{1}{p} + \frac{1}{q} > 0$. If $(r, p) \neq (1, 1)$, and if $(r, p) = (1, 1)$ and $1 \leq q \leq 2$, then*

$$d_n(W_{p,\alpha}^r)_{L_q} \asymp n^{-r+(\max\{\frac{1}{p},\frac{1}{2}\}-\max\{\frac{1}{q},\frac{1}{2}\})_+}, \quad n \geq r. \tag{1}$$

If on the other hand, $(r,p) = (1,1)$ *and* $2 < q < \infty$, *then*

$$c_1 n^{-\frac{1}{2}} \le d_n(W^1_{1,\alpha})_{L_q} \le c_2 n^{-\frac{1}{2}} \big(\log(n+1)\big)^{\frac{3}{2}}, \quad n \ge 1, \tag{2}$$

where $c_1 > 0$ *and* c_2 *do not depend on* n.

Thus we see that the asymptotic order of the Kolmogorov widths is independent of α although the classes $W^r_{p,\alpha}$ become bigger as α increases. The smallest of course is $W^r_{p,0}$ for which the above asymptotics are well known. In view of this it is clear that the lower estimates in (1) and (2) need no proof. The proof of the upper bounds is too long to be given here and we refer the reader to [2].

The exact orders of the widths of the classes $W^1_{1,\alpha}$ in L_q, $2 < q < \infty$, are not known even when $\alpha = 0$.

It turns out that the Kolmogorov widths of the smaller classes $\Delta^s_+ W^r_{p,\alpha}$, $0 \le s \le r$, are, in general, of the same order of magnitude as those of the classes $W^r_{p,\alpha}$. However, they are significantly smaller for the class $\Delta^{r+1}_+ W^r_{p,\alpha}$.

What we have if $0 \le s \le r$, is (see [3]),

Theorem 2.2 *Let* $r \in \mathbb{N}$, $1 \le p, q \le \infty$ *and* $0 \le \alpha < \infty$, *be such that* $r - \alpha - \frac{1}{p} + \frac{1}{q} > 0$. *If* $(r,p) \ne (1,1)$ *and if* $(r,p) = (1,1)$ *and* $1 \le q \le 2$, *then for each* $s = 0, 1, \ldots, r$,

$$d_n(\Delta^s_+ W^r_{p,\alpha})_{L_q} \asymp n^{-r+(\max\{\frac{1}{p},\frac{1}{2}\}-\max\{\frac{1}{q},\frac{1}{2}\})_+}, \quad n \ge r. \tag{3}$$

If on the other hand, $(r,p) = (1,1)$ *and* $2 < q < \infty$, *then for* $s = 0, 1$,

$$c_1 n^{-\frac{1}{2}} \le d_n(\Delta^s_+ W^1_{1,\alpha})_{L_q} \le c_2 n^{-\frac{1}{2}} \big(\log(n+1)\big)^{\frac{3}{2}}, \quad n \ge 1, \tag{4}$$

where $c_1 > 0$ *and* c_2 *do not depend on* n.

But in case $s = r + 1$, we have [3],

Theorem 2.3 *Let* $r \in \mathbb{N}$, $1 \le p, q \le \infty$ *and* $0 \le \alpha < \infty$, *be such that* $r - \alpha - \frac{1}{p} + \frac{1}{q} > 0$. *Then*

$$d_n(\Delta^{r+1}_+ W^r_{p,\alpha})_{L_q} \asymp n^{-r-\max\{\frac{1}{q},\frac{1}{2}\}}, \quad n > r. \tag{5}$$

In view of (1) and (2), we don't have to prove the upper bounds in (3) and (4). The proof of the lower bounds is too long to be included in this paper and the interested reader should consult [3]. On the other hand, it is interesting to note that

$$\begin{aligned}(b-a)^{1/p}\Delta^{r+1}_+ W^r_{p,\alpha} &\supseteq \Delta^{r+1}_+ W^r_{\infty,\alpha} \\ &\supseteq \Delta^{r+1}_+ W^r_{\infty} \\ &\supseteq \Delta^{r+1}_+ W^{r+1}_1 \cap \{x \mid x^{(i)}(a) = 0, 0 \le i \le r\},\end{aligned} \tag{6}$$

where $I = [a,b]$, and that the latter set differs from $\Delta^{r+1}_+ W^{r+1}_1$ by a linear subspace of dimension $r+1$. Hence, the lower bound in (5) for $n > 2r+1$ follows from (3) taking $r+1$ instead of r there and applying it to $s = r+1$ and $p = 1$.

For linear widths we can show [2] that

Theorem 2.4 *Let $r \in \mathbb{N}$, $1 \le p, q \le \infty$ and $0 < \alpha < \infty$ be such that $r - \alpha - \frac{1}{p} + \frac{1}{q} > 0$. If $(r,p) \ne (1,1)$, or if $(r,p) = (1,1)$ and $1 \le q \le 2$, and if $(r,q) = (1,\infty)$ and $2 \le p \le \infty$, then*

$$d_n(W^r_{p,\alpha})^{lin}_{L_q} \asymp n^{-r+(\frac{1}{p}-\frac{1}{q})_+ - \min\{(\frac{1}{p}-\frac{1}{2})_+,(\frac{1}{2}-\frac{1}{q})_+\}}, \quad n \ge r. \tag{7}$$

If on the other hand, $(r,p) = (1,1)$ and $2 < q < \infty$, then

$$c_1 n^{-\frac{1}{2}} \le d_n(W^1_{1,\alpha})^{lin}_{L_q} \le c_2 n^{-\frac{1}{2}} (\log(n+1))^{\frac{3}{2}}, \quad n \ge 1, \tag{8}$$

and if $(r,q) = (1,\infty)$ and $1 < p < 2$, then

$$c_1 n^{-\frac{1}{2}} \le d_n(W^1_{p,\alpha})^{lin}_{L_\infty} \le c_2 n^{-\frac{1}{2}} (\log(n+1))^{\frac{3}{2}}, \quad n \ge 1, \tag{9}$$

where $c_1 > 0$ and c_2 do not depend on n.

The exact orders of the linear widths of the classes $W^1_{1,\alpha}$ in L_q, $2 < q < \infty$, and those of the classes $W^1_{p,\alpha}$ in L_∞, are not known even when $\alpha = 0$.

We see the same phenomenon that the asymptotic order of the linear widths is independent of α although the classes $W^r_{p,\alpha}$ become bigger as α increases. The smallest of course is $W^r_{p,0}$ for which the above asymptotics are well known. In view of this it is clear that the lower estimates in (7), (8) and (9) need no proof. For the proof of the upper bounds we refer the reader to [2].

We have here the same phenomenon as for the Kolmogorov widths, namely, the linear widths of the smaller classes $\Delta^s_+ W^r_{p,\alpha}$, $0 \le s \le r$, are, in general, of the same order of magnitude as those of the classes $W^r_{p,\alpha}$. However, they are significantly smaller for the class $\Delta^{r+1}_+ W^r_{p,\alpha}$. Here we have for $0 \le s \le r$ (see [3]),

Theorem 2.5 *Let $r \in \mathbb{N}$, $1 \le p, q \le \infty$ and $0 \le \alpha < \infty$, be such that $r - \alpha - \frac{1}{p} + \frac{1}{q} > 0$. If $(r,p) \ne (1,1)$, or if $(r,p) = (1,1)$ and $1 \le q \le 2$ and if $(r,q) = (1,\infty)$ and $2 \le p \le \infty$, then for each $s = 0, 1, \ldots, r$,*

$$d_n(\Delta^s_+ W^r_{p,\alpha})^{lin}_{L_q} \asymp n^{-r+(\frac{1}{p}-\frac{1}{q})_+ - \min\{(\frac{1}{p}-\frac{1}{2})_+,(\frac{1}{2}-\frac{1}{q})_+\}}, \quad n \ge r. \tag{10}$$

If on the other hand, $(r,p) = (1,1)$ and $2 < q < \infty$ then for $s = 0, 1$,

$$c_1 n^{-\frac{1}{2}} \le d_n(\Delta^s_+ W^1_{1,\alpha})^{lin}_{L_q} \le c_2 n^{-\frac{1}{2}} (\log(n+1))^{\frac{3}{2}}, \quad n \ge 1, \tag{11}$$

and if $(r,q) = (1,\infty)$ and $1 < p < 2$, then for $s = 0, 1$,

$$c_1 n^{-\frac{1}{2}} \le d_n(\Delta^s_+ W^1_{p,\alpha})^{lin}_{L_\infty} \le c_2 n^{-\frac{1}{2}} (\log(n+1))^{\frac{3}{2}}, \quad n \ge 1, \tag{12}$$

where $c_1 > 0$ and c_2 do not depend on n.

In the case $s = r + 1$ we show in [3] that

Theorem 2.6 *Let $r \in \mathbb{N}$, $1 \le p, q \le \infty$ and $0 \le \alpha < \infty$, be such that $r - \alpha - \frac{1}{p} + \frac{1}{q} > 0$. Then*

$$d_n(\Delta_+^{r+1} W_{p,\alpha}^r)_{L_q}^{lin} \asymp n^{-r-\max\{\frac{1}{q},\frac{1}{2}\}}, \quad n > r. \tag{13}$$

Again, in view of (7), (8) and (9), we do not have to prove the upper bounds in (10), (11) and (12). Also as we have noted after Theorem 2.3, the lower bound in (13) follows from (10).

Remark. Note that for each fixed q and all p such that $r - \alpha - \frac{1}{p} + \frac{1}{q} > 0$, the Kolmogorov and linear widths of the classes $\Delta_+^{r+1} W_{p,\alpha}^r$ are of the same order of magnitude.

3 Shape preserving widths

The most important shapes that we normally wish to preserve are positivity, monotonicity and convexity. For positivity preserving widths we have, [4],

Theorem 3.1 *Let $r \in \mathbb{N}$, $1 \le p, q \le \infty$ and $0 \le \alpha < \infty$, be such that $r - \alpha - \frac{1}{p} + \frac{1}{q} > 0$. Then*

$$\begin{aligned} c_1 n^{-r+(\max\{\frac{1}{p},\frac{1}{2}\}-\max\{\frac{1}{q},\frac{1}{2}\})_+} &\le d_n(\Delta_+^0 W_{p,\alpha}^r, \Delta_+^0 L_q)_{L_q} \\ &\le c_2 n^{-r+(\frac{1}{p}-\frac{1}{q})_+}, \quad n \ge r, \end{aligned} \tag{14}$$

and in particular if $1 \le q \le p \le \infty$, and if $1 \le p \le q \le 2$, then this implies

$$d_n(\Delta_+^0 W_{p,\alpha}^r, \Delta_+^0 L_q)_{L_q} \asymp n^{-r+(\max\{\frac{1}{p},\frac{1}{2}\}-\max\{\frac{1}{q},\frac{1}{2}\})_+}, \quad n \ge r. \tag{15}$$

Furthermore, (15) *holds for all other cases of p and q, if we actually have the (stronger) inequality $r - \alpha - \frac{1}{p} > 0$. (Note that under our assumptions, the latter always holds when $q = \infty$.) Finally, if $(r, \alpha, p) = (1, 0, 1)$ and $2 < q < \infty$, then*

$$c_1 n^{-\frac{1}{2}} \le d_n(\Delta_+^0 W_{1,0}^1, \Delta_+^0 L_q)_{L_q} \le c_2 n^{-\frac{1}{2}} (\ln(n+1))^{\frac{3}{2}}, \quad n \ge 1, \tag{16}$$

where $c_1 > 0$ and c_2 do not depend on n.

Here too, the lower bounds in (14) and (16) follow by virtue of (3) and (4), respectively.

One may be tempted to conjecture that (15) gives the correct asymptotics in all missing cases as well. This however is not clear at all in view of the following asymptotics for monotonicity and for convexity preserving widths that agree instead with the right-hand side of (14). For monotonicity preserving widths we show in [4],

Theorem 3.2 *Let $r \in \mathbb{N}$, $1 \le p, q \le \infty$ and $0 \le \alpha < \infty$, be such that $r - \alpha - \frac{1}{p} + \frac{1}{q} > 0$. Then*

$$d_n(\Delta_+^1 W_{p,\alpha}^r, \Delta_+^1 L_q)_{L_q} \asymp n^{-r+(\frac{1}{p}-\frac{1}{q})_+}, \quad n \ge r.$$

And for convexity preserving widths we obtain (see [4]),

Theorem 3.3 *Let $r \in \mathbb{N}$, $1 \le p, q \le \infty$ and $0 \le \alpha < \infty$, be such that $r - \alpha - \frac{1}{p} + \frac{1}{q} > 0$. If $r > 1$, then*

$$d_n(\Delta_+^2 W_{p,\alpha}^r, \Delta_+^2 L_q)_{L_q} \asymp n^{-r+(\frac{1}{p}-\frac{1}{q})_+}, \quad n \ge r, \tag{17}$$

and if $r = 1$, then

$$d_n(\Delta_+^2 W_{p,\alpha}^1, \Delta_+^2 L_q)_{L_q} \asymp n^{-1-\frac{1}{q}}, \quad n \ge 1. \tag{18}$$

Note that by virtue of (6), the lower bound in (18) for $n > 2$, can be obtained from (17) with $r = 2$ and $p = 1$.

Finally we deal with s-monotone functions with $s \ge 3$. We encounter here a completely different behavior (see [6]), namely,

Theorem 3.4 *Let $r \in \mathbb{N}$, $s \in \mathbb{N}$ and $1 \le p, q \le \infty$. For $3 \le s \le r$, we have*

$$d_n\left(\Delta_+^s W_p^r, \Delta_+^s L_q\right)_{L_q} \asymp n^{-r+s+\frac{1}{p}-3}, \quad n \ge r. \tag{19}$$

Also if $s = r + 1$, $r \ge 2$, then

$$d_n\left(\Delta_+^{r+1} W_p^r, \Delta_+^{r+1} L_q\right)_{L_q} \asymp n^{-2}, \quad n \ge r. \tag{20}$$

Here too, the lower bound in (20) for $n \ge 2r+1$, follows by virtue of (6) from (20) with r replaced by $r+1$ and taking $s = r+1$ and $p = 1$.

Remarks. (i) Note that the asymptotic relations are independent of q and if $s = r + 1$ also independent of p. Note also that they become worse as s increases (inside the range $3 \le s \le r + 1$).

(ii) It is worthwhile noting that as a byproduct we may conclude that the lower bound in (2) with $s = r > 3$, excludes the possibility of Jackson-type estimates involving the fourth modulus of smoothness of x evaluated at $1/n$, in s-monotone approximation of x, by piecewise polynomials or splines with n equidistant knots and thus also not by polynomials of degree $\le n$. Moreover, it even excludes Jackson-type estimates involving the generally bigger $Cn^{-3}\omega(x^{(3)}, n^{-1})_p$.

(iii) Recall that up until now we knew that Shvedov [7] had shown that Jackson-type estimates of s-monotone approximation of an s-monotone x, by polynomials of degree $\le n$, cannot be had with $C\omega_{s+2}(x, n^{-1})_p$. Thus the above is somewhat unexpected to us in view of what seemed to have been a pattern that we have Jackson-type estimates involving $C\omega_2(x, n^{-1})_p$ for monotone approximation, and by Shvedov [7], it is impossible to have such estimates with $\omega_3(x, n^{-1})_p$, and we have Jackson-type estimates for convex approximation involving $\omega_3(x, n^{-1})_p$, while again by Shvedov [7], it is impossible to have such estimates with $\omega_4(x, n^{-1})_p$.

4 Outline of proofs

Proof of Theorems 2.1 *and* 2.4. To prove upper bounds in Theorems 2.1 and 2.4 we divide the generic interval $I = [-1, 1]$ by the partition points

$$t_{\beta,n,i} := \begin{cases} 1 - (\frac{n-i}{n})^\beta, & i = 0, 1, \ldots, n, \\ -1 + (\frac{n+i}{n})^\beta, & i = -1, \ldots, -n, \end{cases}$$

where $\beta = \beta(r, \alpha, p, q) \geq 1$ is to be prescribed. Given a function $x \in W^r_{p,\alpha}(I)$, we define on each subinterval $I_{n,i} := I_{\beta,n,i} := [t_{\beta,n,i-1}, t_{\beta,n,i}]$, $i = 1, \ldots, n$, and $I_{n,i} := I_{\beta,n,i} := [t_{\beta,n,i}, t_{\beta,n,i+1}]$, $i = -1, \ldots, -n$, polynomial splines $\sigma_i(\cdot; x)$, $i = \pm 1, \ldots, \pm n$, of degree $\leq r + 1$, with three fixed knots, yielding the estimates

$$\|x(\cdot) - \sigma_i(x; \cdot)\|_{L_q(I_{n,i})} \leq c|I_{n,i}|^{r-\frac{1}{p}+\frac{1}{q}} \Big(\sum_{j=i}^{n} |I_{n,j}|\Big)^{-\alpha} \|x^{(r)}\rho^\alpha\|_{L_p(I_{n,i})}, \tag{21}$$

$i = 1, \ldots, n$, and similar estimates for $i = -1, \ldots, -n$. It is easy to check that the length $|I_{n,i}|$ of the interval I_{ni}, $i = \pm 1, \ldots, \pm n$, satisfies the inequalities

$$c_1 n^{-\beta}(n - |i| + 1)^{\beta-1} \leq |I_{n,i}| \leq c_2 n^{-\beta}(n - |i| + 1)^{\beta-1}, \tag{22}$$

where $c_1 = c_1(\beta)$, $c_2 = c_2(\beta)$.

The combined spline $\sigma_{r,n}(t; x) := \sigma_i(t; x)$, $t \in I_{n,i}$, $i = \pm 1, \ldots, \pm n$ is so constructed that it belongs to $C^{(1)}(I)$ and it depends linearly on x. It follows from (21) and (22) that for

$$\beta = \Big(r - \frac{1}{p} + \frac{1}{q}\Big)\Big(r - \alpha - \frac{1}{p} + \frac{1}{q}\Big)^{-1},$$

we obtain

$$\|x(\cdot) - \sigma_{r,n}(x; \cdot)\|_{L_q(I)} \leq cn^{-r+(\frac{1}{p}-\frac{1}{q})_+}, \quad 1 \leq p, q \leq \infty, \tag{23}$$

where $c = c(r, \alpha, p, q)$.

The upper bounds (23) already yield the required upper estimates in Theorem 2.1 for a partial range of p and q. We have to improve them when $2 < q \leq \infty$ and $p < q$. This we do by applying discretization techniques and except for the case $r = p = 1$, we obtain

$$d_n\left(W^r_{p,\alpha}\right)_{L_q} \leq \begin{cases} cn^{-r+\frac{1}{p}-\frac{1}{2}}, & 1 \leq p \leq 2 < q \leq \infty, \\ cn^{-r}, & 2 \leq p < q \leq \infty, \end{cases}$$

where $c = c(r, \alpha, p, q)$, for all $n \geq r$. If $r = p = 1$ and $2 < q < \infty$, then we can only prove

$$d_n\left(W^1_{1,\alpha}\right)_{L_q} \leq cn^{-\frac{1}{2}}\left(\log(n+1)\right)^{\frac{3}{2}}.$$

We obtain similar inequalities for the linear widths in Theorem 2.4. The lower bounds in Theorems 2.1 and 2.4 are immediate consequences of the well-known lower bounds for the smaller classes W_p^r. □

Proof of Theorems 2.2 *and* 2.5. As was noted above, the upper bounds in Theorems 2.2 and 2.5 follow from the upper bounds in Theorems 2.1 and 2.4, respectively. Thus, we only need to prove the lower bounds. To this end we choose a system $\Psi_{r,p}^{2n} = \{\psi_{r,n,p,i}\}_{i=1}^{2n}$ of functions

$$\psi_{r,n,p,i}(t) := \frac{n^{\frac{1}{p}}}{r!}\left(\left(t+\frac{n+i-1}{n}\right)_+^r - \left(t+\frac{n+i-2}{n}\right)_+^r\right), \tag{24}$$
$$t \in I,\ i = 1,\ldots,2n.$$

It is easy to see that $\psi_{r,n,p,i} \in \Delta_+^s W_p^r(I)$, $i = 1,\ldots,2n$. Let

$$S_p^+(\Psi_{r,p}^{2n})) := \{\psi \mid \psi(t) = \sum_{i=1}^{2n} a_i\psi_{r,n,p,i}(t), \quad t \in I\}$$

over this system, where $a = (a_1,\ldots,a_{2n}) \in \mathbb{R}_+^{2n}$ and $\|a\|_{l_p^{2n}} \le 1$, be the positive p-sector. Clearly, $S_p^+(\Psi_{r,p}^{2n})) \subseteq \Delta_+^s W_p^r(I)$. Hence

$$d_n(\Delta_+^s W_p^r)_{L_q} \ge d_n(S_p^+(\Psi_{r,p}^{2n}))_{L_q} \quad \text{and} \quad d_n(\Delta_+^s W_p^r)_{L_q}^{lin} \ge d_n(S_p^+(\Psi_{r,p}^{2n}))_{L_q}^{lin}.$$

Using discretization techniques we reduce the problem of estimating the widths of the positive p-sector $S_p^+(\Psi_{r,p}^{2n})$ in the function space L_q, to that of estimating the widths of the positive p-sector in the space l_q^{2n}. More precisely, we obtain

$$d_n(S_p^+(\Psi_{r,p}^{2n}))_{L_q} \ge cn^{-r+\frac{1}{p}-\frac{1}{q}} d_n(S_p^+(E^{2n}))_{l_q^{2n}}, \tag{25}$$

$$d_n(S_p^+(\Psi_{r,p}^{2n}))_{L_q}^{lin} \ge cn^{-r+\frac{1}{p}-\frac{1}{q}} d_n(S_p^+(E^{2n}))_{l_q^{2n}}^{lin}, \tag{26}$$

where

$$S_p^+(E^{2n}) := \{e \mid e = \sum_{i=1}^{2n} a_i e^{(i)},\ a = (a_1,\ldots,a_{2n}) \in \mathbb{R}_+^{2n},\ \|a\|_{l_p^{2n}} \le 1\}$$

is the positive p-sector in $\mathbb{R}^{2n}$ over the system E^{2n} of the standard orthonormal vectors $e^{(i)}$, $i = 1,\ldots,2n$.

Since $S_p^+(E^{2n})$ contains the cube $n^{-\frac{1}{p}} S_\infty^+(E^{2n})$ and the simplex $S_1^+(E^{2n})$, we apply well-known lower estimates of the Kolmogorov $2n$-widths of $S_\infty^+(E^{2n})$ and $S_1^+(E^{2n})$ in l_q^{2n}, and get from (25) the needed lower bounds in Theorem 2.2. Similarly, we use (26) to prove the lower bounds in Theorem 2.5. □

Proof of Theorems 2.3 *and* 2.6. In order to prove the upper bounds in Theorems 2.3 and 2.6 we have to proceed in another way. We apply some ideas of V.M.

Tikhomirov for obtaining the Kolmogorov widths of the classes $\Delta^2_+ W^1_p$ in L_q. Following Tikhomirov we reduce our problem to that of the isoperimetric problem in $\mathbb{R}^n$. To that end we fix

$$\beta > (r + \frac{1}{q})(r - \alpha - \frac{1}{p} + \frac{1}{q})^{-1}, \tag{27}$$

and for any function $x \in \Delta^{r+1}_+ W^r_{p,\alpha}$ we define the approximating spline $\sigma_{r,n}(\cdot; x)$ on each subinterval $I_{n,i} = I_{\beta,n,i}$, $i = \pm 1, \ldots, \pm(n-1)$, as the Lagrange polynomials $l_{r,n,i}(;x)$ of degree r, which interpolate $x(\cdot)$ at the $r+1$ equidistant points. For the end intervals $I_{n,\pm n}$ we take $\sigma_{r,n}(\cdot; x) := l_{r,n,\pm(n-1)}(\cdot; x)$, respectively. Note that the rth derivative $x^{(r)}$ is nondecreasing on I. Also, without loss of generality, we may assume that $x^{(r)}(t) \geq 0$, $t \in [0,1)$, and $x^{(r)}(t) \leq 0$, $t \in (-1, 0]$.

Using well-known Whitney's theorem we have

$$\|x(\cdot) - \sigma_{r,n}(x; \cdot)\|_{L_\infty(I_{n,i})} \leq c|I_{n,i}|^r \omega_{r,n,i}, \quad i = \pm 1, \ldots, \pm(n-1),$$

where

$$\omega_{r,n,i} := \operatorname{esssup}_{t_1,t_2 \in I_{n,i}} |x(t_1) - x(t_2)|.$$

In view of definition of $\sigma_{r,n}(\cdot; x)$ we conclude that for all $1 \leq q \leq \infty$

$$\|x(\cdot) - \sigma_{r,n}(x; \cdot)\|_{L_q(I_{n,i})} \leq c|I_{ni}|^{r+\frac{1}{q}} \omega_{r,n,i}, \quad i = \pm 1, \ldots, \pm(n-1). \tag{28}$$

Also it is easy to verify that

$$\|x(\cdot) - \sigma_{r,n}(x; \cdot)\|_{L_q(I_{n,\pm n})} \leq c|I_{n,\pm n}|^{r-\alpha-\frac{1}{p}+\frac{1}{q}}. \tag{29}$$

So combining (28) and (29) we get

$$\|x(\cdot) - \sigma_{r,n}(x; \cdot)\|_{L_q(0,1)} \leq c \left(\sum_{i=1}^{n-1} |I_{n,i}|^{rq+1} \omega^q_{r,n,i} \right)^{\frac{1}{q}} + c|I_{n,n}|^{r-\alpha-\frac{1}{p}+\frac{1}{q}}, \tag{30}$$

where $c = c(r, \alpha, p, q)$.

On the other hand, we have

$$\|x^{(r)} \rho^\alpha\|_{L_p(0.1)} \geq \left(\sum_{i=1}^{n-1} \|x^{(r)} \rho^\alpha\|^p_{L_p(I_{n.i+1})} \right)^{\frac{1}{p}}.$$

Note that a.e. in $I_{n,i+1}$, $i = 1, \ldots, n-1$,

$$\sum_{j=1}^{i} \omega_{r,n,i} \leq x^{(r)}(t),$$

so that for each $1 \le p \le \infty$

$$\|\rho^\alpha\|_{L_p(I_{n,i+1})} \sum_{j=1}^{i} \omega_{r,n,i} \le \|x^{(r)}\rho^\alpha\|_{L_p(I_{n,i+1})}.$$

Hence for all $1 \le p \le \infty$ we have

$$\left(\sum_{i=1}^{n-1} \|\rho^\alpha\|^p_{L_p(I_{n,i+1})} \left(\sum_{j=1}^{i} \omega_{r,n,i}\right)^p\right)^{\frac{1}{p}} \le 1. \tag{31}$$

If we put $a_i := c|I_{n,i}|$, $b_i := \|\rho^\alpha\|_{L_p(I_{n,i+1})}$, and replace $\omega_{r,n,i}$ by τ_i, then by (30) and (31) we are in the setup of the following

Lemma 4.1 *Let $n \in \mathbb{N}$ and $1 \le p, q \le \infty$. Given $a_i > 0$ and $b_i > 0$, $i = 1, \ldots, n$, let $\tau := (\tau_1, \ldots, \tau_n)$ belong to the set*

$$T_p = \{\tau \mid \tau_i \ge 0,\ 1 \le i \le n,\ (\sum_{i=1}^{n} b_i^p (\sum_{j=1}^{i} \tau_i)^p)^{\frac{1}{p}} \le 1\}, \quad 1 \le p \le \infty. \tag{32}$$

Set

$$f_q(\tau) := (\sum_{i=1}^{n} a_i^q \tau_i^q)^{\frac{1}{q}}, \quad 1 \le q \le \infty. \tag{33}$$

Then setting $a_{n+1} := 0$ and $\frac{1}{p} + \frac{1}{p'} = 1$,

$$\max_{\tau \in T_p} f_q(\tau) \le \left(\sum_{i=1}^{n} \left(|a_i - a_{i+1}| b_i^{-1}\right)^{p'}\right)^{\frac{1}{p'}}, \quad 1 \le p \le \infty. \tag{34}$$

Note that the right-hand side of (34) is independent of q. Hence by (22) and (27),

$$|a_i - a_{i+1}| \le cn^{-\beta(r+\frac{1}{q})}(n - i + 1)^{(\beta-1)(r+\frac{1}{q})-1}, \quad i = 1, \ldots, n-1,$$

and

$$b_i \ge cn^{-\beta(\alpha+\frac{1}{p})}(n - i + 1)^{(\beta-1)(\alpha+\frac{1}{p})-\frac{1}{p}}, \quad i = 1, \ldots, n-1,$$

which in turn imply

$$\|x(\cdot) - \sigma_{r,n}(x;\cdot)\|_{L_q(0,1)} \le cn^{-r-\frac{1}{q}}, \quad 1 \le q \le \infty. \tag{35}$$

where $c = c(r, \alpha, p, q)$. We have similar inequalities for the interval $(-1, 0)$. This completes the proof of upper bound in Theorem 2.3 for $1 \le q \le 2$.

To improve this upper bound for $2 < q \le \infty$, we first show that for each $x \in \Delta_+^{r+1} W_{p,\alpha}^r$,

$$\left\|(x(\cdot) - \sigma_{r,n}(x;\cdot)) w_n(\cdot)^{-1+\frac{1}{q}}\right\|_{L_1(I)} \le cn^{-r-\frac{1}{q}}, \tag{36}$$

where $c = c(r, \alpha, p, q)$ and

$$w_n(t) := n^{-1}(1 - |t| + n^{-\beta})^{\frac{\beta-1}{\beta}}, \quad t \in I.$$

This we obtain by virtue of Lemma 4.1 since

$$c_1|I_{ni}| \le \max_{t \in I_{ni}} w_n(t) \le c_2|I_{ni}|, \quad i = \pm 1, \dots, \pm n.$$

We proceed, using discretization techniques, to prove that there exist splines $\tilde{\sigma}_{r,n}(x;\cdot)$ from n-dimensional subspaces for which the estimate

$$\|x(\cdot) - \tilde{\sigma}_{r,n}(x;\cdot)\|_{L_q(I)} \le cn^{-r-\frac{1}{2}}, \quad 2 < q \le \infty, \tag{37}$$

holds, where $c = c(r, \alpha, p, q)$. Combining (35) and (37) we get the upper bound in Theorem 2.3. Since the splines $\sigma_{r,n}(\cdot;x)$, of (35) and $\tilde{\sigma}_{r,n}(x;\cdot)$ of (37) are linear, the proof of the upper bound in Theorem 2.6 is similar.

In order to prove the lower bounds in Theorems 2.3 and 2.6, we consider the system $\Psi^{2n}_{r+1,1} := \{\psi_{r+1,n,1,i}\}_{i=1}^{2n}$, where the functions $\psi_{r+1,n,1,i}(\cdot)$ are defined by (24) for $p = 1$, and replacing r by $r+1$. Since $\|\psi^{(r)}_{r+1,n,1,i}\|_{L_\infty(I)} = 1$, we have $\psi_{r+1,n,1,i} \in \Delta^{r+1}_+ W^r_\infty$. If $S^+_1(\Psi^{2n}_{r+1,1})$ denotes the positive 1-sector over the system $\Psi^{2n}_{r+1,1}$, then $S^+_1(\Psi^{2n}_{r+1,1}) \subset \Delta^{r+1}_+ W^r_\infty$, whence

$$d_n\left(\Delta^{r+1}_+ W^r_\infty\right)_{L_q} \ge d_n\left(S^+_1(\Psi^{2n}_{r+1,1})\right)_{L_q}.$$

Using discretization techniques, we obtain

$$d_n\left(S^+_1(\Psi^{2n}_{r+1,1})\right)_{L_q} \ge cn^{-r-\frac{1}{q}} d_n\left(S^+_1(E^{2n})\right)_{l_q^{2n}},$$

where $S^+_1(E^{2n})$ is the positive 1-sector over the system $E^{2n} := \{e^{(i)}\}_{i=1}^{2n}$ of the standard orthonormal vectors $e^{(i)}$, $i = 1, \dots, 2n$ in $\mathbb{R}^{2n}$. The lower bounds in Theorem 2.3 and 2.6 now follow from the well-known lower estimates of the Kolmogorov widths. □

Proofs of Theorems 3.1, 3.2 *and* 3.3. In order to obtain upper bounds in Theorems 3.1–3.3, we take the polynomial splines $\sigma_{r,n}(x;\cdot)$ of the proof of the upper bounds in Theorem 2.2. These splines yield good approximation but, in general, they do not preserve the shape. We use correcting splines of small norm on each subinterval $I_{n,i}$, $i = \pm 1, \dots, \pm n$, to modify $\sigma_{r,n}(x;\cdot)$ in a way that the resulting splines preserves the shape. Due to the small norm of the correction we still obtain the same rate of approximation.

More interesting is the proof of the lower bounds for monotone and convex functions when $2 < q \le \infty$ and $p < q$, namely, when the lower bounds in (3) are too small.

To this end, we construct the system $\tilde{\Phi}^n_{r,p} := \{\tilde{\phi}_{r,n,p,i}(\cdot)\}_{i=1}^n$ of monotone functions such that $\|\tilde{\phi}^{(r)}_{r,n,p,i}\|_{L_p(I)} = 1$, their supports are subintervals $\tilde{I}_{n,i} \subset I$ of length $|\tilde{I}_{n,i}| \asymp n^{-1}$, which do not intersect, $\tilde{\phi}_{r,n,p,i}$ vanish to the left of $\tilde{I}_{n,i}$, and $\tilde{\phi}_{r,n,p,i}(\cdot) \equiv cn^{-r+\frac{1}{p}}$ to the right of $\tilde{I}_{n,i}$. For the positive p-sector $S_p^+(\tilde{\Phi}_{r,p}^n)$ over the system we have $S_p^+(\tilde{\Phi}_{r,p}^n) \subseteq \Delta_+^1 W_p^r$. Hence

$$d_{n-2}(\Delta_+^1 W_p^r, \Delta_+^1 L_q)_{L_q} \geq d_{n-2}(S_p^+(\tilde{\Phi}_{r,p}^n), \Delta_+^1 L_q)_{L_q}. \tag{38}$$

Using a discretization operator, we show that

$$d_{n-2}(S_p^+(\tilde{\Phi}_{r,p}^n), \Delta_+^1 L_q)_{L_q} \geq cn^{-r+\frac{1}{p}-\frac{1}{q}} d_{n-2}(S_p^+(\tilde{E}^n), \Delta_+^1)_{l_q^n}, \tag{39}$$

where $S_p^+(\tilde{E}^n)$ is the positive p-sector in $\mathbb{R}^n$ over the system $\tilde{E}^n := \{\tilde{e}^{(i)}\}_{i=1}^n$ of vectors $\tilde{e}^{(1)} := (1, \dots, 1)$, $\tilde{e}^{(2)} := (0, 1, \dots, 1)$, ..., $\tilde{e}^{(n)} := (0, \dots, 0, 1)$, and

$$\Delta_+^1 := \{x \mid x = (x_1, \dots, x_n) \in \mathbb{R}^n,\ x_1 \leq x_2 \leq \dots \leq x_n\},$$

is the cone of vectors $x = (x_1, \dots, x_n)$ with monotone coordinates.

All we need now is the estimate

$$d_m(S_p^+(\tilde{E}^n), \Delta_+^1)_{l_q^n} \geq \frac{1}{8}, \quad m+1 < n, \quad 1 \leq p \leq q \leq \infty, \tag{40}$$

which readily follows from

Lemma 4.2 *Let $n \in \mathbb{N}$, $n > 1$ and denote $\delta B_1^n := \{x \mid x \in \mathbb{R}^n, \|x\|_{l_1^n} \leq \delta\}$. Then for any $\delta_*, \delta^* > 0$ one has*

$$d_{n-1}(\delta_* B_1^n, \delta^* B_1^n)_{l_\infty^n} = \max\left\{\delta_* - \frac{\delta^*}{2}, \frac{\delta_*}{n}\right\}.$$

From (38) through (40) we obtain

$$d_n(\Delta_+^1 W_p^r, \Delta_+^1 L_q)_{L_q} \geq cn^{-r+\frac{1}{p}-\frac{1}{q}},$$

where $c = c(r, p, q)$.

The proof of lower bounds for convex functions in Theorem 3.3 is similar. We construct a system $\breve{\Phi}_{r,p}^n := \{\breve{\phi}_{r,n,p,i}(\cdot)\}_{i=1}^n$ of convex functions $\breve{\phi}_{r,n,p,i}(\cdot)$, such that the positive p-sector over this system, $S_p^+(\breve{\Phi}_{r,p}^n) \subseteq \Delta_+^2 W_p^r$. Hence

$$d_{n-2}(\Delta_+^2 W_p^r, \Delta_+^2 L_q)_{L_q} \geq d_{n-2}(S_p^+(\breve{\Phi}_{r,p}^n), \Delta_+^2 L_q)_{L_q}. \tag{41}$$

Using a discretization operator we show

$$d_{n-2}(S_p^+(\breve{\Phi}_{r.p}^n), \Delta_+^2 L_q)_{L_q} \geq cn^{-r+\frac{1}{p}-\frac{1}{q}} d_{n-2}(S_p^+(\breve{E}^n), \Delta_+^2)_{l_q^n}, \tag{42}$$

where $S_p^+(\breve{E}^n)$ is the positive p-sector in $\mathbb{R}^n$ over the system $\breve{E}^n := \{\breve{e}^{(i)}\}_{i=1}^n$ of vectors $\breve{e}^{(1)} := (1, 2, \dots, n)$, $\breve{e}^{(2)} := (0, 1, \dots, n-1)$, ..., $\breve{e}^{(n)} := (0, \dots, 0, 1)$, and

$$\Delta_+^2 := \{x = (x \mid x = (x_1, \dots, x_n) \in \mathbb{R}^n,\ x_2 - x_1 \leq \dots \leq x_n - x_{n-1}\},$$

is the cone of vectors $x = (x_1, \dots, x_n)$ with convex coordinates. It follows by Lemma 4.2 that

$$d_m\big(S_p^+(\breve{E}^n), \Delta_+^2\big)_{l_q^n} \geq \frac{1}{26}, \quad m+1 < n, \quad 1 \leq p \leq q \leq \infty,$$

which together with (41), (42) implies

$$d_n(\Delta_+^2 W_p^r, \Delta_+^2 L_q)_{L_q} \geq c n^{-r+\frac{1}{p}-\frac{1}{q}},$$

where $c = c(r, p, q)$. □

Proof of Theorem 3.4. For the upper bounds we require the following lemma which is interesting in its own right (see [5]).

Lemma 4.3 *Let $x \in C^2[a,b]$, be 3-monotone, and for $m \in \mathbb{N}$, set $t_i = t_{m,i} := a + im^{-1}|J|$, $i = 0, 1, \dots, m$, where $J := [a,b]$. Then there exists a 3-monotone quadratic spline $\sigma_{2,m}(x;\cdot)$ with knots t_i, $i = 1, \dots, m-1$, such that*

$$x''(t_{i-1}) \leq \sigma''_{2,m}(x;t) \leq x''(t_i), \quad t \in (t_{i-1}, t_i), \quad i = 1, \dots, m,$$

and

$$\|x(\cdot) - \sigma_{2,m}(x;\cdot)\|_{L_\infty(J)} \leq \frac{3}{2} m^{-2} |I|^2 \omega(x''; m^{-1}|J|),$$

$$\big\|x'(\cdot) - \sigma'_{2,m}(x;\cdot)\big\|_{L_\infty(J)} \leq \frac{7}{2} m^{-1} |J| \omega(x''; m^{-1}|J|),$$

and

$$\big\|x''(\cdot) - \sigma''_{2,m}(x;\cdot)\big\|_{L_\infty(J)} \leq \omega(x''; m^{-1}|J|).$$

Here $\omega(x;h)$ is the ordinary modulus of continuity of x.

In particular we have

Corollary. *Let x be 3-monotone and assume that $x \in W_p^3$, $1 \leq p \leq \infty$. For $m \in \mathbb{N}$, let t_i, $i = 0, 1, \dots, m$, as in Lemma 4.3. Then there exists a 3-monotone quadratic spline $\sigma_{2,m}(x;\cdot)$ with knots t_i, $i = 1, \dots, m-1$, such that*

$$x''(t_{i-1}) \leq \sigma''_{2,m}(x;t) \leq x''(t_i), \quad t \in (t_{i-1}, t_i), \quad i = 1, \dots, m,$$

and

$$\|x(\cdot) - \sigma_{2,m}(x;\cdot)\|_{L_\infty(J)} \leq \frac{3}{2} m^{-3+\frac{1}{p}} |I|^{3-\frac{1}{p}},$$

$$\big\|x'(\cdot) - \sigma'_{2,m}(x;\cdot)\big\|_{L_\infty(J)} \leq \frac{7}{2} m^{-2+\frac{1}{p}} |J|^{2-\frac{1}{p}},$$

and

$$\big\|x''(\cdot) - \sigma''_{2,m}(x;\cdot)\big\|_{L_\infty(J)} \leq m^{-1+\frac{1}{p}} |J|^{1-\frac{1}{p}}$$

Given $x \in \Delta_+^s W_p^r(I)$, $3 \le s \le r$, we then construct a spline $\sigma_{r,n}(x^{(s-3)};.)$ with $\asymp n$ equidistant knots, which is 3-monotone and approximate well $x^{(s-3)}$

$$\|x^{(s-3)}(\cdot) - \sigma_{r,n}(x^{(s-3)};\cdot)\|_{L_\infty(I)} \le cn^{-r+s+\frac{1}{p}-3},$$

where $c = c(r,s,p)$. Then by $s-3$ integrations and appropriate corrections yield an s-monotone spline $\sigma_{r,n}(x;\cdot)$, which is close enough to x. Thus proving the required upper bounds.

In order to prove the lower bounds, we construct a collection of functions $\{\psi_{r,s,n,i}\}_{i=1}^{2n}$ from $\Delta_+^s W_p^r(I)$, that behave very much like the truncated powers of degree $s-1$, such that the distance of any linear manifold M^n in $L_1(I)$, to at least one of them is no better than $cn^{-r+s+\frac{1}{p}-3}$. To this end we have estimates from below on the distance of the $(s-1)$st derivatives are from the same derivatives of arbitrary elements of the manifold, and we need to translate it to distance between from the functions themselves. Instead, we replace the functions by the truncated powers and apply the following lemma which is again interesting in its own right.

Lemma 4.4 *For $\tau \in \mathbb{R}$, $b > 0$ denote*

$$\chi_{s,\tau,b}(t) := \frac{b}{s!}(t-\tau)_+^s, \quad t \in \mathbb{R}, \quad s \in \mathbb{N}.$$

Let $s > 1$ and $\psi \in C^s[\tau - a, \tau + a]$, $a > 0$, and assume that $\psi^{(s)}$ is nondecreasing and $0 \le \psi^{(s)}(t) \le b$, in $[\tau - a, \tau + a]$. Then, if

$$\|\chi_{s,\tau,b}^{(s)} - \psi^{(s)}\|_{L_1[\tau-a,\tau+a]} \ge A,$$

where $0 < A \le ab$, then

$$\|\chi_{s,\tau,b} - \psi\|_{L_1[\tau-a,\tau+a]} \ge 2^{-s^2-4s-3}a^{s-1}b^{-1}A^2.$$

Note that since we apply Lemma 4.4 for the $(s-1)$st derivative we must have $s-1 > 1$, i.e., $s \ge 3$. This is the main reason why the case $s \ge 3$ behaves so much different than the case $1 \le s \le 2$. □

References

[1] V.N. Konovalov *Estimates of diameters of Kolmogorov type for classes of differentiable periodic functions*, Mat. Zametki, **35** (1984), 369–380.

[2] V.N. Konovalov and D. Leviatan, *Kolmogorov and linear widths of weighted Sobolev-type classes on a finite interval*, Analysis Math., to appear.

[3] V.N. Konovalov and D. Leviatan, *Kolmogorov and linear widths of weighted Sobolev-type classes on a finite interval II*, J. Approx. Theory, **113** (2001), 266–297.

[4] V.N. Konovalov and D. Leviatan, *Shape preserving widths of weighted Sobolev-type classes of positive, monotone and convex functions on a finite interval*, Constr. Approx., to appear.

[5] V.N. Konovalov and D. Leviatan, *Estimates on the approximation of 3-monotone functions by 3-monotone quadratic splines*, East J. Approx., **7** (2001), 333–349.

[6] V.N. Konovalov and D. Leviatan, *Shape preserving widths of Sobolev-type classes of s-monotone functions on a finite interval*, Israel J. Math., to appear.

[7] A.S. Shvedov, *Orders of coapproximation of functions by algebraic polynomials*, Mat. Zametki, **29** (1981), 117–130. English transl. in Math. Notes, **29** (1981), 63–70.

V.N. Konovalov
Institute of Mathematics
National Academy of Sciences of Ukraine
Kyiv 01601, Ukraine
Email address: vikono@imath.kiev.ua

D. Leviatan
School of Mathematical Sciences
Sackler Faculty of Exact Sciences
Tel Aviv University
Tel Aviv 69978, Israel
Email address: leviatan@post.tau.ac.il

Advanced Problems in Constructive Approximation
(Eds.) M.D. Buhmann and D.H. Mache

International Series of Numerical Mathematics Vol. 142, 95–107

On Approximation Methods by Using Orthogonal Polynomial Expansions

Rupert Lasser, Detlef H. Mache & Josef Obermaier

1 Introduction and basic facts

The following investigations start from a general point of view. Therefore let $(P_n)_{n\in\mathbf{N}_0}$ be an orthogonal polynomial sequence (OPS) on the real line with respect to a probability measure π with compact support S and $\text{card}(S) = \infty$. The polynomials P_n are assumed to be real valued with $\deg(P_n) = n$.
Then the sequence $(P_n)_{n\in\mathbf{N}_0}$ satisfies a three term recurrence relation of the following type

$$P_1(x)P_n(x) = a_n P_{n+1}(x) + b_n P_n(x) + c_n P_{n-1}(x), \quad n \geq 1, \tag{1}$$

with $P_0(x) = q_0$ and $P_1(x) = q_0(x - b_0)/a_0$, where the coefficients are real numbers with $c_1 q_0 > 0$, $c_n a_{n-1} > 0$, $n > 1$, and $(c_n a_{n-1})_{n\in\mathbf{N}}$, $(b_n)_{n\in\mathbf{N}}$ are bounded sequences. On the contrary, if we define $(P_n)_{n\in\mathbf{N}_0}$ by (1) we get an OPS with the assumed properties, see [3].
With

$$h(k) = \left(\int_S P_k^2(x)\, d\pi(x)\right)^{-1} = (< P_k, P_k >)^{-1} \tag{2}$$

the corresponding orthonormal polynomials are given by $p_n(x) = \sqrt{h(n)}P_n(x)$. By the Christoffel-Darboux formula, see [3], we have

$$\sum_{k=0}^{n} P_k(x)P_k(y)h(k) = \frac{a_0}{q_0} a_n h(n) \frac{P_{n+1}(x)P_n(y) - P_n(x)P_{n+1}(y)}{x - y}, \quad n \geq 1. \tag{3}$$

For $f \in L_1(S,\pi)$ one may form orthogonal expansion with respect to the OPS by

$$f \sim \sum_{k=0}^{\infty} \hat{f}(k) P_k h(k), \tag{4}$$

where the Fourier coefficients are defined by

$$\hat{f}(k) = \int_S f(x) P_k(x)\, d\pi(x) = < f, P_k > . \tag{5}$$

In this paper we will focus on weighted expansions

$$\sum_{k=0}^{n} a_{n,k} \hat{f}(k) P_k h(k), \ n \to \infty, \tag{6}$$

and study convergence properties in various norms, e.g. for $1 \leq p < \infty$ the L_p-norm when $f \in L_p(S,\pi) \subseteq L_1(S,\pi)$ or for $p = \infty$ the sup-norm when $f \in C(S)$.

For essential parts of our investigations we make the additional assumption that there exists a point $x_0 \in S$ such that

$$|P_n(x)| \leq P_n(x_0) = 1 \quad \text{for all} \quad x \in S,\ n \in \mathbf{N}_0. \tag{7}$$

Property (7) implies that the coefficients in (1) fulfill $q_0 = 1$, $a_0 + b_0 = x_0$ and $a_n + b_n + c_n = 1$, $n > 0$.
If the linearization coefficients g in

$$P_i P_j = \sum_{k=|i-j|}^{i+j} g(i,j,k) P_k \tag{8}$$

are non-negative, then there exists a normalized version of $(P_n)_{n \in \mathbf{N}_0}$ with property (7). Those polynomials are associated with a so-called hypergroup structure on $\mathbf{N}_0$ and there exist a lot of examples which are well studied, see [6] and [1]. Further on we denote an OPS with property (7) by $(R_n)_{n \in \mathbf{N}_0}$.

2 Approximate identities

Denote by B one of the Banach spaces C(S) or $L_p(S,\pi)$, $1 \leq p < \infty$, with respect to the orthogonalization measure π and by $\|\cdot\|_B$ the actual norm.
Let $(a_{n,k})_{0 \leq n < \infty, 0 \leq k \leq n}$ be a triangular matrix of complex numbers. Then we define the generating polynomial A_n by

$$A_n(x) = \sum_{k=0}^{n} a_{n,k} h(k) P_k(x) \tag{9}$$

and call the sequence $(A_n)_{n\in\mathbf{N}_0}$ a kernel. We also may identify A_n with a continuous linear operator from B into B by

$$A_nf(x) = \sum_{k=0}^{n} a_{n,k}\hat{f}(k)P_k(x)h(k) = \sum_{k=0}^{n} a_{n,k}\frac{< f, P_k >}{< P_k, P_k >}P_k(x). \qquad (10)$$

The weight coefficients $a_{n,k}$ have to be chosen appropriately to guarantee concrete features of the approximation process.
Before one goes into details let us give the following definition.

Definition 2.1 *We say that the sequence $(A_n)_{n\in\mathbf{N}_0}$ is an approximate identity with respect to B, if*

$$\lim_{n\to\infty} \|A_nf - f\|_B = \lim_{n\to\infty} \|\sum_{k=0}^{n} a_{n,k}\hat{f}(k)P_kh(k) - f\|_B = 0 \quad \textit{for all } f \in B. \quad (11)$$

The Banach-Steinhaus theorem yields necessary and sufficient conditions for $(A_n)_{n\in\mathbf{N}_0}$ to be an approximate identity, see [9].

Theorem 2.2 *The sequence $(A_n)_{n\in\mathbf{N}_0}$ is an approximate identity with respect to B if and only if the following two conditions hold.*

(i) $\lim_{n\to\infty} a_{n,k} = 1$ *for all* $k \in \mathbf{N}_0$.

(ii) *There exists a constant $C > 0$ with $\|A_nf\|_B \leq C\|f\|_B$ for all $f \in B$ and for all $n \in \mathbf{N}_0$.*

In [9] it is also shown that $(A_n)_{n\in\mathbf{N}_0}$ is an approximate identity with respect to $L_1(S,\pi)$ if and only if it is an approximate identity with respect to $C(S)$. Moreover, if $(A_n)_{n\in\mathbf{N}_0}$ is an approximate identity with respect to $C(S)$ then also with respect to $L_p(S,\pi)$, $1 < p < \infty$. Of course, the opposite direction is not true. For this reason one may focus on approximate identities with respect to $C(S)$.

Many classical approximation processes concerning trigonometric polynomials are performed by a sequence of convolution operators, see [2] and [7]. In some special cases of algebraic polynomial systems, there also does exist a proper convolution structure on $C(S)$.

Definition 2.3 *If for all $x, y \in S$ there exists a complex Borel measure $\mu_{x,y}$ with $\|\mu_{x,y}\| \leq M$, where $M > 0$ is independent of x and y, such that*

$$P_n(x)P_n(y) = \int_S P_n(z)\, d\mu_{x,y}(z) \quad \textit{for all} \quad n \in \mathbf{N}_0, \qquad (12)$$

then we say that for the OPS $(P_n)_{n\in\mathbf{N}_0}$ a product formula holds.

If a product formula holds, then we are able to define a convolution on $C(S)$ by

$$\varphi * \psi(y) = \int_S \int_S \varphi(x)\psi(z)\, d\mu_{x,y}(z)\, d\pi(x) = \int_S \int_S \varphi(z)\psi(x)\, d\mu_{x,y}(z)\, d\pi(x), \quad (13)$$

see [10], where $\varphi * \psi \in C(S)$ and

$$\|\varphi * \psi\|_\infty \leq M\|\varphi\|_1\|\psi\|_\infty. \quad (14)$$

Then the operator A_n acts as a convolution operator, that is

$$A_n f = \sum_{k=0}^{n} a_{n,k}\hat{f}(k)P_k h(k) = A_n \star f. \quad (15)$$

Now, by (14), the uniform boundedness of $\|A_n\|_1$ implies (ii) of Theorem 2.2. So one may derive great benefit from the existence of a convolution structure.

3 Positive approximate identities

The situation becomes more handsome, if we assume the operators A_n to be positive. For the remainder of this section we suppose $(R_n)_{n\in\mathbf{N}_0}$ to be an OPS with property (7).

Definition 3.1 *An operator G from $C(S)$ into $C(S)$ is called positive, if $f \geq 0$ implies $Gf \geq 0$.*

In case of positive operators there is a simplification of Theorem 2.2.

Theorem 3.2 *Let $(R_n)_{n\in\mathbf{N}_0}$ be an OPS with property (7) and $(A_n)_{n\in\mathbf{N}_0}$ be a sequence of positive operators.*
Then $(A_n)_{n\in\mathbf{N}_0}$ is an approximate identity with respect to B if and only if the following two conditions hold.

(i) $\lim_{n\to\infty} a_{n,0} = \lim_{n\to\infty} a_{n,1} = 1$.

(ii) *There exists a constant $C > 0$ with $\|A_n f\|_B \leq C\|f\|_B$ for all $f \in B$ and for all $n \in \mathbf{N}_0$.*

Proof. Define by $D_k(x) = \sum_{i=0}^{k} R_i(x)h(i)$ the so-called Dirichlet kernel. By the Christoffel-Darboux formula (3) and (7) we derive

$$(1 - R_1(x))D_k(x) = a_k h(k)(R_k(x) - R_{k+1}(x)), \quad k \geq 1.$$

Let $x \in S$. By (7) again it holds $|D_k(x)| \leq D_k(x_0)$ and therefore

$$-(1 - R_1(x))\frac{D_k(x_0)}{|a_k|h(k)} \leq R_k(x) - R_{k+1}(x) \leq (1 - R_1(x))\frac{D_k(x_0)}{|a_k|h(k)}.$$

In case $n \geq m$ it is simple to deduce that $A_n R_m = a_{n,m} R_m$.

Hence, if $n > k$, then the stated positivity of the operators implies

$$\begin{aligned} -(a_{n,0} - a_{n,1}R_1(x))\frac{D_k(x_0)}{|a_k|h(k)} &\leq a_{n,k}R_k(x) - a_{n,k+1}R_{k+1}(x) \\ &\leq (a_{n,0} - a_{n,1}R_1(x))\frac{D_k(x_0)}{|a_k|h(k)}. \end{aligned}$$

With $x = x_0$ we get

$$|a_{n,k} - a_{n,k+1}| \leq \frac{D_k(x_0)}{|a_k|h(k)}(a_{n,0} - a_{n,1}) \quad \text{for all} \quad n > k \geq 1.$$

Now condition (i) yields $\lim_{n\to\infty} |a_{n,k} - a_{n,k+1}| = 0$.
By induction we get $\lim_{n\to\infty} a_{n,k} = 1$ for all $k \in \mathbf{N}_0$ and according to Theorem 2.2 the proof is complete. □

In case of a product formula with corresponding positive measures we may achieve positive operators by a simple procedure.

Theorem 3.3 *Let $(R_n)_{n\in\mathbf{N}_0}$ be an OPS with property* (7) *and suppose that a product formula holds, where $\mu_{x,y}$ is a positive measure for all $x, y \in S$.*
If the generating polynomials A_n are non-negative, i.e., $A_n(x) \geq 0$ for all $x \in S$, then $(A_n)_{n\in\mathbf{N}_0}$ is a sequence of positive operators.
Additionally, if $\lim_{n\to\infty} a_{n,0} = \lim_{n\to\infty} a_{n,1} = 1$, then $(A_n)_{n\in\mathbf{N}_0}$ is an approximate identity with respect to any B.

Proof. Since $A_n(z) \geq 0$ on S and $\mu_{x,y} \geq 0$, the positivity of the operators is shown by $A_n f(y) = \int_S \int_S A_n(z) f(x)\, \mu_{x,y}(z)\, \pi(x)$. Moreover, we have $\|A_n\|_1 = a_{n,0}$ and therefore $\lim_{n\to\infty} a_{n,0} = 1$ implies the uniform boundedness of $\|A_n\|_1$. Now, by (14) and Theorem 3.2, $(A_n)_{n\in\mathbf{N}_0}$ is an approximate identity with respect to $C(S)$ and according to the remark after Theorem 2.2 with respect to any B. □

4 Local convergence behaviour of A_n

At first we give a local error estimate for the approximation of functions $f \in B = L_p(S)$ or $C(S)$, $S = [0, 1]$, by $A_n f$ defined in (10). In the following we assume that A_n is a positive linear operator with $a_{n,0} = 1$, i.e., that the operator A_n preserves constant functions ($A_n R_0 = 1$).

Let us note that the estimates in the following theorem will show that the order of approximation depends on the first weight coefficient $a_{n,1}$. To do this, recall that the Lipschitz type maximal function of order σ introduced by B. Lenze [11] is defined as

$$f_\sigma^{\sim}(x) = \sup_{t\neq x, t\in S} \frac{|f(x) - f(t)|}{|x - t|^\sigma}, \qquad x \in S,\ \sigma \in]0, 1].$$

Then we have

Theorem 4.1 *There exists a constant $C > 0$ such that for each bounded function $f \in B$ and for all $x \in S$*

$$| f(x) - (A_n f)(x) | \leq C \, f_\sigma^\sim(x) \left(\max\{|g_1(x)|, |g_2(x)|\}(1 - a_{n,1}) \right)^{\frac{\sigma}{2}},$$

for $\sigma \in]0, 1]$ by using the functions

$$g_1(x) := a_0^2 c_1 - (b_0 - x)(a_0 b_1 + (b_0 - x)) \qquad g_2(x) := (b_0 - x)(a_0 b_1 + 2(b_0 - x)). \quad (16)$$

Proof. For the sequence $(R_n)_{n\in\mathbf{N}_0}$ we have the recurrence relation (1) with property (7)

$$R_1(x) R_n(x) = a_n R_{n+1}(x) + b_n R_n(x) + c_n R_{n-1}(x), \quad n \geq 1,$$

with $R_0(x) = 1$ and $R_1(x) = (x - b_0)/a_0$. By using the eigenstructure of A_n, i.e., $A_n R_k = a_{n,k} R_k$ and the known inequality

$$|f(t) - f(x)| \leq |t - x|^\sigma \, f_\sigma^\sim(x), \qquad x, t \in S,$$

for $0 < \sigma \leq 1$, one obtains with $A_n((t-x)^2; x) = (1 - a_{n,2}) g_1(x) + (1 - a_{n,1}) g_2(x)$ and Hölder's inequality

$$\begin{aligned} | f(x) - (A_n f)(x) | &\leq f_\sigma^\sim(x) \, A_n(|t - x|^\sigma; x) \leq f_\sigma^\sim(x) \, A_n((t-x)^2; x)^{\frac{\sigma}{2}} \\ &\leq C \, f_\sigma^\sim(x) \left(\max\{|g_1(x)|, |g_2(x)|\}(1 - a_{n,1}) \right)^{\frac{\sigma}{2}}, \end{aligned}$$

which concludes the proof. □

At this point we mention that the constant C denotes a positive constant which can be different at each occurrence.

Now if $f \in B := C(S)$, for every $x \in S$, the k-th difference $\Delta_h^k f(x)$ of f with the step $h \in \mathbf{R}, h \neq 0$ at the point x is given by $\Delta_h^f(x) := \sum_{m=0}^{k} (-1)^{m+k} \binom{k}{m} f(x + mh)$, provided that the arguments $x + kh \in S$. For the sake of brevity one sets $\Delta_h f(x) := \Delta_h^1 f(x) = f(x + h) - f(x)$. Now, if $f : S \to \mathbf{R}$ is a bounded real function and if $\delta > 0$, the k-th modulus of continuity $\omega_k(f; \delta)$ of f is defined by

$$\omega_k(f; \delta) := \sup_{|h| \leq \delta, \, x, x+h \in S} |\Delta_h^k f(x)|,$$

where for $k = 1$ we have the well-known modulus of continuity $\omega(f \; \delta) := \omega_1(f; \delta)$.

Now one can formulate the following

Theorem 4.2 *(Local Direct Results)*
For $f \in B := C(S)$, $x \in S$, we have

$$| f(x) - (A_n f)(x) | \leq C \omega(f; \sqrt{1 - a_{n,1}}) \quad (17)$$

and further in addition with the second modulus

$$|f(x) - (A_n f)(x)| \leq C \left(\omega_2(f; \sqrt{1-a_{n,1}}) + |b_0 - x|\sqrt{1-a_{n,1}}\ \omega(f; \sqrt{1-a_{n,1}})\right), \tag{18}$$

where C is a positive constant.

Proof. Following the known arguments in [13] and [16] we have that the method A_n defined as

$$(A_n f)(x) = \sum_{k=0}^{n} a_{n,k}\hat{f}(k)R_k(x)h(k) = \sum_{k=0}^{n} a_{n,k}\frac{< f, R_k >}{< R_k, R_k >}R_k(x)$$

satisfies the inequality

$$|f(x) - (A_n f)(x)| \leq 2\ \omega(f; A_n(|x-t|; x)),$$

which proves (17). The second inequality (18) follows by the estimates used in [13], i.e., for $h \in (0, 2]$ and $x \in S$

$$\begin{aligned} |f(x) - (A_n f)(x)| &\leq \left(3 + \frac{1}{h^2}A_n((t-x)^2; x)\right)\omega_2(f; h) \\ &\quad + \frac{2}{h}|A_n(x-t; x)|\omega(f; h) \\ &\leq \left(3 + \frac{1}{h^2}\max\{g_1(x), g_2(x)\}(1-a_{n,1})\right)\omega_2(f; h) \\ &\quad + \frac{2}{h}|b_0 - x|(1-a_{n,1})\ \omega(f; h). \end{aligned}$$

Therefore, with $h = \sqrt{1-a_{n,1}}$

$$\begin{aligned} |f(x) - (A_n f)(x)| &\leq (3+C)\omega_2(f; \sqrt{1-a_{n,1}}) \\ &\quad + 2|b_0 - x|\sqrt{1-a_{n,1}}\ \omega(f; \sqrt{1-a_{n,1}}), \end{aligned}$$

which proves (18). □

With the first part of the above theorem we can prove the following equivalence result

Theorem 4.3 *(Local Characterization Result)*
Let A_n, be a positive linear operator defined as in (10). Under the assumption that $1 - a_{n,1} = \mathcal{O}(n^{-2})$ we have for $f \in B(S)$ and $0 < \sigma < 1$

$$\begin{aligned} |f(x) - (A_n f)(x)| &\leq C\left(\tfrac{1}{n}\right)^{\sigma} && (n \in \mathbf{N}) \\ \Longleftrightarrow \quad \omega(f; t) &= \mathcal{O}(t^{\sigma}) && (t \to 0). \end{aligned}$$

Some approximation operators in a similar form like in (10) were investigated in [13] and [14].

5 Examples of kernels

Having in mind Theorem 3.3 let us now derive some important kernels, which are associated with a sequence of positive operators and investigate there convergence behaviour. We also lay stress on the relationship to the corresponding trigonometric kernels.
The most outstanding examples of OPS are the Jacobi polynomials $(J_n^{(\alpha,\beta)})_{n\in\mathbf{N}_0}$, $\alpha,\beta>-1$, which are orthogonal with respect to $(1-x)^\alpha(1+x)^\beta dx$ and normalized by $J_n^{(\alpha,\beta)}(1)=1$. They exactly fit the conditions of Theorem 3.3, if $\alpha\geq\beta$ and either $\alpha+\beta\geq 0$ or $\beta\geq-\frac{1}{2}$, see [4] and [6]. For Jacobi polynomials $(J_n^{(\alpha,\beta)})_{n\in\mathbf{N}_0}$ we have $a_0^{(\alpha,\beta)}=\frac{2(\alpha+1)}{\alpha+\beta+2}$, $b_0^{(\alpha,\beta)}=\frac{\beta-\alpha}{\alpha+\beta+2}$, $a_n^{(\alpha,\beta)}=\frac{(\alpha+\beta+2)(n+\alpha+\beta+1)(n+\alpha+1)}{(\alpha+1)(2n+\alpha+\beta+1)(2n+\alpha+\beta)}$, $b_n^{(\alpha,\beta)}=\frac{\alpha-\beta}{2(\alpha+1)}\left(1-\frac{(\alpha+\beta)(\alpha+\beta+2)}{(2n+\alpha+\beta)(2n+\alpha+\beta+2)}\right)$, $c_n^{(\alpha,\beta)}=\frac{(\alpha+\beta+2)n(n+\beta}{(\alpha+1)(2n+\alpha+\beta)(2n+\alpha+\beta+1)}$ und $h^{(\alpha,\beta)}(n)=\frac{(2n+\alpha+\beta+1)\Gamma(\beta+1)\Gamma(n+\alpha+1)\Gamma(n+\alpha+\beta+1)}{\Gamma(\alpha+1)\Gamma(\alpha+\beta+2)\Gamma(n+1)\Gamma(n+\beta+1)}$, $n\geq 1$.
Another example are the generalized Chebyshev polynomials $(C_n^{(\alpha,\beta)})_{n\in\mathbf{N}_0}$, $\alpha,\beta>-1$, which are orthogonal with respect to $(1-x^2)^\alpha|x|^{2\beta+1}dx$ and normalized by $C_n^{(\alpha,\beta)}(1)=1$. If ($\alpha\geq\beta$ and $\alpha+\beta>-1$) or $\beta\geq-\frac{1}{2}$, then they are fitting our conditions, too, see [5] and [6].

5.1 Fejér kernel

In the trigonometric case the Fejér kernel $(F_n)_{n\in\mathbf{N}_0}$ is defined as (C,1)-series of the Dirichlet kernel $(D_n)_{n\in\mathbf{N}_0}$ by

$$F_n(t)=\frac{1}{n+1}\sum_{k=0}^{n}D_k(t)=\sum_{k=-n}^{n}(1-\frac{|k|}{n+1})e^{ikt},\quad t\in[0,2\pi[,\tag{19}$$

see [2, Sec. 1.2.2]. Moreover, in the even case $n=2p$ we have the representation

$$F_{2p}(t)=\frac{D_p^2(t)}{D_p(0)}.\tag{20}$$

Following the even trigonometric case we define a general Fejér kernel $(F_{2p})_{p\in\mathbf{N}_0}$ for OPS $(R_n)_{n\in\mathbf{N}_0}$ by

$$F_{2p}(x)=\frac{D_p^2(x)}{\chi_{2p,0}}=\sum_{k=0}^{2p}\varphi_{2p,k}R_k(x)h(k),\quad x\in S,\tag{21}$$

where $D_p(x)=\sum_{k=0}^{p}R_k(x)h(k)$, see [8] and [9]. The coefficients $\chi_{2p,k}$ are uniquely defined by $D_p^2(x)=\sum_{k=0}^{2p}\chi_{2p,k}R_k(x)h(k)$. More explicitly we get

$$\chi_{2p,k}=\sum_{j=0}^{p}\sum_{i=|k-j|}^{p}g(k,j,i)h(j).\tag{22}$$

Hence, the Fejér weights are determined by

$$\varphi_{2p,k} = \frac{\chi_{2p,k}}{\chi_{2p,0}} = \frac{\sum_{j=0}^{p} \sum_{i=|k-j|}^{p} g(k,j,i)h(j)}{\sum_{j=0}^{p} h(j)}. \tag{23}$$

Obviously, our definition yields $F_{2p}(x) \geq 0$ and $\|F_{2p}\|_1 = \varphi_{2p,0} = 1$.
In case $p \geq 1$ we derive

$$\varphi_{2p,1} = 1 - \frac{a_p h(p)}{\sum_{j=0}^{p} h(j)}. \tag{24}$$

In [8] we have shown that $\lim_{n\to\infty} c_n/a_{n-1} = 1$ implies $\lim_{n\to\infty} \varphi_{2p,1} = 1$.
For Jacobi polynomial systems $(J_n^{(\alpha,\beta)})_{n\in\mathbf{N}_0}$ we get

$$F_{2p}^{(\alpha,\beta)}(x) = \frac{\Gamma(p+\alpha+\beta+2)\Gamma(\beta+1)\Gamma(p+\alpha+2)}{\Gamma(\alpha+\beta+2)\Gamma(p+\beta+1)\Gamma(\alpha+2)\Gamma(p+1)} (J_p^{(\alpha+1,\beta)}(x))^2 \tag{25}$$

and

$$\varphi_{2p,1} = 1 - \frac{\alpha+\beta+2}{2p+\alpha+\beta+2}. \tag{26}$$

Thus $1 - \varphi_{2p,1} = \mathcal{O}(p^{-1})$.
Especially for Chebyshev polynomials of the first kind ($\alpha = \beta = -1/2$) our Fejér kernel coincides with the trigonometric one and for Chebyshev polynomials of the second kind ($\alpha = \beta = 1/2$) the weights are given by

$$\varphi_{2p,k} = \begin{cases} 1 - \frac{q(1+q)(6p^2+18p+13-2q-2q^2)}{(1+2q)(1+p)(2+p)(3+2p)} & \text{if} \quad k = 2q, \\ \\ 1 - \frac{(1+q)(3p^2+6+9p-2q-q^2)}{(1+p)(2+p)(3+2p)} & \text{if} \quad k = 2q+1. \end{cases} \tag{27}$$

5.2 Fejér-Korovkin kernel

In the trigonometric case the Fejér-Korovkin kernel $(FK_n)_{n\in\mathbf{N}_0}$ is defined by

$$FK_n(t) = \frac{2\sin^2(\pi/(n+2))}{n+2} \left(\frac{\cos((n+2)t/2)}{\cos t - \cos(\pi/(n+2))} \right)^2, \quad t \in [0, 2\pi[, \tag{28}$$

see [2, Sec. 1.6.1]. By substitution $x = \cos t$ we get

$$\frac{\sin^2(\pi/(n+2))}{n+2} \frac{1 + T_{n+2}(x)}{(x - \cos(\pi/(n+2)))^2}, \tag{29}$$

where $T_n(x) = \cos(n \arccos x)$ are the Chebyshev polynomials of the first kind. Whereas in the even case $n = 2p$ we achieve

$$\frac{\sin^2(\pi/(2(p+1)))}{p+1} \left(\frac{T_{p+1}(x)}{x - \cos(\pi/(2(p+1)))} \right)^2. \tag{30}$$

Following the even case we define a general Fejér-Korovkin kernel $(F_{2p})_{p\in\mathbf{N}_0}$ for OPS $(R_n)_{n\in\mathbf{N}_0}$ by

$$FK_{2p}(x) = \frac{1}{\chi_{2p,0}}\left(\frac{R_{p+1}(x)}{x-z_{p+1}}\right)^2 = \sum_{k=0}^{2p}\kappa_{2p,k}R_k(x)h(k), \quad x\in S, \tag{31}$$

where z_{p+1} is that zero of R_{p+1}, which is as close to x_0 as possible.
The coefficients $\chi_{2p,k}$ are uniquely defined by $\left(\frac{R_{p+1}(x)}{x-z_{p+1}}\right)^2 = \sum_{k=0}^{2p}\chi_{2p,k}R_k(x)h(k)$.
It holds

$$\chi_{2p,k} = \frac{1}{(R_{p+2}(z_{p+1})h(p+1)a_0a_{p+1})^2}\sum_{j=0}^{p}R_j(z_{p+1})h(j)\sum_{i=|k-j|}^{p}g(k,j,i)R_i(z_{p+1}). \tag{32}$$

The Fejér-Korovkin weights are given by

$$\kappa_{2p,k} = \frac{\chi_{2p,k}}{\chi_{2p,0}} = \frac{\sum_{j=0}^{p}R_j(z_{p+1})h(j)\sum_{i=|k-j|}^{p}g(k,j,i)R_i(z_{p+1})}{\sum_{j=0}^{p}R_j^2(z_{p+1})h(j)}. \tag{33}$$

Obviously, our definition yields $FK_{2p}(x)\geq 0$ and $\|FK_{2p}\|_1 = \kappa_{2p,0} = 1$.
In case $p\geq 1$ we derive

$$\kappa_{2p,1} = R_1(z_{p+1}). \tag{34}$$

Since it is well known that $\lim_{p\to\infty} z_p = x_0$ we get $\lim_{p\to\infty}\kappa_{2p,1} = 1$.
For Jacobi polynomial systems $(J_n^{(\alpha,\beta)})_{n\in\mathbf{N}_0}$ we also define kernel polynomials of odd degree by

$$FK_n^{(\alpha,\beta)}(x) = \gamma_n^{(\alpha,\beta)}(x+1)^r\left(\frac{J_{\lfloor\frac{n}{2}\rfloor+1}^{(\alpha,\beta+r)}(x)}{x-z_{\lfloor\frac{n}{2}\rfloor+1}^{(\alpha,\beta+r)}}\right)^2, \tag{35}$$

where $r = n \mod 2$, $z_{\lfloor\frac{n}{2}\rfloor+1}^{(\alpha,\beta+r)}$ is that zero of $J_{\lfloor\frac{n}{2}\rfloor+1}^{(\alpha,\beta+r)}$ which is as close to 1 as possible,

$$\gamma_n^{(\alpha,\beta)} = \left(\frac{\alpha+\beta+2}{2(\beta+1)}\right)^{1-r}\frac{(J_{\lfloor\frac{n}{2}\rfloor+2}^{(\alpha,\beta+r)}(z_{\lfloor\frac{n}{2}\rfloor+1}^{(\alpha,\beta+r)})h^{(\alpha,\beta+r)}(\lfloor\frac{n}{2}\rfloor+1)a_0^{(\alpha,\beta+r)}a_{\lfloor\frac{n}{2}\rfloor+1}^{(\alpha,\beta+r)})^2}{\sum_{j=0}^{\lfloor\frac{n}{2}\rfloor}h^{(\alpha,\beta+r)}(j)(J_j^{(\alpha,\beta+r)}(z_{\lfloor\frac{n}{2}\rfloor+1}^{(\alpha,\beta+r)}))^2}$$

and

$$\kappa_{n,1}^{(\alpha,\beta)} = J_1^{(\alpha,\beta)}(z_{\lfloor\frac{n}{2}\rfloor+1}^{(\alpha,\beta+r)}) = 1-\frac{\alpha+\beta+2}{2(\alpha+1)}(1-z_{\lfloor\frac{n}{2}\rfloor+1}^{(\alpha,\beta+r)}). \tag{36}$$

This kernel is also known as general Jacobi kernel. It holds $1-\kappa_{n,1}^{(\alpha,\beta)} = \mathcal{O}(n^{-2})$, which for positive kernels is the best achievable rate of convergence, see [16].

Especially for Chebyshev polynomials of the first kind ($\alpha = \beta = -1/2$) our Fejér-Korovkin kernel coincides with the well-known kernel (29) and we have

$$\kappa_{n,k}^{(-\frac{1}{2},-\frac{1}{2})} = \frac{n-k+2}{n+2}\cos\frac{k\pi}{n+2} + \frac{\cos(\pi/(n+2))}{(n+2)\sin(\pi/(n+2))}\sin\frac{k\pi}{n+2}. \qquad (37)$$

For Chebyshev polynomials of the first kind we may also define a nearby Fejér-Korovkin kernel $(NFK_n^{(-\frac{1}{2},-\frac{1}{2})})_{n\in\mathbf{N}_0}$ by

$$NFK_n^{(-\frac{1}{2},-\frac{1}{2})}(x) = \cos\frac{k\pi}{n+2}FK_n^{(-\frac{1}{2},-\frac{1}{2})}(x) = 1 + 2\sum_{k=1}^{n}\nu_{n,k}^{(-\frac{1}{2},-\frac{1}{2})}T_k(x), \qquad (38)$$

with $\nu_{n,k}^{(-\frac{1}{2},-\frac{1}{2})} = \cos\frac{k\pi}{n+2}\kappa_{n,k}^{(-\frac{1}{2},-\frac{1}{2})}$. Of course, this nearby Fejér-Korovkin kernel is also a positive approximate identity and $1 - \nu_{n,1}^{(-\frac{1}{2},-\frac{1}{2})} = \sin^2\frac{\pi}{n+2} = \mathcal{O}(n^{-2})$.

5.3 De la Vallée-Poussin kernel

For trigonometric polynomials the de la Vallée-Poussin kernel $(V_n)_{n\in\mathbf{N}_0}$ is defined by

$$V_n(t) = 1 + \sum_{k=1}^{n} v_{n,k}\cos kt = \frac{(n!)^2}{(2n)!}(\cos t + 1)^n, \quad v_{n,k} = \frac{(n!)^2}{(n-k)!(n+k)!}, \qquad (39)$$

$t \in [-\pi, \pi)$, see [2, Sec. 2.5.2]. We are able to give the definition of a general de la Vallée-Poussin kernel for OPS $(R_n)_{n\in\mathbf{N}_0}$, see [18], where the assumptions on the coefficients in (1) are changed slightly.
Let us fix $s = -\min_{x\in S} R_1(x) > 0$. The de la Vallée-Poussin kernel $(V_n)_{n\in\mathbf{N}_0}$ and the de la Vallée-Poussin weights $v_{n,k}$ are defined by

$$V_n(x) = \frac{(R_1(x)+s)^n}{\chi_{n,0}} = \sum_{k=0}^{n} v_{n,k}R_k(x)h(k), \qquad (40)$$

where the coefficients $\chi_{n,k}$ are uniquely determined by

$$(R_1(x)+s)^n = \sum_{k=0}^{n}\chi_{n,k}R_k(x)h(k).$$

If we define the coefficients $\delta_{i,j}$ by

$$R_1^i(x) = \sum_{j=0}^{\infty}\delta_{i,j}R_j(x) = \sum_{j=0}^{i}\delta_{i,j}R_j(x), \qquad (41)$$

then we get $\chi_{n,k} = \sum_{i=0}^{n} \binom{n}{i} s^{n-i}\delta_{i,k}/h(k)$ and

$$v_{n,k} = \frac{\chi_{n,k}}{\chi_{n,0}} = \frac{\sum_{i=0}^{n} \binom{n}{i} s^{n-i}\delta_{i,k}}{\sum_{i=0}^{n} \binom{n}{i} s^{n-i}\delta_{i,0}} \frac{1}{h(k)}. \tag{42}$$

Obviously, our definition yields $V_n(x) \geq 0$, $x \in S$ and $\|V_n\|_1 = v_{n,0} = 1$. Moreover, if in the non-symmetric case, i.e., $\exists n \in \mathbf{N} : b_n \neq 0$, holds $\lim_{i\to\infty} \delta_{i+1,0}/\delta_{i,0} = 1$ and in the symmetric case, i.e., $\forall n \in \mathbf{N} : b_n = 0$, holds $\lim_{i\to\infty} \delta_{2i+2,0}/\delta_{2i,0} = 1$, then $\lim_{n\to\infty} v_{n,1} = 1$.
Especially for Jacobi polynomials we have

$$V_n(x) = \sum_{k=0}^{n} v_{n,k} R_k(x) h(k) = \frac{\Gamma(\beta+1)\Gamma(n+\alpha+\beta+2)}{\Gamma(n+\beta+1)\Gamma(\alpha+\beta+2)} \left(\frac{1+x}{2}\right)^n \tag{43}$$

with de la Vallée Poussin weights

$$v_{n,k} = \frac{n!\Gamma(n+\alpha+\beta+2)}{(n-k)!\Gamma(n+k+\alpha+\beta+2)}. \tag{44}$$

Thus $1 - v_{n,1} = \frac{\alpha+\beta+2}{n+\alpha+\beta+2} = \mathcal{O}(n^{-1})$.

References

[1] Bloom, W.R., Heyer, H., *Harmonic Analysis of Probability Measures on Hypergroups*, de Gruyter, Berlin–New York, (1995).

[2] Butzer, P.L., Nessel, R.J., *Fourier analysis and approximation*, Birkhäuser, Basel–Stuttgart, (1971).

[3] Chihara, T.S., *An introduction to orthogonal polynomials*, Gordon and Breach, New York, (1978).

[4] Gasper, G., *Banach algebras for Jacobi series and positivity of a kernel*, *Ann. of Math.*, Vol. 95, (1972), 261–280.

[5] Laine, T.P., *The product formula and convolution structure for the generalized Chebyshev polynomials*, *SIAM J. Math. Anal.*, Vol. 11, (1980), 133–146.

[6] Lasser, R., *Orthogonal polynomials and hypergroups*, *Rend. Mat.*, Vol. 3, (1983), 185–209.

[7] Lasser, R., *Introduction to Fourier Series*, Marcel Dekker, New York, (1996).

[8] Lasser, R., Obermaier, J., *On Fejér means with respect to orthogonal polynomials: A hypergroup theoretic approach*, in: Progress of Approximation Theorie, Academic Press, Boston, (1991), 551–565.

[9] Lasser, R., Obermaier, J., *On the convergence of weighted Fourier expansions*, *Acta. Sci. Math.*, VOL. 61, (1995), 345–355.

[10] Lasser, R., Obermaier, J., *Orthogonal Expansions for L^p and C-spaces*, in: Special Functions, Proceedings of the International Workshop, World Scientific Publishing, Singapore – London, (2000), 194–206.

[11] Lenze, B., *On Lipschitz-type maximal functions and their smoothness spaces*, Proc. Netherl. Acad. Sci. A 91 (1988), 53–63.

[12] Lubinsky, D.S. and Mache, D.H. $(C, 1)$ *Means of Orthonormal Expansions for Exponential Weights, Journal of Approximation Theory*, VOL. 103, No. 1, (2000), 151–182.

[13] Lupas, A. & Mache, D.H.; *The Θ-Transformation of Certain Positive Linear Operators, in: International Journal of Mathematics and Mathematical Sciences*, VOL. 19, No. 4, (1996), 667–678.

[14] Mache, D.H.; *Generalized Methods using Integral Transforms and Convolution Structures with Jacobi Orthogonal Polynomials, Journal Rendiconti del Circolo Matematico di Palermo*, Serie II, Suppl. 68 (2002), 625–640.

[15] Mache, D.H.; *Summation of Durrmeyer Operators with Chebyshev Weights, in: Approximation and Optimization, Proceedings of the Int. Conf. on Approximation and Optimization – ICAOR*, Cluj – Napoca (1996), VOL. I, 307–318.

[16] Mache, D.H.; *Optimale Konvergenzordnung positiver Summationsverfahren der Durrmeyer Operatoren mit Jacobi-Gewichtungen*; *Habilitationsschrift*, Dortmund (1997).

[17] Mache, D.H. & Zhou, D.X.; *Characterization Theorems for the Approximation by a Family of Operators, in: Journal of Approximation Theory*, VOL. 84, No. 2, (1996), 145–161.

[18] Obermaier, J., *The de la Vallée Poussin Kernel for Orthogonal Polynomial Systems*, *Analaysis*, VOL. 21, (2001), 277–288.

Rupert Lasser
Josef Obermaier
GSF-Forschungszentrum
für Umwelt und Gesundheit
Institut f. Biomathematik und Biometrie
Ingolstädter Landstrasse 1
D-85764 Neuherberg
lasser@gsf.de
josef.obermaier@gsf.de

Detlef H. Mache

Universität Dortmund
Institut f. Angewandte Mathematik
Vogelpothsweg 87
D-44221 Dortmund
mache@math.uni-dortmund.de

Advanced Problems in Constructive Approximation
(Eds.) M.D. Buhmann and D.H. Mache
International Series of Numerical Mathematics Vol. 142, 109–128

Curious q-Series as Counterexamples in Padé Approximation

D.S. Lubinsky

Abstract

Basic hypergeometric, or q-series, are usually investigated when $|q| < 1$. Less common is the case $|q| > 1$, and the case where q is on the unit circle is extremely rare. It is the latter curious, exotic, choice of q that has yielded a number of interesting examples and counterexamples in Padé approximation, including a counterexample to the Baker-Gammel-Wills Conjecture. We survey some of these, and also pose a number of problems involving q-series for q on the unit circle.

1 Introduction

Let

$$f(z) = \sum_{j=0}^{\infty} a_j z^j \tag{1}$$

be a formal power series, with complex coefficients. Given integers $m, n \geq 0$, the (m,n) Padé approximant to f is a rational function

$$[m/n] = P/Q$$

where P, Q are polynomials of degree at most m, n respectively, such that Q is not identically 0, and such that

$$(fQ - P)(z) = O\left(z^{m+n+1}\right).$$

By this last relation, we mean that the coefficients of $1, z, z^2, \ldots, z^{m+n}$ in the formal power series on the left-hand side vanish. The basic idea is that $[m/n]$ is a rational function with given upper bounds on its numerator and denominator degrees, chosen in such a way that its Maclaurin series reproduces as many terms as possible in the power series f. It is not difficult to see that $[m/n]$ exists and is unique.

Because there are two parameters m and n, it is natural to form the array or Padé table

$$\begin{array}{ccccc} [0/0] & [0/1] & [0/2] & [0/3] & \cdots \\ [1/0] & [1/1] & [1/2] & [1/3] & \cdots \\ [2/0] & [2/1] & [2/2] & [2/3] & \cdots \\ [3/0] & [3/1] & [3/2] & [3/3] & \cdots \\ \vdots & \vdots & \vdots & \vdots & \ddots \end{array}$$

and then to investigate convergence of sequences of approximants as we traverse some path in the table.

The path traversed has a dramatic effect on the convergence properties of the sequence. For example, the first column $\{[m/0]\}_{m=1}^{\infty}$ is nothing more than the sequence of partial sums of the MacLaurin series:

$$[m/0](z) = \sum_{j=0}^{m} a_j z^j.$$

So the first column has the convergence properties of a Taylor series.

What about the nth column, where $n \geq 1$? Here $[m/n]$ is a rational function with at most n poles, so cannot be expected to approximate as $m \to \infty$, a function with more than n poles. That it does approximate functions with exactly n poles is the de Montessus de Ballore theorem, the oldest and one of the most widely applied convergence results on Padé approximation. Here is the simplest form of the theorem [6, p. 282]:

Theorem 1.1 (De Montessus de Ballore's Theorem) *Let f be analytic at 0 and in the unit ball $U = \{z : |z| < 1\}$, except for poles of total multiplicity n. Then*

$$\lim_{m \to \infty} [m/n](z) = f(z),$$

uniformly in compact subsets of the unit ball omitting poles of f.

What happens if we try to approximate a function f with $< n$ poles in U, using the sequence $\{[m/n]\}_{m=1}^{\infty}$? Because the approximants have "extra" poles, some of those extra poles do not know where to go. In this case, the sequence may converge or diverge. This is a whole topic on its own, the so-called "intermediate rows." See [29].

Even when the full sequence $\{[m/n]\}_{m=1}^{\infty}$ does not converge in this intermediate row case, is it possible that a *subsequence* converges? A. Beardon proved this true for the case $n = 1$, but G. Baker and P. Graves-Morris observed that a subsequence often converges for any n. They obtained partial results and formulated a general conjecture [5]:

Conjecture 1.2 (Baker-Graves-Morris Conjecture) *Let f be analytic at 0 and in the unit ball $U = \{z : |z| < 1\}$, except for poles of total multiplicity $\ell < n$. Then there exists an increasing sequence $\mathcal{S}$ of positive integers such that*

$$\lim_{\substack{m \to \infty \\ m \in \mathcal{S}}} [m/n](z) = f(z),$$

uniformly in compact subsets of the unit ball omitting poles of f.

The conjecture was finally resolved by Buslaev, Goncar and Suetin [8], after the efforts of many authors. They showed that there is a function analytic in the unit ball for which the conjecture is false for $n = 2$. Nevertheless, they did prove that there is a constant $\sigma_n > 0$, independent of the function, such that some subsequence converges in $\{z : |z| < \sigma_n\}$ away from the poles. Using scale invariance of Padé approximants, they deduced that the Baker-Graves-Morris Conjecture is true for functions meromorphic in the whole plane, with less than n poles there.

In the next section, we shall discuss how the partial theta function

$$\sum_{j=0}^{\infty} q^{j(j-1)/2} z^j$$

gives an example of a function for which the Baker-Graves-Morris Conjecture fails for every $n \geq 2$.

Traversing a diagonal seems to be the next natural case to study. In fact, surely $[n/n]$ should be the "best" Padé approximant, as it makes full use of its rational nature? The convergence nature of the diagonal sequence is complicated and not yet fully understood. There are power series f with zero radius of convergence, for which $[n/n](z)$ converges as $n \to \infty$ to a function single valued and analytic in the cut-plane $\mathbf{C} \backslash [0, \infty)$. On the other hand, Hans Wallin constructed in the early 1970's [33] an entire function f for which

$$\limsup_{n \to \infty} |[n/n](z)| = \infty$$

for all $z \in \mathbf{C} \backslash \{0\}$. The problem in Wallin's example is that each point in the plane is a limit point of poles of $\{[n/n]\}_{n=1}^{\infty}$. These poles are called *spurious poles*, because they do not reflect the analytic properties of the underlying function.

About the same time, John Nuttall made a seminal discovery: the spurious poles only affect a small area. More precisely, he proved [25]:

Theorem 1.3 (Nuttall's Theorem) *Let f be analytic at 0 and meromorphic in $\mathbf{C}$. Then $\{[n/n]\}_{n=1}^{\infty}$ converges in measure to f in compact subsets of the plane. More precisely, let $r, \varepsilon > 0$ and meas denote the planar Lebesgue measure. Then*

$$meas\, \{z : |z| \leq r \text{ and } |f - [n/n]|(z) > \varepsilon\} \to 0, \quad n \to \infty.$$

Subsequently Pommerenke [27] showed that one may replace ε by ε^n, planar measure by logarithmic capacity, and allow f to have singularities of logarithmic capacity 0. In particular f, can have essential singularities – but not branchpoints. There are far deeper analogues of the Nuttall-Pommerenke theorem for functions with branchpoints, due to H. Stahl [30], [31]. In essence, the branchpoints determine a so-called extremal set in the plane and the approximants converge in capacity inside that extremal set.

For functions meromorphic only in U, there is no analogue of the Nuttall-Pommerenke theorem: $\{[n/n]\}_{n=1}^{\infty}$ need not converge in measure or capacity in any open set within U [20], [28]. Nevertheless, there are still attempts to make some positive statement in this case [22].

Even these brief remarks convey to the reader the complexity of the convergence theory, due to spurious poles. Despite this inherent problem, George Baker and his collaborators found Padé approximants to be an invaluable tool in analysing singularities of series in a variety of physical problems. They also noted that in the situations where spurious poles did arise, it nevertheless affected only a subsequence of approximants. This led them to formulate a now famous conjecture [3], [4]. We shall concentrate on the following form of it:

Conjecture 1.4 (Baker-Gammel-Wills Conjecture (1961)) *Let f be meromorphic in the unit ball, and analytic at 0. There is an infinite subsequence $\{[n/n]\}_{n\in S}$ of the diagonal sequence $\{[n/n]\}_{n=1}^{\infty}$ that converges uniformly in all compact subsets of the unit ball omitting poles of f.*

The conjecture was generally disbelieved from the early 1970's, at least in the above form. It was thought to be possibly true for entire functions, or functions meromorphic in the whole plane. While the latter is still unresolved, the author recently proved a counterexample to the stated form of the Baker-Gammel-Wills Conjecture. For q not a root of unity, let

$$G_q(z) := \sum_{j=0}^{\infty} \frac{q^{j^2}}{(1-q)(1-q^2)\cdots(1-q^j)} z^j$$

denote the Rogers-Ramanujan function. Moreover, let

$$H_q(z) := G_q(z)/G_q(qz).$$

For appropriate q on the unit circle, the author showed [23] that H_q provides a counterexample. This is discussed in Section 4.

Of course, the comments above provide only a small glimpse into the Padé forest. For various perspectives on the convergence theory, including the important converse results of the Russian school, see [15], [21], [32], [34].

This paper is organised as follows: in Section 2, we discuss the partial theta function. In Section 3, we discuss the work of K. Driver on Wynn's series. In Section 4, we discuss the Rogers-Ramanujan continued fraction. Finally in Section 5, we discuss a number of unresolved questions and problems that we believe are worthwhile.

2 The partial theta function

Basic hypergeometric, or q-series, is a vast topic [1], [13], [14] and I cannot pretend that I am competent to survey even parts of it. Essentially, I am a user of a tiny part of the theory. My own interest began when the Rogers-Szegö polynomials turned up in describing the behaviour of $[m/n]$ with $m \to \infty$ and n fixed, when the coefficients of the underlying power series f are "smooth." Let us recall some of q-notation: for $n \geq j \geq 0$, the Gaussian binomial coefficient is

$$\begin{bmatrix} n \\ j \end{bmatrix} = \frac{(1-q^n)(1-q^{n-1})(1-q^{n-2})\cdots(1-q^{n-j+1})}{(1-q^j)(1-q^{j-1})(1-q^{j-2})\cdots(1-q)}.$$

Here if q is a root of unity, it must be interpreted in a limiting sense. In particular, as $q \to 1$,

$$\begin{bmatrix} n \\ j \end{bmatrix} \to \binom{n}{j}.$$

The Rogers-Szegö polynomial of degree n is

$$\mathcal{G}_n(z) = \sum_{j=0}^{n} \begin{bmatrix} n \\ j \end{bmatrix} z^j.$$

It is closely related to polynomials appearing in the q-binomial theorem, which has the form

$$\mathcal{H}_n(z) = \sum_{j=0}^{n} \begin{bmatrix} n \\ j \end{bmatrix} q^{j(j+1)/2} z^j = \prod_{j=1}^{n} \left(1 + q^j z\right).$$

The Rogers-Szegö polynomial $\mathcal{G}_n$ also turns up in the Padé denominators for the *partial theta function*

$$h_q(z) = \sum_{j=0}^{\infty} q^{j(j-1)/2} z^j.$$

These functions bear this name because they are essentially part of the theta function

$$\sum_{j=-\infty}^{\infty} q^{j^2} z^j.$$

The partial theta function satisfies a very simple functional relation, namely,

$$z h_q(qz) = h_q(z) - 1.$$

The following identity was established by amongst others, Wynn [35].

Lemma 2.1 *Let $n \geq 1$ and let q not be a j th root of unity for $1 \leq j \leq n$. Let $m \geq n-1$ and let $[m/n] = P_{mn}/Q_{mn}$ denote the (m,n) Padé approximant for h_q, normalized by $Q_{mn}(0) = 1$. Then*

$$Q_{mn}(z) = \mathcal{G}_n(-zq^m).$$

One proof is given in [24, p. 354 ff.]. In particular when $|q| = 1$, this lemma implies that if we fix n, and let m increase to ∞, all the zeroes of Q_{mn} will lie on circles centre 0 that contain a zero of the fixed polynomial $\mathcal{G}_n$. Even more, if q is not a root of unity, $\{q^m\}_{m=1}^{\infty}$ is dense on the unit circle, and so the zeroes of $\{Q_{mn}\}_{m=1}^{\infty}$ will have as their limit points precisely the circles centre 0 containing zeroes of $\mathcal{G}_n$.

This suggests a possible counterexample to the Baker-Graves-Morris Conjecture: if $|q| = 1$, then h_q is analytic inside the unit ball. If for some $|q| = 1$, $\mathcal{G}_n$ has a zero, α say, inside the unit ball, then every $[m/n]$ with $m \geq n-1$ will have a pole on the circle $\{z : |z| = \alpha\}$, and so no subsequence of $\{[m/n]\}_{m=1}^{\infty}$ can converge to h_q uniformly in all compact subsets of the unit ball. It turns out that for all such q and for *every* $n \geq 2$, $\mathcal{G}_n$ has such a zero α, and so h_q provides a counterexample to the Baker-Graves-Morris Conjecture for *every* $n \geq 2$ [24]:

Theorem 2.2 *Let $|q| = 1$ and q not be a root of unity. Then for $n \geq 2$, $\mathcal{G}_n$ has at least one zero in the unit ball. Consequently, there does not exist a subsequence of $\{[m/n]\}_{m=1}^{\infty}$ that converges to h_q uniformly in all compact subsets of the unit ball.*

We remind the reader, as mentioned in the previous section, that this was not the first counterexample. The first was given by Buslaev, Goncar, and Suetin [8]. They showed that the function

$$f(z) = \frac{1 + 2^{1/3}z}{1 - z^3}$$

has no subsequence of $\{[m/2]\}_{m=1}^{\infty}$ converging uniformly in $\{z : |z| \leq 2^{-1/3}\}$. We also recall that they did show that a subsequence converges in some ball centre 0, with the radius being independent of the underlying function.

Where is the smallest zero of $\mathcal{G}_n$? Using numerical computation, we showed [24] that for $n = 2$, there exists a q such that $\mathcal{G}_2$ has a zero with absolute value $0.58\ldots$ and that this is smallest possible as q ranges over the unit circle. As n increases, the size of the smallest zero of $\mathcal{G}_n$ as q ranges over the unit circle decreases, and reaches $0.24\ldots$ for $n = 17$.

This suggests an interesting problem, when taken in conjunction with Buslaev-Goncar-Suetin's positive result:

Problem 2.3 *Let f be analytic in U and $n \geq 2$. Let $\sigma_n(f)$ denote the radius of the largest disc centre 0 for which some subsequence of $\{[m/n]\}_{m=1}^{\infty}$ converges uniformly in each compact subset of that disc. Compute*

$$\sigma_n^* := \inf\{\sigma_n(f) : f \text{ analytic in } U\}.$$

The only known lower bound for σ_n^* is due to Buslaev, Goncar and Suetin. From the examples mentioned above, we know

$$\sigma_2^* \leq 0.58\ldots$$

and

$$\sigma_{17}^{*} \leq 0.24 \ldots$$

There has been virtually no work on this problem, maybe because it is so difficult. But surely, it can be resolved for $n = 2$ for example?

In actual fact, Ed Saff and the author were fishing for something far bigger with the partial-theta function: we had initially hoped that it would provide a counterexample to the Baker-Gammel-Wills Conjecture. As it turned out, for every q, the Baker-Gammel-Wills Conjecture is true for h_q. In retrospect, I understand why it cannot provide a counterexample, and this is best explained using the concept of a continued fraction.

Given a formal power series (1), we may also formally write

$$f(z) = c_0 + \cfrac{c_1 z^{k_1}}{1 + \cfrac{c_2 z^{k_2}}{1 + \cfrac{c_3 z^{k_3}}{1 + \ddots}}},$$

or more compactly,

$$f(z) = c_0 + \frac{c_1 z^{k_1} \mid}{\mid 1} + \frac{c_2 z^{k_2} \mid}{\mid 1} + \frac{c_3 z^{k_3} \mid}{\mid 1} + \cdots$$

where $\{c_j\}$ are complex numbers, and $\{k_j\}$ are positive integers. This is called the C-fraction corresponding to f. In the case that all $k_j = 1$ (a most desirable phenomenon), the C-fraction is said to be normal.

Just as we define the value of an infinite series to be the limit of the sequence of partial sums (each of which is a truncation of the series), so we define the value of a continued fraction to be the limit of the sequence formed by successive truncations of it. For $n \geq 1$, let

$$\frac{\mu_n}{\nu_n}(z) = c_0 + \frac{c_1 z^{k_1} \mid}{\mid 1} + \frac{c_2 z^{k_2} \mid}{\mid 1} + \frac{c_3 z^{k_3} \mid}{\mid 1} + \cdots + \frac{c_n z^{k_n} \mid}{\mid 1}.$$

This is called the nth *convergent* of the continued fraction; it is a rational function of z. We define the value of the continued fraction to be

$$\lim_{n \to \infty} \frac{\mu_n}{\nu_n}(z),$$

if this limit exists.

There is a close relationship between continued fractions and Padé approximants [19]. In particular, in the normal case, where all $k_j = 1$, the sequence of convergents $\{\mu_n/\nu_n\}_{n=1}^{\infty}$ comprises the main diagonal $\{[n/n]\}_{n=1}^{\infty}$ and the super-diagonal $\{[n+1/n]\}_{n=1}^{\infty}$. Thus we can use continued fraction techniques to study Padé approximants, and conversely, Padé methods give some insight into continued fractions.

For the partial theta function h_q, the continued fraction has the form

$$h_q(z) = 1 + \frac{z\ |}{|\ 1} + \frac{-qz\ |}{|\ 1} + \frac{q(1-q)z\ |}{|\ 1} + \frac{-q^3z\ |}{|\ 1} + \frac{q^2(1-q^2)z\ |}{|\ 1} + \cdots$$

So the continued fraction is normal. For $|q| = 1$, we see that the coefficients of z are bounded in absolute value by 2. An old theorem of Worpitzky [19] then ensures that the continued fraction converges at least for $|z| \leq \frac{1}{8}$. However, the coefficients are also oscillatory, some taking the form $-q^n$ and others taking the form $q^n(1-q^n)$, $n \geq 1$. This prevents application of standard convergence theorems for continued fractions beyond the range of z covered by Worpitzky's theorem. It also suggests that convergence of the full sequence of convergents, throughout the unit ball U, may not take place.

Indeed, for a given q, let Δ_q denote the inf of the absolute values of the zeroes of $\mathcal{G}_n$, so that

$$\Delta_q = \inf\{|z| : \mathcal{G}_n(z) = 0, \text{ some } n \geq 1\}.$$

We know that $\Delta_q < 1$. It was shown in [24] that the continued fraction converges in $\{z : |z| < \Delta_q\}$ but not in any larger disk. Nevertheless, as Ed Saff and I found to our disappointment, some subsequence of the convergents does converge throughout the unit ball to h_q, and then (with a little more work) also some subsequence of $\{[n/n]\}_{n=1}^{\infty}$. That subsequence corresponded to an infinite sequence $\mathcal{S}$ of integers for which

$$q^n \to 1, \quad n \to \infty, \quad n \in \mathcal{S}. \tag{2}$$

For this subsequence the coefficients

$$q^n(1-q^n) \to 0,$$

which was sufficient to guarantee convergence. In retrospect, it should not be surprising that such a subsequence yields good convergents: recall the theorem that when the full sequence of continued fraction coefficients converges to 0, the continued fraction converges to a meromorphic function in the whole complex plane (except at the poles).

There were many other fascinating features of the partial theta function, and of its Padé approximants. But undoubtedly the most significant is that for each $|q| = 1$ with q not a root of unity, h_q provides a counterexample to the Baker-Graves-Morris Conjecture for every $n \geq 2$.

3 Wynn's series

While the Padé approximants for the partial theta function failed to provide a counterexample to the Baker-Gammel-Wills Conjecture, their curious and irregular behaviour suggested that a counterexample might well be found in some close cousin. Certainly it seems a good idea to use formulas for q-series not in the usual setting, namely $|q| < 1$, but in the exotic domain $|q| = 1$, with the hope that something truly pathological might arise.

Of course one needs explicit formulas if any analysis is to be possible, and it was the 1967 work of P. Wynn [35] that suggested the next candidates for study. To some extent, Wynn's work overlapped with earlier work of Heine, Balk, Gragg. In that paper, Wynn considered three classes of series.

$$\text{(I)} \qquad f_1(z) = \sum_{j=0}^{\infty} \left[\prod_{k=0}^{j-1} \left(A - q^{k+\alpha}\right) \right] z^j = {}_2\Phi_1\left(\begin{matrix} A^{-1}q^{\alpha}, q; q, Az \\ 0 \end{matrix} \right),$$

in the language of basic hypergeometric series. Here $A, q \in \mathbf{C}, \alpha \in \mathbf{R}$ and we assume that

$$A \neq q^{k+\alpha}, \quad k \geq 0, \tag{3}$$

so that the series does not terminate. The functional relation is [10]

$$f_1(z)(1 - zA) = 1 - f_1(qz)\, zq^{\alpha}.$$

Note that if $A = \alpha = 0$, f_1 reduces essentially to the partial theta function.

$$\text{(II)} \qquad f_2(z) = \sum_{j=0}^{\infty} \left[\prod_{k=0}^{j-1} \left(\frac{1}{C - q^{k+\gamma}} \right) \right] z^j = {}_2\Phi_1\left(\begin{matrix} q, 0; q, C^{-1}z \\ C^{-1}q^{\gamma} \end{matrix} \right),$$

where $C, q \in \mathbf{C}, \gamma \in \mathbf{R}$ and we assume that

$$C \neq q^{k+\gamma}, \quad k \geq 0. \tag{4}$$

The functional relation is [11]

$$f_2(z)(C - z) = C - q^{\gamma-1} + q^{\gamma-1} f_2(qz).$$

$$\text{(III)} \qquad f_3(z) = \sum_{j=0}^{\infty} \left[\prod_{k=0}^{j-1} \left(\frac{A - q^{k+\alpha}}{C - q^{k+\gamma}} \right) \right] z^j = {}_2\Phi_1\left(\begin{matrix} A^{-1}q^{\alpha}, q; q, AC^{-1}z \\ C^{-1}q^{\gamma} \end{matrix} \right),$$

where $A, C, q \in \mathbf{C}; \alpha, \gamma \in \mathbf{R}$ and we assume that (3) and (4) hold. The functional relation is [12]

$$f_3(z)(C - zA) = C - q^{\gamma-1} + f_3(qz)\left(q^{\gamma-1} - zq^{\alpha}\right).$$

In all three cases, the functional relation is a useful tool in investigating the analytic properties of the function. Let us look at f_3. In her thesis, K.A. Driver [9] proved, amongst other things, the following: let $A, C \neq 0$, $|A|, |C| \neq 1$, let $|q| = 1$ and q not be a root of unity. Then f_3 has radius of convergence

$$R = \left[\max\{1, |A|\} \min\left\{1, |C|^{-1}\right\}\right]^{-1},$$

and if

$$C \neq Aq^{\gamma-\alpha+j}, \quad j = 0, 1, 2, \ldots,$$

f_3 has a natural boundary on its circle of convergence. (If this last condition fails, f_3 is a rational function.) The continued fraction has the form

$$f_3(z) = 1 + \frac{c_1 z \mid}{\mid 1} + \frac{c_2 z \mid}{\mid 1} + \frac{c_3 z \mid}{\mid 1} + \cdots,$$

where

$$c_{2n+1} = \frac{q^{2n-1}(1-q^n)\left(Aq^\gamma - Cq^{1-n+\alpha}\right)}{(C - q^{2n+\gamma})(C - q^{2n+\gamma-1})}$$

and

$$c_{2n} = \frac{-q^{n-1}\left(A - q^{n+\alpha}\right)\left(C - q^{n+\gamma-1}\right)}{\left(C - q^{2n+\gamma-1}\right)\left(C - q^{2n+\gamma-2}\right)}.$$

A detailed analysis was provided of the continued fraction. In particular, Driver proved that the full sequence of convergents (and hence $\{[n/n]\}_{n=1}^{\infty}$) converges in measure and in capacity in compact subsets of $\{z : |z| < R\}$. Moreover, some subsequence does converge uniformly in compact subsets of that ball. In fact such a subsequence corresponds to the infinite sequence of integers $\mathcal{S}$ satisfying (2), just as for the partial theta function.

Thus again hopes of a counterexample to the Baker-Gammel-Wills Conjecture dissipated, although there were a host of other interesting features.

4 The Rogers-Ramanujan continued fraction

For q not a root of unity, let

$$G_q(z) := \sum_{j=0}^{\infty} \frac{q^{j^2}}{(1-q)(1-q^2)\cdots(1-q^j)} z^j$$

denote the Rogers-Ramanujan function. (At this stage, it is a formal power series.) It admits the functional relation

$$G_q(z) = G_q(qz) + qzG_q\left(q^2 z\right).$$

Moreover, let

$$H_q(z) := G_q(z)/G_q(qz). \tag{5}$$

From the functional relation for G_q, it is easy to derive one for H_q :

$$H_q(z) = 1 + \frac{qz}{H_q(qz)}.$$

Iterating this leads to

$$H_q(z) = 1 + \cfrac{qz}{1 + \cfrac{q^2 z}{1 + \ddots \, \cfrac{q^n z}{H_q(q^n z)}}}$$

and hence to the formal infinite continued fraction

$$H_q(z) = 1 + \frac{qz\,|}{|\,1} + \frac{q^2 z|}{|\,1} + \frac{q^3 z|}{|\,1} + \cdots . \tag{6}$$

For $|q| < 1$, the continued fraction was considered independently by L.J. Rogers and S. Ramanujan in the early part of the twentieth century.

There are several differences between the Rogers-Ramanujan continued fraction (c.f.), and those from Wynn's series. Firstly, if $|q| = 1$, all the coefficients in the Rogers-Ramanujan c.f. have modulus 1, whereas a subsequence of the coefficients in the c.f. for Wynn's series converges to 0. Moreover the latter subsequence is associated with a subsequence of the convergents to the c.f. that converges throughout the region of analyticity. This already suggests that there may not be a uniformly convergent subsequence of the convergents for the Rogers-Ramanujan c.f. Secondly, in the case where q is a root of unity, all of the Wynn's series reduce to rational functions, while the Rogers-Ramanujan c.f. corresponds to a function with branchpoints.

We see that the radius of convergence of G_q is

$$R(q) := \liminf_{j \to \infty} \left| \prod_{k=0}^{j-1} (1 - q^k) \right|^{1/j}.$$

It was essentially proved in [16] that

$$R(q) = \liminf_{j \to \infty} |1 - q^j|^{1/j}.$$

If we write $q = e^{2\pi i \tau}$, this is readily reformulated in terms of the diophantine approximation properties of τ. Since $|1 - q^j| = 2\,|\sin[\pi(j\tau - k)]|$ for any integer k, we see that

$$R(q) = \liminf_{j \to \infty} \; \|j\tau\|^{1/j},$$

where $\|x\|$ denotes the distance from x to the nearest integer. In particular, elementary diophantine approximation theory shows that for a.e. q on the unit circle, $R(q) = 1$.

Using the functional relation, one can show that G_q has a natural boundary on its circle of convergence. Then H_q is meromorphic inside this ball. One can also show that H_q has a natural boundary on its circle of meromorphy, that is on the largest circle centre 0, inside which it is meromorphic. This does not follow from the fact that G_q has a natural boundary on $\{z : |z| = R(q)\}$, and must be proved independently from the functional relation for H_q.

Far more curious, is the fact that the natural boundary of H_q need not coincide with that of G_q: somehow in the division in (5), the natural boundary of G_q "cancels out" [23]:

Theorem 4.1 *Let $0 < \sigma < \frac{1}{4}$. Then there exists $|q| = 1$, with q not a root of unity, such that G_q is analytic in $\{z : |z| < \sigma\}$and has a natural boundary on $\{z : |z| = \sigma\}$. However, if we define H_q by (5), then it may be continued meromorphically to $\{z : |z| < \rho\}$, where $\rho \geq \frac{1}{4} > \sigma$. Thus H_q is meromorphic in $\{z : |z| < \rho\}$, and has a natural boundary on $\{z : |z| = \rho\}$.*

This is the first time that this author has seen a natural boundary cancel out: we are all familiar with poles that cancel, but natural boundaries?

Problem 4.2 *Explain this cancellation.*

This phenomenon is unusual. Indeed, for a.e. q on the unit circle, G_q is analytic in the unit ball U with natural boundary on the unit circle, and H_q is meromorphic in the unit ball, with natural boundary on the unit circle.

Proofs of the explicit formulae for the numerator and denominator polynomials μ_n and ν_n in the convergent

$$\frac{\mu_n(z)}{\nu_n(z)} = 1 + \frac{qz \mid}{\mid 1} + \frac{q^2 z\mid}{\mid 1} + \cdots + \frac{q^3 z\mid}{\mid 1}$$

were first published by M. Hirschorn in 1972 [17]:

$$\mu_n(z) = \sum_{k=0}^{[\frac{n+1}{2}]} z^k q^{k^2} \left[\begin{array}{c} n+1-k \\ k \end{array} \right]$$

and

$$\nu_n(z) = \mu_{n-1}(qz) = \sum_{k=0}^{[\frac{n}{2}]} z^k q^{k(k+1)} \left[\begin{array}{c} n-k \\ k \end{array} \right],$$

where $[x]$ is the greatest integer $\leq x$. The author met Hirschhorn in Sydney in 2000, and was intrigued by the story of these identities. Like so many other q-identities, they appeared in notes of Ramanujan, but without proof. Moreover,

Hirschhorn was not aware of Ramanujan's notes at the time he wrote his papers – again a common occurrence.

In describing the behaviour of $\{\mu_n\}$ and $\{\nu_n\}$, we need an elementary observation from number theory: if q is not a root of unity, then $\{q^n\}_{n=1}^{\infty}$ is dense on the unit circle, and one may extract a subsequence converging to an arbitrary β on the unit circle. This helps to introduce our main convergence theorem for the convergents (recall that $R(q)$ is the radius of convergence of G_q) :

Theorem 4.3 *Let $|q| = 1$, and q not be a root of unity. Let $|\beta| = 1$ and $\mathcal{S}$ be any infinite sequence of positive integers with*

$$\lim_{n\to\infty, n\in\mathcal{S}} q^n = \beta. \tag{7}$$

Then uniformly in compact subsets of $\{z : |z| < R(q)\}$,

$$\lim_{n\to\infty, n\in\mathcal{S}} \mu_n(z) = \overline{G_q(\overline{\beta q z})} G_q(z); \tag{8}$$

$$\lim_{n\to\infty, n\in\mathcal{S}} \nu_n(z) = \overline{G_q(\overline{\beta q z})} G_q(qz); \tag{9}$$

and uniformly in compact subsets of $\{z : |z| < R(q)\}$ omitting zeroes of $G_q(\overline{\beta q z})$ and $G_q(qz)$,

$$\lim_{n\to\infty, n\in\mathcal{S}} \frac{H_q(z) - \frac{\mu_n(z)}{\nu_n(z)}}{(-1)^n z^{n+1} q^{(n+1)(n+2)/2}} = \frac{G_q(\beta q^2 z)}{G_q(qz)^2 \overline{G_q(\overline{\beta q z})}}$$

and so in such sets omitting these zeroes,

$$\lim_{n\to\infty, n\in\mathcal{S}} \frac{\mu_n(z)}{\nu_n(z)} = H_q(z).$$

The crucial point in the last line is that the convergence takes place away from the zeroes of both $G_q(z)$ and $G_q(\overline{\beta q z})$. The zeroes of $G_q(\overline{\beta q z})$ need not be poles of H_q, and yet (9) shows that they attract poles of the convergents. Moreover, because $|\beta| = 1$, both $G_q(z)$ and $\overline{G_q(\overline{\beta q z})}$ have the same number of zeroes on any circle centre 0, and this is true of every such β. Hence:

Corollary 4.4 *Let $|q| = 1$, not a root of unity. Assume that $r < R(q)$ and H_q has poles of total multiplicity ℓ on $\{z : |z| = r\}$. Let $\mathcal{O}$ be an open set containing this circle. Then there exists n_0 such that for $n \geq n_0$, μ_n/ν_n has poles of total multiplicity $\geq 2\ell$ in $\mathcal{O}$.*

This is the first such example in the literature, in which *all* approximants of large order have more poles than the approximated function in a region of meromorphy. If we could show that there does not exist β for which the zero sets of $G_q(qz)$ and $G_q(\overline{\beta qz})$ are the same, then it establishes a counterexample to the Baker-Gammel-Wills conjecture. For then, given any subsequence of the convergents, we can extract a further subsequence for which (7) holds for some β; that subsequence cannot converge uniformly in a compact set containing zeroes of $G_q(\overline{\beta qz})$ that are not zeroes of $G_q(z)$.

A little thought shows that the zero sets of $G_q(qz)$ and $G_q(\overline{\beta qz})$ are not the same for any $|\beta| = 1$, iff the zeroes of G_q are not symmetric about any line through 0. Thus:

Corollary 4.5 *Let $|q| = 1$, and q not be a root of unity. Assume that the zeroes of G_q inside its circle of convergence are not symmetric about any line through* 0. *Then H_q provides a counterexample to the Baker-Gammel-Wills Conjecture.*

Intuitively, there was a lot of reason to believe in the desired asymmetry, at least from the following standpoint: recall that a Maclaurin series with real coefficients has zeroes symmetric about the real axis, that is, they occur in conjugate pairs. Conversely, one might hope that for special functions, zeroes that occur in conjugate pairs are associated with Maclaurin series with real coefficients. After a rotation of the variable, symmetry of zeroes of G_q about some line through 0, would become symmetry about the real axis. Yet the arguments of the coefficients of $G_q(\gamma z)$ are highly oscillatory for any γ on the unit circle, and there is no reason to expect symmetry.

For a long time, the author tried to prove this asymmetry property, but failed. Since numerical computation might provide some insight, the author was fortunate to be able to ask A. Knopfmacher (who is, amongst other things, a Mathematica expert) to plot some zeroes a few years ago. Of course, we cannot easily compute G_q itself, but we can with reasonable accuracy, plot the zeroes of the partial sums

$$S_{m,q}(z) = \sum_{j=0}^{m} \frac{q^{j^2}}{(1-q)(1-q^2)\cdots(1-q^j)} z^j.$$

In the case when the radius of convergence is $R(q) = 1$, the partial sums converge rapidly within the unit ball to G_q as $m \to \infty$, and so their zeroes should approximate the zeroes of G_q well inside the unit circle.

We typically chose [18]

$$q = \exp\left(2\pi i/\sqrt{\ell}\right),$$

where ℓ is some positive integer. For almost all the choices of ℓ, $S_{m,q}$ had zeroes asymmetric with respect to any line through 0. Moreover, as we increased m from 10 through to 100, the zeroes well within the unit circle remained the same. In fact, in almost all cases, the two zeroes closest to the origin had distinct modulus and distinct argument, which is already enough to establish asymmetry. This very strongly suggested that we do have a counterexample. However, it took some time to find a choice of q for which I could obtain a sufficiently fine estimate of

$$G_q - S_{m,q}$$

to turn this into a proof. This was finally done for a special q in January 2001 [23]:

Theorem 4.6 *Let*

$$q := \exp(2\pi i\tau)$$

where

$$\tau := \frac{2}{99+\sqrt{5}}.$$

Then H_q is meromorphic in the unit ball and analytic at 0. There does not exist any subsequence of $\{\mu_n/\nu_n\}_{n=1}^{\infty}$ that converges uniformly in all compact subsets of

$$\mathcal{A} := \{z : |z| < 0.46\}$$

omitting poles of H_q. In particular no subsequence of $\{[n/n]\}_{n=1}^{\infty}$ or $\{[n+1/n]\}_{n=1}^{\infty}$ can converge uniformly in all compact subsets of $\mathcal{A}$ omitting poles of H_q.

After this counterexample was announced, it was discussed in the seminar of A. Gonchar at the Steklov Institute of Mathematics in Moscow. This inspired V. Buslaev to construct a simpler counterexample to the Baker-Gammel-Wills Conjecture, namely the algebraic function:

$$f(z) = \frac{-27 + 6z^2 + 3(9+\zeta)z^3 + \sqrt{81(3-(3+\xi)z^3)^2 + 4z^6}}{2z(9+9z+(9+\xi)z^2)},$$

where

$$\xi = -\frac{1}{2} + \frac{\sqrt{3}}{2}i.$$

He shows that this function is analytic in the unit ball, but no subsequence of $\{[n/n]\}_{n=1}^{\infty}$ can converge at one of three special points inside the unit ball. See the announcement [7].

5 Suggested problems

In a complicated subject like Padé convergence theory, the resolution of the Baker-Gammel-Wills Conjecture raises many problems about weaker forms of the conjecture. Some of those are discussed in [23], [32]. Here we shall discuss a few problems specifically relating to q-series, that we believe are interesting.

We have already mentioned the strange cancellation of the natural boundary of G_q (Problem 4.2). But there are many others. We know, thanks to work of G. Petruska [26], that given $R \in [0,1]$, we may find q such that G_q has radius of convergence R, that is $R = R(q)$. But what values can the radius of meromorphy of H_q assume? Thanks to a theorem of Worpitzky, which ensures that the continued fraction (6) converges for $|z| \leq \frac{1}{4}$, we know this radius is $\geq \frac{1}{4}$. Moreover, for almost every q, it is 1, but are there any exceptional values?

Problem 5.1 *Let $|q| = 1$, and q not be a root of unity. Let $\rho(q)$ denote the largest circle centre 0 inside which H_q may be meromorphically continued. Can $\rho(q)$ assume any value other than 1? If so, what is its range of values?*

Another interesting problem, is to investigate the zeroes of G_q without the use of any numerical package:

Problem 5.2

(i) *Investigate the structure of zeroes of G_q when $|q| = 1$ and $R(q) > 0$.*

(ii) *Moreover, investigate whether for every such q, the zeroes of G_q are not symmetric about any line through 0.*

(iii) *Investigate the behaviour of the zero of G_q closest to the origin as q traverses the unit circle.*

It is instructive here to recall that the q-exponential functions are

$$e_q(z) = \sum_{j=0}^{\infty} \frac{z^j}{(1-q)(1-q^2)(1-q^3)\cdots(1-q^j)};$$

$$E_q(z) = \sum_{j=0}^{\infty} \frac{q^{j(j-1)/2} z^j}{(1-q)(1-q^2)(1-q^3)\cdots(1-q^j)}.$$

Now in $\{z : |z| < R(q)\}$, a direct calculation shows that

$$e_q(z) E_q(-z) = 1$$

and hence e_q and E_q have no zeroes in that ball. In contrast, we know that

$$G_q(z) = \sum_{j=0}^{\infty} \frac{q^{j^2} z^j}{(1-q)(1-q^2)(1-q^3)\cdots(1-q^j)}$$

may have zeroes. This suggests:

Problem 5.3 *Let $|q| = 1$ and $R(q) > 0$. For $a \geq 0$, let*

$$G_{a,q}(z) = \sum_{j=0}^{\infty} \frac{q^{aj^2} z^j}{(1-q)(1-q^2)(1-q^3)\cdots(1-q^j)}.$$

For which $a > 0$ does $G_{a,q}$ have zeroes?

Of course these are very specific problems. But there are also some quite general ones, involving the mere definition of q-special functions. We have seen above that there is no problem with defining q-exponential functions for q not a root of unity. But a q-gamma function is far more challenging. The q-gamma function $\Gamma_q(x)$ is the unique solution of the difference equation

$$\Gamma_q(x+1) = \frac{q^x - 1}{q - 1}\Gamma_q(x), \tag{10}$$

with normalisation

$$\Gamma_q(1) = 1,$$

and with $\log \Gamma_q(x)$ convex for $x > 0$. As $q \to 1$, we see that this becomes the classical relation

$$\Gamma(x+1) = x\Gamma(x).$$

For $q \in (0,1)$, we have [2]

$$\Gamma_q(x) = (1-q)^{1-x} \prod_{n=0}^{\infty} \frac{1-q^{n+1}}{1-q^{n+x}}.$$

It is easily seen that this also defines a function analytic in the upper-half plane ($\mathrm{Re}\, x > 0$).

Problem 5.4 *Define a q-gamma function for $|q| = 1$.*

The problem seems to be that any solution of the functional relation (10) with $|q| = 1$, q not a root of unity, cannot be defined on the real axis. Of course there are other q-special functions for which similar problems might arise. The definition of q-special functions for q on the unit circle might seem like the sort of subject one invents for the sake of "creating a gap in the literature." However, the author believes that q-series with $|q| = 1$, and q not a root of unity, will have a growing number of applications. Moreover, not all of these will be in the search for pathological examples.

Acknowledgement

It was a privilege to take part in IDOMAT2001. The author thanks the organisers for all their hard work, the kind invitation, and the stimulating and well run conference.

References

[1] G. Andrews, R. Askey, R. Roy, *Special Functions*, Cambridge University Press, Cambridge, 1999.

[2] R. Askey, *The q-Gamma and q-Beta Functions*, Applicable Analysis, **8**(1978), 125–141.

[3] G.A. Baker, *Essentials of Padé Approximants*, Academic Press, New York, 1975.

[4] G.A. Baker, J.L. Gammel, J.G. Wills, *An Investigation of the Applicability of the Padé Approximant Method*, J. Math. Anal. Appl., **2**(1961), 405–418.

[5] G.A. Baker and P. R. Graves-Morris, *Convergence of Rows of the Padé Table*, J. Math. Anal. Appl., **57**(1977), 323–339.

[6] G.A. Baker and P. R. Graves-Morris, *Padé Approximants*, 2nd edn., Encyclopaedia of Mathematics and its Applications, Vol. 59, Cambridge University Press, Cambridge, 1996.

[7] V. I. Buslaev, *Simple Counterexample to the Baker-Gammel-Wills Conjecture*, East Journal on Approximations, **4**(2001), 515–517.

[8] V. I. Buslaev, A.A. Goncar, and S.P. Suetin, *On Convergence of Subsequences of the mth Row of a Padé Table*, Math. USSR. Sbornik, **48**(1984), 535–540.

[9] K.A. Driver, Convergence of Padé Approximants for some q-Hypergeometric Series (Wynn's Power Series I, II, III), Thesis, University of the Witwatersrand, 1991.

[10] K.A. Driver and D.S. Lubinsky, *Convergence of Padé Approximants for a q-Hypergeometric Series* (Wynn's Power Series I), Aequationes Mathematicae, **42**(1991), 85–106.

[11] K.A. Driver and D.S. Lubinsky, *Convergence of Padé Approximants for a q-Hypergeometric Series* (Wynn's Power Series II), Colloquia Mathematica Societatis Janos Bolyai, Vol. 58, pp. 221–239, Janos Bolyai Math Society, 1990.

[12] K.A. Driver and D.S. Lubinsky, *Convergence of Padé Approximants for a q-Hypergeometric Series* (Wynn's Power Series III), Aequationes Mathematicae, **45**(1993), 1–23.

[13] N.J. Fine, *Basic Hypergeometric Series and Their Applications*, American Mathematical Society Mathematical Surveys and Monographs No. 27, American Mathematical Society, Providence, Rhode Island, 1988.

[14] G. Gasper and M. Rahman, *Basic Hypergeometric Series*, Cambridge University Press, Cambridge, 1990.

[15] A.A. Goncar, *On Uniform Convergence of Diagonal Padé Approximants*, Math. USSR. Sbornik, **46**(1982), 539–559.

[16] G.H. Hardy and J.E. Littlewood, *Notes on the Theory of Series (XXIV): A Curious Power Series*, Math. Proc. Cambridge Philosophical Soc., **42**(1946), 85–90.

[17] M.D. Hirschhorn, *Partitions and Ramanujan's Continued Fraction*, Duke Math. Journal, **39**(1972), 789–791.

[18] A. Knopfmacher and D.S. Lubinsky, *Mathematica Evidence that Ramanujan kills Baker-Gammel-Wills*, Applied Mathematics and Computation, **128**(2002), 289–302.

[19] L. Lorentzen and H. Waadeland, *Continued Fractions with Applications*, North Holland, 1992.

[20] D.S. Lubinsky, *Diagonal Padé Approximants and Capacity*, J. Math. Anal. Applns., **78**(1980), 58–67.

[21] D.S. Lubinsky, *Will Ramanujan kill Baker-Gammel-Wills? (A Selective Survey of Padé Approximation)*, (in) Proceedings of IDOMAT98, International Series in Numerical Mathematics, Vol. 132, Birkhäuser, Basel, 1999, pp. 159–174.

[22] D.S. Lubinsky, *Weighted Maximum over Minimum Modulus of Polynomials, with Application to Ray Sequences of Padé Approximants*, Constructive Approximation, **18**(2002), 285–308.

[23] D.S. Lubinsky, *Rogers-Ramanujan and the Baker-Gammel-Wills (Padé) Conjecture*, to appear in Annals of Mathematics.

[24] D.S. Lubinsky and E.B. Saff, *Convergence of Padé Approximants of Partial Theta Functions and the Rogers-Szegö Polynomials*, Constr. Approx., **3**(1987), 331–361.

[25] J. Nuttall, *Convergence of Padé Approximants of Meromorphic Functions*, J. Math. Anal. Appl., **31**(1970), 147–153.

[26] G. Petruska, *On the Radius of Convergence of q-Series*, Indagationes Mathematicae, **3**(3) (1992), 353–364.

[27] C. Pommerenke, *Padé Approximants and Convergence in Capacity*, J. Math. Anal. Appl., **41**(1973), 775–780.

[28] E.A. Rakhmanov, *On the Convergence of Padé Approximants in Classes of Holomorphic Functions*, Math. Ussr. Sbornik, **40**(1981), 149–155.

[29] A. Sidi, *Quantitative and Constructive Aspects of the Generalized Koenig's and de Montessus's Theorems for Padé Approximants*, J. Comp. Appl. Math., **29**(1990), 257-291.

[30] H. Stahl, *General Convergence Results for Padé Approximants*, (in) Approximation Theory VI, (eds. C.K. Chui, L.L. Schumaker, J.D. Ward), Academic Press, San Diego, 1989, pp. 605–634.

[31] H. Stahl, *The Convergence of Padé Approximants to Functions with Branch Points*, J. Approx. Theory, **91**(1997), 139–204.

[32] H. Stahl, *Conjectures Around the Baker-Gammel-Wills Conjecture: Research Problems* 97-2, Constr. Approx., **13**(1997), 287–292.

[33] H. Wallin, *The Convergence of Padé Approximants and the Size of the Power Series Coefficients*, Applicable Anal., **4**(1974), 235–251.

[34] H. Wallin, *Potential Theory and Approximation of Analytic Functions by Rational Interpolation*, (in) Proc. Colloquium on Complex Analysis, Joensuu, Finland, 1978, Springer Lecture Notes in Mathematics, Vol. 747, Springer-Verlag, Berlin, 1979, pp. 434–450.

[35] P. Wynn, *A General System of Orthogonal Polynomials*, Quart. J. Math. Oxford Ser., **18**(1967), 81–96.

School of Mathematics
Georgia Institute of Technology
Atlanta, GA 30332-0160 USA
Email address: lubinsky@math.gatech.edu

Advanced Problems in Constructive Approximation
(Eds.) M.D. Buhmann and D.H. Mache
International Series of Numerical Mathematics Vol. 142, 129–141

On the Degree of Approximation in Multivariate Weighted Approximation

H.N. Mhaskar

Abstract

Let $s \geq 1$ be an integer, $f \in L^p(\mathbf{R}^s)$ for some p, $1 \leq p < \infty$ or be a continuous function on $\mathbf{R}^s$, vanishing at infinity. We consider the degree of approximation of f by expressions of the form $\exp(-\sum_{k=1}^{s} Q_k(x_k))P(x_1, \ldots, x_s)$ where each $\exp(-Q_k(\cdot))$ is a Freud type weight function, and P is a polynomial of specific degrees in each coordinate. Direct and converse theorems are stated. In particular, it is shown that if each $Q_k = |x|^\alpha$ for some even, positive integer α, and the degree of approximation has a power decay, then the same property holds when the weight function is replaced by $\exp(-\sum_{k=1}^{s} a_k Q_k(x_k))$ for any positive constants a_k.

1 Introduction

Let $C_0(\mathbf{R})$ be the class of all continuous real functions on $\mathbf{R}$, vanishing at infinity. The classical Bernstein approximation problem seeks conditions on a weight function w such that expressions of the form wP, P a polynomial, are dense in $C_0(\mathbf{R})$. Such expressions will be called (w-)weighted polynomials. In the 1970's, G. Freud initiated a detailed study of the degree of approximation of a function $f \in C_0(\mathbf{R})$ by weighted polynomials of a given degree, when the weight function satisfies certain technical conditions, somewhat more stringent than those necessary to ensure density. Freud found it convenient to write $f = wg$ for some g, and formulated his results as those on approximation of g in a weighted norm. This theme has been studied in great generality in the past twentyfive years (cf. [5, 6, 7, 2, 1].)

Nevertheless, very little is known regarding the connection between the degrees of approximation of g with respect to different weights ([8]).

The purpose of this paper is to record the observation that in the case when w is a Freud weight, there exists an equivalent weight $\overline{w}$ such that a polynomial rate of decay for the degree of approximation of $f \in C_0(\mathbf{R})$ by $\overline{w}$-weighted polynomials implies the same rate for the degree of approximation by $\overline{w}^\lambda$-weighted polynomials for any positive λ. We will formulate our results in the multivariate, tensor product case. To the best of our knowledge, the only previously published work on the degree of weighted approximation in the multivariate setting is by Dzrbasyan and Tavadyan [3], where "direct theorem"s are obtained under very strong assumptions about the target function, f. For example, in the univariate context, it is required that $w^{-1}f \in C_0(\mathbf{R})$.

In the next section, we recall the definition of Freud weights and a construction of equivalent, smoother weights. Our main results are stated in Section 3. The proof of these results involves developing the analogues of both the direct and converse theorems of weighted approximation in multivariate setting, as well as evaluating a new K-functional. The general approach is the same as in the univariate case. We will only sketch the proof, supplying details when they are significantly different from those in [7]. In Section 4, we define and discuss some properties of the shifted average operators and use these to prove the direct, converse, and equivalence theorems in terms of the K-functional. In Section 5, we evaluate the order of magnitude of this K-functional in terms of the forward differences of the target function.

2 Freud weights

Let $w : \mathbf{R} \to [0, \infty)$. We say that w is a Freud(-type) weight (function) if

$$Q(x) := \log\left\{\frac{1}{w(x)}\right\} \tag{1}$$

is an even, convex function on $\mathbf{R}$, Q is twice differentiable on $(0, \infty)$, and there are constants $c_1, c_2 > 0$ such that

$$0 < c_1 \le \frac{xQ''(x)}{Q'(x)} \le c_2 < \infty, \qquad x \in (0, \infty). \tag{2}$$

In general, if $Q : \mathbf{R} \to \mathbf{R}$, we will write $w_Q := \exp(-Q)$. It is clear that if w_Q is a Freud weight, then $Q'(A) > 0$ for some $A > 0$, and hence, that $xQ'(x)$ is a strictly increasing function of $x \in [A, \infty)$. We choose and fix such constant A. The Freud numbers, q_x, are defined by

$$q_x Q'(q_x) = x, \qquad x \ge A, \tag{3}$$

and $q_x = 1$ for $0 \le x < A$. The prototypical Freud weights are $\exp(-|x|^\alpha)$, $\alpha > 1$; the corresponding Freud numbers being $(x/\alpha)^{1/\alpha}$.

In the sequel, we adopt the following convention regarding constants. The symbols $c, c_1, \ldots$ will denote positive constants depending only on the fixed parameters such as the weight functions and norms involved. The values of these constants may be different at different occurrences, even within the same formula. The notation $D \sim B$ will mean that $c_1 D \leq B \leq c_2 D$.

In [7, Theorem 5.1.1], we proved the following theorem regarding the smoothing of Freud weights.

Theorem 2.1 *Let w_Q be a Freud weight, $r \geq 2$ be an integer, and*

$$\omega_r^*(x) := \max_{x-r-1 \leq u \leq x+r+1} \max_{-r \leq t \leq r} \left| \sum_{\nu=0}^{r-1} (-1)^{r-\nu} \binom{r-1}{\nu} Q'(u+\nu t) \right| \to 0 \quad (4)$$

as $|x| \to \infty$. Then there exists an even, r times continuously differentiable function $\overline{Q} : \mathbf{R} \to \mathbf{R}$ such that

$$|Q^{(j)}(x) - \overline{Q}^{(j)}(x)| \leq c\omega_r^*(x), \qquad x \in \mathbf{R}, \ j = 0, 1, \quad (5)$$

and for $j = 2, \ldots, r$,

$$\lim_{|x| \to \infty} \left| \frac{\overline{Q}^{(j)}(x)}{\overline{Q}'(x)^j} \right| = 0. \quad (6)$$

In particular, $w_Q(x) \sim \exp(-\overline{Q}(x))$ and $1 + Q'(x)^2 \sim 1 + \overline{Q}'(x)^2$ for $x \in \mathbf{R}$.

In general, the function $\overline{Q}$ will depend upon the choice of r, although we prefer to treat r as a fixed parameter, and write $\overline{Q}$ rather than $\overline{Q}_r$. In the case when $r = 1$, we find it convenient to let (4) be an "empty condition", and write $\overline{Q} := Q$. In the case of the prototypical Freud weights $\exp(-|x|^\alpha)$, the condition (4) is satisfied for all $r > \alpha$, and we may write $\overline{Q} = Q$ if $r < \alpha$. In the case when α is an even integer, we may choose $\overline{Q} = Q$ for all integer $r \geq 1$, even though (4) is not satisfied.

3 Main results

In this section, $s \geq 1$ will denote a fixed integer, and all constants will depend upon s as well. Bold face letters will denote vectors, with components indicated by subscripts, for example, $\mathbf{x} = (x_1, \ldots, x_s) \in \mathbf{R}^s$. Similarly, if $f_k : \mathbf{R} \to \mathbf{R}$, $1 \leq k \leq s$, $\mathbf{a} \in \mathbf{R}^s$, we will write $\mathbf{f}(\mathbf{x}) := (f_1(x_1), \ldots, f_s(x_s))$, $\mathbf{a} \odot \mathbf{f}(\mathbf{x}) := (a_1 f_1(x_1), \ldots, a_s f_s(x_s))$, and $w_{\mathbf{f}}(\mathbf{x}) := \exp(-\sum_{k=1}^{s} f_k(x_k))$.

If $1 \leq k \leq s$, $A \subseteq \mathbf{R}^k$ is Lebesgue measurable with respect to the k-dimensional Lebesgue measure, and $f : A \to \mathbf{R}$ is Lebsegue measurable, we define

$$\|f\|_{p,A} := \left\{ \int_A |f(\mathbf{x})|^p d\mathbf{x} \right\}^{1/p},$$

if $0 < p < \infty$,

$$\|f\|_{p,A} := \operatorname*{ess\,sup}_{\mathbf{x}\in A} |f(\mathbf{x})|,$$

if $p = \infty$, where the measure is the k-dimensional Lebesgue measure. The space of all f for which $\|f\|_{p,A} < \infty$ will be denoted by $L^p(A)$, where two functions are considered equal if they are equal almost everywhere. The subspace of $L^\infty(A)$ consisting of all continuous functions vanishing at infinity will be denoted by $C_0(A)$. The notation $\|\cdot\|_p$ will mean $\|\cdot\|_{p,\mathbf{R}^s}$.

In the remainder of this section, we assume that $w_{Q_k} = \exp(-Q_k), 1 \le k \le s$, are Freud weights satisfying (4) with some $r \ge 1$, and $\overline{Q}_k$ are the corresponding smooth functions as in Theorem 2.1. In the sequel, r will be a fixed parameter, and all constants may depend upon r as well. For $1 \le k \le s$, let $q_{x,k}$ denote the Freud number for w_{Q_k}, and p_k denote an inverse function of $x \mapsto x/q_{x,k}$. For $\mathbf{y} \ge 0, \mathbf{y} \in \mathbf{R}^s$, we write $\Pi_{\mathbf{y}}$ to be the class of all polynomials in s variables, having degree at most y_k in the k-th variable. For $1 \le p \le \infty$, $f \in L^p(\mathbf{R}^s)$, and $y \ge 0$, we are interested in

$$\tilde{\epsilon}_{p,y}(\mathbf{Q}; f) := \inf\{\|f - w_{\overline{\mathbf{Q}}}P\|_p \ : \ P \in \Pi_{\mathbf{p}(y)}\}. \tag{7}$$

Let D_k denote the partial derivative with respect to the k-th variable, and $w_{\mathbf{Q}}g \in L^p(\mathbf{R}^s)$. We define a K-functional of order r by

$$K_r(\mathbf{Q}, p; g, \delta) := \inf\left\{\|(g-h)w_{\mathbf{Q}}\|_p + \delta^r\left(\|w_{\mathbf{Q}}h\|_p + \sum_{k=1}^{s}\|w_{\mathbf{Q}}D_k^r h\|_p\right)\right\}, \tag{8}$$

where the infimum is taken over all $h : \mathbf{R}^s \to \mathbf{R}$ such that h is $r-1$ times continuously differentiable in each variable, $D_k^{r-1}h$ is absolutely continuous, and $w_{\mathbf{Q}}D_k^r h \in L^p(\mathbf{R}^s)$ for $1 \le k \le s$.

Our main theorem in this section is the following.

Theorem 3.1 *Let $r \ge 1$ be an integer. For $k = 1, \ldots, s$, let w_{Q_k} be Freud weights satisfying* (4)*, and $\overline{Q}_k$ be the smooth functions as in Theorem 2.1. Let $1 \le p \le \infty$, $f \in L^p(\mathbf{R}^s)$. Let $a_1, \ldots, a_s$ be positive numbers, and $0 < \beta < r$. The following are equivalent.*

$$\tilde{\epsilon}_{p,y}(\mathbf{Q}; f) = \mathcal{O}(y^{-\beta}), \qquad y > 0. \tag{9}$$

$$\tilde{\epsilon}_{p,y}(\mathbf{a} \odot \mathbf{Q}; f) = \mathcal{O}(y^{-\beta}), \qquad y > 0. \tag{10}$$

$$K_r(\mathbf{Q}, p; w_{\overline{\mathbf{Q}}}^{-1} f, \delta) = \mathcal{O}(\delta^\beta), \qquad 0 < \delta < c. \tag{11}$$

It is worthwhile to formulate a corollary of this theorem in the case when all the weights are equal. We define, for $1 \le p \le \infty$, $f \in L^p(\mathbf{R}^s)$, and $y \ge 0$,

$$E_{p,y}^{[s]}(\mathbf{Q}; f) := \inf\{\|f - w_{\overline{\mathbf{Q}}}P\|_p \ : \ P \in \Pi_{(y,\ldots,y)}\}, \tag{12}$$

where, in contrast to the definition of $\tilde{\epsilon}_{p,y}(\mathbf{Q}; f)$, the infimum is taken over all polynomials in s variables with coordinatewise degree at most y.

Corollary 3.2 *Let $r \geq 1$ be an integer, w_Q be a Freud weight satisfying* (4), $\mathbf{Q} = (Q, \ldots, Q)$, $1 \leq p \leq \infty$, $f \in L^p(\mathbf{R}^s)$, $a_1, \ldots, a_s$ *be positive numbers, and* $0 < \beta < r$. *Then*

$$E_{p,y}^{[s]}(\mathbf{Q}; f) = \mathcal{O}\left(\left(\frac{q_y}{y}\right)^{\beta}\right), \qquad y > 0,$$

if and only if

$$E_{p,y}^{[s]}(\mathbf{a} \odot \mathbf{Q}; f) = \mathcal{O}\left(\left(\frac{q_y}{y}\right)^{\beta}\right), \qquad y > 0.$$

In particular, if $Q(x) = |x|^\alpha$ for an even, positive integer α, then for any $\gamma > 0$, $E_{p,y}^{[s]}(\mathbf{Q}; f) = \mathcal{O}(y^{-\gamma})$ if and only if $E_{p,y}^{[s]}(\mathbf{a} \odot \mathbf{Q}; f) = \mathcal{O}(y^{-\gamma})$.

4 Shifted average operators

4.1 The univariate case

If w_Q is a weight function (*i.e.*, $\int_{\mathbf{R}} w_Q(t)|t|^m dt < \infty$ for all integer $m \geq 0$), $g : \mathbf{R} \to \mathbf{R}$, and $w_Q g \in L^p(\mathbf{R})$, we write

$$E_{p,x}(Q; g) := \inf_{P \in \Pi_x} \|(g - P)w_Q\|_{p,\mathbf{R}}.$$

While the best approximation operator is not linear, Freud developed in a series of papers (cf. [7]) a sequence of linear operators which are "near best" approximants.

There exists a unique system of orthonormal polynomials $\{p_k(Q) \in \Pi_k\}_{k=0}^{\infty}$ with positive leading coefficients such that

$$\int_{\mathbf{R}} p_k(Q; x) p_j(Q; x) w_Q^2(x) dx = 1,$$

if $k = j$, and 0 otherwise.

The shifted average kernel is defined for integer $n \geq 1$ and $x, t \in \mathbf{R}$ by

$$V_n(Q; x, t) := \sum_{k=0}^{n} p_k(Q; x) p_k(Q; t) + \sum_{k=n+1}^{2n-1} \left(2 - \frac{k}{n}\right) p_k(Q; x) p_k(Q; t), \qquad (13)$$

and the shifted average operator is defined by

$$v_n(Q; g, x) := \int_{\mathbf{R}} g(t) V_n(Q; x, t) w^2(t) dt, \qquad (14)$$

when the integral on the right-hand side is well defined. The following theorem ([7]) summarizes some of the important properties of the operators v_n.

Theorem 4.1 *Let w_Q be a Freud weight, and $n \geq 1$ be an integer. For all $P \in \Pi_n$, we have $v_n(Q;P) = P$. Let $1 \leq p \leq \infty$, and $w_Q g \in L^p(\mathbf{R})$. Then $v_n(Q;g) \in \Pi_{2n-1}$, and*

$$\|v_n(Q;g)w_Q\|_{p,\mathbf{R}} \leq c\|w_Q g\|_{p,\mathbf{R}}. \tag{15}$$

In particular,

$$E_{p,2n-1}(Q;g) \leq \|(g - v_n(Q;g))w_Q\|_{p,\mathbf{R}} \leq cE_{p,n}(Q;g). \tag{16}$$

If $r \geq 1$ is an integer, and g is an r times iterated integral of a function $g^{(r)}$ with $w_Q g^{(r)} \in L^p(\mathbf{R})$, then

$$\|(g^{(r)} - v_n^{(r)}(Q;g))w_Q\|_{p,\mathbf{R}} \leq cE_{p,n-r}(Q;g^{(r)}), \tag{17}$$

and

$$\|(g - v_n(Q;g))w_Q\|_{p,\mathbf{R}} \leq c\left(\frac{q_n}{n+1}\right)^r \|w_Q g^{(r)}\|_{p,\mathbf{R}}. \tag{18}$$

4.2 Multivariate case

Let w_{Q_k}, $1 \leq k \leq s$, be Freud weights. We introduce the shifted average operators coordinatewise as follows.

$$\begin{aligned} v_{n,k}(g,\mathbf{x}) &:= v_{n,k}(Q_k; g, \mathbf{x}) \\ &:= \int_{\mathbf{R}} V_n(Q_k; x_k, t) g(x_1, \ldots, x_{k-1}, t, x_{k+1}, \ldots, x_s) w_{Q_k}^2(t)dt. \end{aligned} \tag{19}$$

For a multi-integer $\mathbf{m} \geq 1$ (i.e., $m_k \geq 1$, $k = 1, \ldots, s$), we further define

$$\begin{aligned} &v_{\mathbf{m}}^{[k]}(g;\mathbf{x}) \\ &:= \int_{\mathbf{R}^k} \prod_{j=1}^{k} V_{m_j}(Q_j; x_j, t_j) g(t_1, \ldots, t_k, x_{k+1}, \ldots, x_s) \\ &\qquad \times \exp(-2\sum_{j=1}^{k} Q_j(t_j))dt_1 \ldots dt_k. \end{aligned} \tag{20}$$

We will also find it convenient to define $v_{\mathbf{m}}^{[0]}(g) := g$.

Similarly, analogous to the K-functional defined in (8), we introduce the coordinatewise K-functionals by

$$K_{r,k}(\mathbf{Q}, p; g, \delta) := \inf \{\|(g-h)w_{\mathbf{Q}}\|_p + \delta^r (\|w_{\mathbf{Q}} h\|_p + \|w_{\mathbf{Q}} D_k^r h\|_p)\}, \tag{21}$$

where the infimum is taken over all functions h such that h is $r-1$ times continuously differentiable in the k-th variable, $D_k^{r-1}h$ is absolutely continuous with respect to the k-th variable, and $w_{\mathbf{Q}} D_k^r h \in L^p(\mathbf{R}^s)$.

The direct theorem of multivariate weighted approximation is the following improvement over the corresponding result of Dzrbasyan and Tavadyan [3].

Theorem 4.2 *Let w_{Q_k} be Freud weights for $k = 1, \ldots, s$.*
(a) *Let $1 \le p \le \infty$, and $g : \mathbf{R}^s \to \mathbf{R}$ be such that $w_{\mathbf{Q}} g \in L^p(\mathbf{R}^s)$. Then for multi-integers $\mathbf{m} \ge 1$, $\mathbf{r} \ge 1$,*

$$\|(g - v^{[s]}_{\mathbf{m}}(g))w_{\mathbf{Q}}\|_p \le c \sum_{k=1}^{s} K_{r_k,k}\left(\mathbf{Q}, p; g, \frac{q_{m_k,k}}{m_k+1}\right). \tag{22}$$

(b) *If $1 \le k, \ell \le s$ are integers, h is $r-1$ times continuously differentiable in the k-th variable, $D_k^{r-1}h$ is absolutely continuous with respect to the k-th variable, and $w_{\mathbf{Q}} D_k^r h \in L^p(\mathbf{R}^s)$ then*

$$\|w_{\mathbf{Q}} D_k^r v^{[\ell]}_{\mathbf{m}}(h)\|_p \le c\|w_{\mathbf{Q}} D_k^r h\|_p. \tag{23}$$

Proof. We prove part (b) first. It is clear that

$$v^{[\ell]}_{\mathbf{m}}(h) = v_{m_1,1}\left(\ldots(v_{m_\ell,\ell}(h))\ldots\right). \tag{24}$$

Therefore, if $1 \le \ell < k$, (23) follows trivially from (15).

Next, let $k \le \ell \le s$. In this proof only, for $\mathbf{x} \in \mathbf{R}^s$, and $1 \le j \le s$, let $h_{\mathbf{x},j} : \mathbf{R} \to \mathbf{R}$ be defined by $h_{\mathbf{x},j}(t) := h(x_1, \ldots, x_{j-1}, t, x_{j+1}, \ldots, x_s)$. The estimate (17) implies that

$$\|v^{(r)}_{m_k}(Q_k; h_{\mathbf{x},k})w_{Q_k}\|_{p,\mathbf{R}} \le c\|h^{(r)}_{\mathbf{x},k} w_{Q_k}\|_{p,\mathbf{R}};$$

i.e.,

$$\|D_k^r v_{m_k,k}(h) w_{Q_k}\|_{p,\mathbf{R},k} \le c\|w_{Q_k} D_k^r h\|_{p,\mathbf{R},k}, \tag{25}$$

where $\|\cdot\|_{p,\mathbf{R},k}$ denotes the $L^p(\mathbf{R})$ norm taken with respect to the k-th variable. Now, (24) implies that

$$D_k^r v^{[\ell]}_{\mathbf{m}}(h) = v_{m_1,1}\left(\ldots v_{m_{k-1},k-1}\left(v_{m_{k+1},k+1}\left(\ldots\left(v_{m_\ell,\ell}\left(D_k^r v_{m_k,k}(h)\right)\right)\ldots\right)\right)\right).$$

Therefore, (23) follows from (15) and (25).

Continuing the same notation, we observe next that (18) implies that

$$\|(h_{\mathbf{x},k} - v_{m_k}(Q_k; h_{\mathbf{x},k}))w_{Q_k}\|_{p,\mathbf{R},k} \le c\left(\frac{q_{m_k}}{m_k+1}\right)^r \|w_{Q_k} h^{(r)}_{\mathbf{x},k}\|_{p,\mathbf{R},k},$$

which leads to

$$\|(h - v_{m_k,k}(h))w_{\mathbf{Q}}\|_p \le c\left(\frac{q_{m_k}}{m_k+1}\right)^r \|w_{\mathbf{Q}} D_k^r h\|_p. \tag{26}$$

Now, let $1 \le k \le s$ be any integer, $w_{\mathbf{Q}} g \in L^p(\mathbf{R}^s)$, and h be found satisfying the conditions of part (b) with r_k in place of r, such that

$$\begin{aligned} &\|(g-h)w_{\mathbf{Q}}\|_p + \left(\frac{q_{m_k,k}}{m_k+1}\right)^{r_k} \left(\|w_{\mathbf{Q}} h\|_p + \|w_{\mathbf{Q}} D_k^{r_k} h\|_p\right) \\ &\le\ 2K_{r_k,k}\left(\mathbf{Q}, p; g, \frac{q_{m_k,k}}{m_k+1}\right). \end{aligned} \tag{27}$$

From (24) and (15), we deduce that

$$\left\| \left(v_{\mathbf{m}}^{[k-1]}(g-h) - v_{\mathbf{m}}^{[k]}(g-h) \right) w_{\mathbf{Q}} \right\|_p \le c\|(g-h)w_{\mathbf{Q}}\|_p. \tag{28}$$

Further, using (26) (with r_k in place of r), we see that

$$\begin{aligned}\left\| \left(v_{\mathbf{m}}^{[k-1]}(h) - v_{\mathbf{m}}^{[k]}(h) \right) w_{\mathbf{Q}} \right\|_p &= \left\| v_{\mathbf{m}}^{[k-1]} \left(h - v_{m_k,k}(h) \right) w_{\mathbf{Q}} \right\|_p \\ &\le c\|(h - v_{m_k,k}(h))w_{\mathbf{Q}}\|_p \le \left(\frac{q_{m_k,k}}{m_k+1} \right)^{r_k} \|w_{\mathbf{Q}} D_k^{r_k} h\|_p.\end{aligned}$$

Along with (28) and (27), this leads to

$$\left\| \left(v_{\mathbf{m}}^{[k-1]}(g) - v_{\mathbf{m}}^{[k]}(g) \right) w_{\mathbf{Q}} \right\|_p \le cK_{r_k,k}\left(\mathbf{Q}, p; g, \frac{q_{m_k,k}}{m_k+1} \right). \tag{29}$$

The estimate (22) now follows from the fact that

$$g - v_{\mathbf{m}}^{[s]}(g) = \sum_{k=1}^{s} \left(v_{\mathbf{m}}^{[k-1]}(g) - v_{\mathbf{m}}^{[k]}(g) \right). \qquad \square$$

As a corollary to this theorem, we prove the following relationship between the K-functionals defined in (8) and (21). We observe that in the definition of $K_r(\mathbf{Q}, p; g, \delta)$, the derivatives with respect to each variable are applied to the same smooth function, while in the definition of the different $K_{r,k}(\mathbf{Q}, p; g, \delta)$'s, there may be a different smooth function for each variable. Further, we observe that p_k being an inverse function of $x \mapsto x/q_{x,k}$, $x \sim p_k(x)/q_{p_k(x),k}$ for $x > 0$ and $1 \le k \le s$. Therefore, the Markov-Bernstein inequality [7, Theorem 6.2.9] implies that

$$\|w_{\mathbf{Q}} D_k^r P\|_p \le c\delta^{-r}\|w_{\mathbf{Q}} P\|_p, \qquad P \in \Pi_{\mathbf{p}(\delta^{-1})},\ 1 \le k \le s,\ r \ge 1. \tag{30}$$

Proposition 4.3 *Let $r \ge 1$ be an integer, and for $k = 1, \ldots, s$, w_{Q_k} be a Freud weight. Let $1 \le p \le \infty$, and $w_{\mathbf{Q}} g \in L^p(\mathbf{R}^s)$. Then for $0 < \delta \le 1$,*

$$K_r(\mathbf{Q}, p; g, \delta) \sim \sum_{k=1}^{s} K_{r,k}(\mathbf{Q}, p; g, \delta). \tag{31}$$

Proof. It is easy to see from the relevant definitions that

$$\sum_{k=1}^{s} K_{r,k}(\mathbf{Q}, p; g, \delta) \le cK_r(\mathbf{Q}, p; g, \delta). \tag{32}$$

Let $\mathbf{m}$ be a multi-integer such that $m_k \sim p_k(\delta^{-1})$, and h_k be found such that

$$\|(g-h_k)w_{\mathbf{Q}}\|_p + \delta^r \left(\|w_{\mathbf{Q}} h_k\|_p + \|w_{\mathbf{Q}} D_k^r h_k\|_p \right) \le 2K_{r,k}(\mathbf{Q}, p; g, \delta). \tag{33}$$

We observe that

$$\|w_{\mathbf{Q}}v_{\mathbf{m}}^{[s]}(g)\|_p \le c\|w_{\mathbf{Q}}g\|_p \le c\{\|(g-h_k)w_{\mathbf{Q}}\|_p + \|w_{\mathbf{Q}}h_k\|_p\} \le c\delta^{-r}K_{r,k}(\mathbf{Q},p;g,\delta). \tag{34}$$

In view of (30), (23), and (33), we have

$$\begin{aligned} \|w_{\mathbf{Q}}D_k^r v_{\mathbf{m}}^{[s]}(g)\|_p &\le \|w_{\mathbf{Q}}D_k^r v_{\mathbf{m}}^{[s]}(g-h_k)\|_p + \|w_{\mathbf{Q}}D_k^r v_{\mathbf{m}}^{[s]}(h_k)\|_p \\ &\le c\delta^{-r}\left\{\|w_{\mathbf{Q}}v_{\mathbf{m}}^{[s]}(g-h_k)\|_p + \delta^r\|w_{\mathbf{Q}}D_k^r h_k\|_p\right\} \\ &\le c\delta^{-r}K_{r,k}(\mathbf{Q},p;g,\delta). \end{aligned} \tag{35}$$

From (22), (34), and (35), we conclude that

$$\begin{aligned} K_r(\mathbf{Q},p;g,\delta) &\le \|(g-v_{\mathbf{m}}^{[s]}(g))w_{\mathbf{Q}}\|_p \\ &\quad +\delta^r\left\{\|w_{\mathbf{Q}}v_{\mathbf{m}}^{[s]}(g)\|_p + \sum_{k=1}^{s}\|w_{\mathbf{Q}}D_k^r v_{\mathbf{m}}^{[s]}(g)\|_p\right\} \\ &\le c\sum_{k=1}^{s}K_{r,k}(\mathbf{Q},p;g,\delta). \end{aligned}$$

Along with (32), this completes the proof. □

We end this section with a statement of the converse theorem and an equivalence theorem. Towards this end, we write for $y \ge 0$, and $w_{\mathbf{Q}}g \in L^p(\mathbf{R}^s)$:

$$\epsilon_{p,y}(\mathbf{Q};g) := \inf\{\|(g-P)w_{\mathbf{Q}}\|_p \ : \ P \in \Pi_{\mathbf{p}(y)}\}, \tag{36}$$

Using (30), it is easy to obtain the following analogue of [7, Theorem 4.2.2].

Theorem 4.4 *Let w_{Q_k} be Freud weights for $k = 1,\ldots,s$, $r \ge 1$ be an integer, $1 \le p \le \infty$, and $w_{\mathbf{Q}}g \in L^p(\mathbf{R}^s)$. Then for $0 < \delta \le 1$,*

$$K_r(\mathbf{Q},p;g,\delta) \le c\delta^r\left\{\|w_{\mathbf{Q}}g\|_p + \sum_{0\le m\le\delta^{-1}}(m+1)^{r-1}\epsilon_{p,m}(\mathbf{Q};g)\right\}. \tag{37}$$

The proof of this theorem, being the usual telescoping argument as in the proof of [7, Theorem 4.2.2] or [4, Theorem 3], is ommitted.

In light of Theorems 4.2, 4.4 and Proposition 4.3, we have the following equivalence theorem.

Proposition 4.5 *Let $r \ge 1$ be an integer, w_{Q_k} be Freud weights for $k = 1,\ldots,s$, $1 \le p \le \infty$, $w_{\mathbf{Q}}g \in L^p(\mathbf{R}^s)$, and $0 < \beta < r$. Then the following are equivalent:*

$$K_r(\mathbf{Q},p;g,\delta) = \mathcal{O}(\delta^\beta), \qquad \delta > 0, \tag{38}$$

$$\epsilon_{p,y}(\mathbf{Q};g) = \mathcal{O}(y^{-\beta}), \qquad y > 0. \tag{39}$$

$$\sum_{k=1}^{s}K_{r,k}(\mathbf{Q},p;g,\delta) = \mathcal{O}(\delta^\beta), \qquad \delta > 0. \tag{40}$$

5 A modulus of smoothness

In this section, we relate the coordinatewise K-functionals $K_{r,k}(\mathbf{Q},p;w_{\overline{\mathbf{Q}}}^{-1}f,\delta)$ with $K_{r,k}(\mathbf{a}\odot\mathbf{Q},p;w_{\overline{\mathbf{a}\odot\mathbf{Q}}}^{-1}f,\delta)$ for $f\in L^p(\mathbf{R}^s)$. Since only one coordinate is (essentially) involved here, we find it convenient to do this in the univariate case. Accordingly, we will evaluate the order of magnitude of the univariate K-functional:

$$K_r(Q,p;g,\delta) := \inf\left\{\|(g-h)w_Q\|_{p,\mathbf{R}} + \delta^r\left(\|hw_Q\|_p + \|h^{(r)}w_Q\|_{p,\mathbf{R}}\right)\right\}, \tag{41}$$

where w_Q is a Freud weight, and the infimum is taken over all $r-1$ times continuously differentiable functions h such that $h^{(r-1)}$ is absolutely continuous, and $h^{(r)}w_Q\in L^p(\mathbf{R})$.

In [7], we have studied a slightly different version of the K-functional defined by

$$\mathcal{K}_r(Q,p;g,\delta) := \inf\left\{\|(g-h)w_Q\|_{p,\mathbf{R}} + \delta^r\|h^{(r)}w_Q\|_{p,\mathbf{R}}\right\}, \tag{42}$$

where the infimum is taken over all $r-1$ times continuously differentiable functions h such that $h^{(r-1)}$ is absolutely continuous, and $h^{(r)}w_Q\in L^p(\mathbf{R})$. In [7], we evaluated the order of magnitude of this K-functional in terms of a modified modulus of smoothness as follows.

For $t>0$ and integer $k\ge 0$, the forward difference of a function $f:\mathbf{R}\to\mathbf{R}$ is defined by

$$\Delta_t^k f(x) := \sum_{\nu=0}^{k}(-1)^{k-\nu}\binom{k}{\nu}f(x+\nu t).$$

Let w_Q be a Freud weight satisfying (4), and $\overline{Q}$ be constructed as in Theorem 2.1. With

$$Q'_\delta(x) := \min\left(\delta^{-1},(1+Q'(x)^2)^{1/2}\right), \qquad \delta>0,\ x\in\mathbf{R},$$

we define a pre-modulus of smoothness by the formula

$$\omega_r(Q,p;g,\delta) := \sum_{\nu=0}^{r}\delta^{r-\nu}\sup_{|t|\le\delta}\|{Q'_\delta}^{r-\nu}\Delta_t^\nu(gw_{\overline{Q}})\|_{p,\mathbf{R}}, \tag{43}$$

and the modulus of smoothness by the formula

$$\Omega_r(Q,p;g,\delta) := \inf_{P\in\Pi_{r-1}}\omega_r(Q,p;g-P,\delta). \tag{44}$$

In [7], we proved that

$$\mathcal{K}_r(Q,p;g,\delta) \sim \Omega_r(Q,p;g,\delta), \tag{45}$$

for $r\ge 1$, $0<\delta\le 1$, $1\le p\le\infty$, and $w_Qg\in L^p(\mathbf{R})$. We will prove here the following theorem.

Theorem 5.1 *Let $r \geq 1$ be an integer, and w_Q be a Freud weight satisfying* (4), $1 \leq p \leq \infty$, *and $w_Q g \in L^p(\mathbf{R})$. Then*

$$K_r(Q,p;g,\delta) \sim \omega_r(Q,p;g,\delta) \sim \mathcal{K}_r(Q,p;g,\delta) + \delta^r \|w_Q g\|_{p,\mathbf{R}}, \qquad 0 < \delta \leq 1. \quad (46)$$

Proof. Using the triangle inequality,

$$|\|w_Q h\|_{p,\mathbf{R}} - \|w_Q g\|_{p,\mathbf{R}}| \leq \|(g-h)w_Q\|_{p,\mathbf{R}},$$

it is easy to derive that

$$K_r(Q,p;g,\delta) \sim \mathcal{K}_r(Q,p;g,\delta) + \delta^r \|w_Q g\|_{p,\mathbf{R}}, \qquad 0 < \delta \leq 1. \quad (47)$$

If $0 < \delta \leq 1$, $Q'_\delta(x) \geq 1$ for all $x \in \mathbf{R}$, and $\delta^r \|w_Q g\|_{p,\mathbf{R}} \leq \delta^r \|(Q'_\delta)^r w_Q g\|_{p,\mathbf{R}}$. The last expression being one of the summands in (43), we see that

$$\delta^r \|w_Q g\|_{p,\mathbf{R}} \leq \omega_r(Q,p;g,\delta). \quad (48)$$

In view of (45) and (44), $\mathcal{K}_r(Q,p;g,\delta) \leq c\omega_r(Q,p;g,\delta)$. Thus, we have

$$\mathcal{K}_r(Q,p;g,\delta) + \delta^r \|w_Q g\|_{p,\mathbf{R}} \leq c\omega_r(Q,p;g,\delta). \quad (49)$$

Let h be found such that

$$\|(g-h)w_Q\|_{p,\mathbf{R}} + \delta^r \left(\|hw_Q\|_p + \|h^{(r)}w_Q\|_{p,\mathbf{R}}\right) \leq 2K_r(Q,p;g,\delta), \quad (50)$$

and

$$T(x) := \sum_{j=0}^{r-1} \frac{h^{(j)}(0)}{j!} x^j.$$

Then

$$\omega_r(Q,p;g,\delta) \leq \omega_r(Q,p;g-h,\delta) + \omega_r(Q,p;h-T,\delta) + \omega_r(Q,p;T,\delta). \quad (51)$$

It is clear from the definition that

$$\omega_r(Q,p;g-h,\delta) \leq c\|(g-h)w_Q\|_{p,\mathbf{R}}. \quad (52)$$

During the proof of [7, Theorem 5.2.1], we have shown (estimate (5.2.16)) that

$$\omega_r(Q,p;h-T,\delta) \leq c\delta^r \|(h-T)^{(r)} w_Q\|_{p,\mathbf{R}} = c\delta^r \|h^{(r)} w_Q\|_{p,\mathbf{R}}. \quad (53)$$

(The notation in [7] is different; the function g there is what we are calling $h - T$ here.) We have also proved (cf. the derivation of [7, estimate (5.2.14)]) that for $|t| \leq \delta$ and $0 \leq \nu \leq r$,

$$\|(Q'_\delta)^{r-\nu} \Delta_t^\nu (w_{\overline{Q}} T)\|_{p,\mathbf{R}} \leq \delta^\nu \|(1 + \overline{Q}'^2)^{(r-\nu)/2} (w_{\overline{Q}} T)^{(\nu)}\|_{p,\mathbf{R}}.$$

Therefore, (43) shows that

$$\omega_r(Q,p;T,\delta) \le \delta^r \sum_{\nu=0}^{r} \|(1+\overline{Q}'^2)^{(r-\nu)/2}(w_{\overline{Q}}T)^{(\nu)}\|_{p,\mathbf{R}}.$$

Now, $\sum_{\nu=0}^{r} \|(1+\overline{Q}'^2)^{(r-\nu)/2}((\cdot)w_{\overline{Q}})^{(\nu)}\|_{p,\mathbf{R}}$ is a norm on Π_{r-1}. Since all norms on Π_{r-1} are equivalent, we deduce that

$$\omega_r(Q,p;T,\delta) \le c\delta^r \|w_Q T\|_{p,\mathbf{R}} \le c\delta^r \left\{\|w_Q h\|_{p,\mathbf{R}} + \|(h-T)w_Q\|_{p,\mathbf{R}}\right\}. \tag{54}$$

Using (48) with $h-T$ in place of g, and then using (53), we see that

$$\delta^r \|(h-T)w_Q\|_{p,\mathbf{R}} \le c\omega_r(Q,p;h-T,\delta) \le c\delta^r \|h^{(r)}w_Q\|_{p,\mathbf{R}}.$$

Hence, (54) implies that

$$\omega_r(Q,p;T,\delta) \le c\delta^r \left\{\|hw_Q\|_{p,\mathbf{R}} + \|h^{(r)}w_Q\|_{p,\mathbf{R}}\right\}.$$

Substituting from this estimate and those in (52), (53) in (51), we get

$$\omega_r(Q,p;g,\delta) \le c\left\{\|(g-h)w_Q\|_{p,\mathbf{R}} + \delta^r \left(\|hw_Q\|_p + \|h^{(r)}w_Q\|_{p,\mathbf{R}}\right)\right\}.$$

Our choice of the function h as in (50) now shows that

$$\omega_r(Q,p;g,\delta) \le cK_r(Q,p;g,\delta).$$

This completes the proof. □

It is obvious that for any $\lambda > 0$ and $f \in L^p(\mathbf{R})$,

$$\omega_r(Q,p;w_{\overline{Q}}^{-1}f,\delta) \sim \omega_r(\lambda Q,p;w_{\overline{Q}}^{-\lambda}f,\delta),$$

where the constants may depend upon λ. Transcribing all these univariate results to the coordinatewise K-functionals, we arrive at the following corollary.

Corollary 5.2 *Let $r \ge 1$ be an integer. For $k = 1,\ldots,s$, let w_{Q_k} be a Freud weight satisfying* (4), *and $\overline{Q}_k$ be the smooth function as in Theorem* 2.1. *Let $1 \le p \le \infty$, $f \in L^p(\mathbf{R}^s)$ and $a_1,\ldots,a_s$ be positive numbers. Then for $1 \le k \le s$:*

$$K_{r,k}(\mathbf{Q},p;w_{\overline{\mathbf{Q}}}^{-1}f,\delta) \sim K_{r,k}(\mathbf{a}\odot\mathbf{Q},p;w_{\mathbf{a}\odot\overline{\mathbf{Q}}}^{-1}f,\delta), \qquad 0<\delta\le 1. \tag{55}$$

Hence, also

$$K_r(\mathbf{Q},p;w_{\overline{\mathbf{Q}}}^{-1}f,\delta) \sim K_r(\mathbf{a}\odot\mathbf{Q},p;w_{\mathbf{a}\odot\overline{\mathbf{Q}}}^{-1}f,\delta), \qquad 0<\delta\le 1. \tag{56}$$

We end this paper with the proof of Theorem 3.1.

Proof of Theorem 3.1. In light of Theorem 2.1, we see that

$$\tilde{\epsilon}_{p,y}(\mathbf{Q};f) = \epsilon_{p,y}(\overline{\mathbf{Q}};w_{\overline{\mathbf{Q}}}^{-1}f) \sim \epsilon_{p,y}(\mathbf{Q};w_{\overline{\mathbf{Q}}}^{-1}f), \qquad y>0. \tag{57}$$

Therefore, the equivalence between (9) and (11) follows from Proposition 4.5. In view of (56), Proposition 4.5 further implies that (11) is equivalent to

$$\tilde{\epsilon}_{p,y}(\mathbf{a}\odot\mathbf{Q};f) \sim \epsilon_{p,y}(\mathbf{a}\odot\mathbf{Q};w_{\mathbf{a}\odot\overline{\mathbf{Q}}}^{-1}f) = \mathcal{O}(y^{-\beta}), \qquad y>0. \tag{58}$$

This completes the proof. □

References

[1] S.B. Damelin, *Converse and smoothness theorems for Erdős weights in L^p $(0<p\le\infty)$,* J. Approx. Theory, **93** (1998), 349–398.

[2] Z. Ditzian and D.S. Lubinsky, *Jackson and smoothness theorems for Freud weights in L^p $(0<p\le\infty)$,* Constr. Approx., **11** (1997), 99–152.

[3] M.M. Dzrbasyan and A.B. Tavadyan, *On weighted uniform approximation by polynomials of functions of several variables,* Mat. Sb., N.S., **43** (85) (1957), 227–256.

[4] G. Freud, *Direct and converse theorems in the theory of weighted polynomial approximation,* Math. Zeitschrift, **126** (1972), 123–134.

[5] A. Kroo and J. Szabados, *Weighted polynomial approximation on the real line,* J. Approx. Theory, **83** (1995), 41–64.

[6] D.S. Lubinsky, *Jackson and Bernstein theorems for exponential weights,* BSMS: Approximation theory and function series, **5** (1996), 85–115.

[7] H.N. Mhaskar, "Introduction to the theory of weighted polynomial approximation", World Scientific, Singapore, 1996.

[8] E.B. Saff and V. Totik, "Logarithmic potentials with external fields", Springer Verlag, New York, 1997.

Department of Mathematics
California State University
Los Angeles, California, 90032, U.S.A.
Email address: hmhaska@calstatela.edu

The research of this author was supported, in part, by grant DMS-9971846 from the National Science Foundation and grant DAAD19-01-1-0001 from the U.S. Army Research Office.

Advanced Problems in Constructive Approximation
(Eds.) M.D. Buhmann and D.H. Mache
International Series of Numerical Mathematics Vol. 142, 143–152

Semigroups Associated to Mache Operators

I. Rasa

Abstract

Mache operators are investigated from the point of view of Altomare's theory. We prove the existence of a Feller semigroup representable as a limit of suitable iterates of Mache operators. The preservation properties of Mache operators lead to qualitative properties of the solution of the associated Cauchy problem. A new Chernoff type approach to the semigroup is presented, as well as quantitative results related to it.

1 Introduction

F. Altomare [1] initiated a systematic study of the connection between some approximation processes (in an arbitrary dimensional context) and the solution of suitable evolution problems, the key ingredients being Voronovskaja's formula and Trotter's theorem. Chapter 6 of [2] gives a first account of Altomare's theory; for further developments see [3]–[9],[11]–[13],[25],[27] and the references therein. The general (i.e., arbitrary dimensional) case has discussed in [1],[2],[11],[7],[8],[25],[27]. In the particularly important one-dimensional case many authors considered these problems either using classical approximation processes or introducing new ones satisfying a prescribed Voronovskaja formula. In [3]–[6],[9],[11]–[13] the authors investigate such classical or new processes and prove the existence of corresponding Feller semigroups; in some cases also the associated Markov process is studied.

In this paper Mache operators are considered from the point of view of Altomare's theory. The existence of an associated Feller semigroup is proved and the preservation properties of this semigroup are studied; as in the general Altomare theory, these preservation properties lead to qualitative properties of the solution of the associated Cauchy problem.

Using the technique of [25], a qualitative and a quantitative Chernoff-type approach to the Feller semigroup are presented. The Chernoff-type approaches to the semigroups are studied in [3]–[6],[9],[11]–[13] will be discussed in forthcoming papers.

2 Mache operators

Let $a, b > -1$ and $\alpha \geq 0$ be real numbers. For $n \geq 0$, $k = 0, 1, \ldots, n$ and for $x \in [0,1]$ let us define the Bernstein-Bezier polynomials $p_{n,k}(x) = \binom{n}{k} x^k (1-x)^{n-k}$. Set $c := [n^\alpha]$ and consider the functionals $A_{n,k} : C[0,1] \to \mathbf{R}$ by

$$A_{n,k}(f) = B(ck + a + 1, cn - ck + b + 1)^{-1} \int_0^1 t^{ck+a}(1-t)^{cn-ck+b} f(t)dt$$

where B is Euler's Beta function.

Now let $P_n : C[0,1] \to C[0,1]$, $n \geq 0$,

$$P_n f = \sum_{k=0}^{n} A_{n,k}(f) p_{n,k}\,. \tag{1}$$

The linear positive operators P_n have been introduced, even in a more general form, by D.H. Mache [21], [22]; by specializing the parameters a, b, α one obtains several known families of (Durrmeyer-type) operators (see [15], [20], [21]–[23]).

The approximation properties and characterization results of Mache operators have been studied in [21]–[23], [26], [27]. Now P_n transforms the polynomials of the degree $\leq m$ into polynomials of degree $\leq m$; see [26], (1). Here is Voronovskaja's formula for Mache operators, established independently in [21]and [26],[27].

Theorem 2.1 *Let $f \in C^2[0,1]$. If $\alpha > 0$ we have*

$$\lim_{n\to\infty} n(P_n f(x) - f(x)) = \frac{x(1-x)}{2} f''(x) \tag{2}$$

uniformly on $[0,1]$.

If $\alpha = 0$ then

$$\lim_{n\to\infty} n(P_n f - f) = uf'' + vf' \tag{3}$$

uniformly on $[0,1]$, where

$$u(x) := x(1-x), \quad v(x) := a + 1 - (a+b+2)x, \qquad x \in [0,1]. \tag{4}$$

For $\alpha = a = b = 0$, (3) was proved in [16], Theorem 2, p. 10. The preservation properties of Mache operators have been investigated in [19],[26],[27]. They are described as follows.

Theorem 2.2 ([26],[27]). *Let $0 \leq m \leq n$ and $\{f_0, \ldots, f_m\} \subset C[0,1]$ be a Chebyshev system. Then $\{P_n f_0, \ldots, P_n f_m\}$ is also a Chebyshev system. If $f \in C[0,1]$ is convex with respect to $\{f_0, \ldots, f_m\}$, then $P_n f$ is also convex with respect to $\{P_n f_0, \ldots, P_n f_m\}$.*

Corollary 2.3 ([10],[19],[26],[27]). *If $0 \leq m \leq n$ and $f \in C[0,1]$ is convex of order m (i.e., convex with respect to $\{e_0, \ldots, e_m\}$ where $e_j(x) = x^j$, $j \in \mathbf{N}_0$), then $P_n f$ is convex of order m.*

Theorem 2.4 ([26],[27]). *Let $Lip(\beta, M)$ be the Hölder class on $[0,1]$ with the exponent $\beta \in (0,1]$ and constant $M > 0$. Then*

$$P_n(Lip(\beta, 1)) \subset Lip(\beta, N^\beta) \tag{5}$$

and

$$\omega(P_n f, t) \leq \frac{2cn + a + b + 2}{cn + a + b + 2}\omega(f, t), \tag{6}$$

where ω is the usual first-order modulus of continuity and $N = \frac{cn}{cn+a+b+2}$.

For $\alpha = a = b = 0$, from (5) we obtain the result expressed in Theorem 1(c) of [10].

3 The Feller semigroup in the case $\alpha = 0$

Let $\alpha = 0$, $a, b \geq 0$. Consider the operator

$$Af(x) = u(x)f''(x) + v(x)f'(x), \qquad 0 < x < 1, \qquad f(x) \in D(A) \tag{7}$$

where u and v are defined by (4), and

$$D(A) := \{f \in C^1[0,1] \cap C^2(0,1) \mid \lim_{x \to 0,1} u(x)f''(x) = 0\}. \tag{8}$$

Theorem 3.1 *Under the above mentioned hypotheses, $(A, D(A))$ generates a semi group $(T(t))_{t \geq 0}$ of positive contractions on the Banach space $C[0,1]$ with the uniform norm. Moreover,*

$$T(t) = \lim_{n \to \infty} P_n^{[nt]} \qquad \textit{strongly on} \qquad C[0,1]. \tag{9}$$

Proof. Since $a \geq 0$ and $b \geq 0$, the condition (4.1) in [13] is satisfied, hence Lemma 4.2, Prop. 4.3 and Theorem 4.4 of [13] show that $(A, D(A))$ generates a semigroup $(T(t))$ of positive contractions on $C[0,1]$ and $C^2[0,1]$ is a core for A.

By Theorem 2.1 we have

$$\lim_{n \to \infty} n(P_n f - f) = Af, \qquad f \in C^2[0,1],$$

and so it is possible to apply Chernoff's product formula ([18], Cor. III.5.4) to the family of contractions

$$V(t) := P_n, \qquad \frac{1}{n} \leq t < \frac{1}{n-1}, \qquad n \geq 2.$$

It follows that

$$T(t) = \lim_{n\to\infty} V(t_n)^{k_n} \qquad \text{strongly on} \qquad C[0,1], \tag{10}$$

whenever $k_n t_n \to t > 0$ with k_n integers tending to $+\infty$.

For $t_n = \frac{1}{n}$ and $k_n = [nt]$, (10) reduces to (9) and the proof is finished. □

Corollary 3.2 *We have the following two statements:*

(i) *If $f \in C[0,1]$ is convex of order m, then $T(t)f$ is convex of order m for $t \geq 0$.*
(ii) $T(t)(Lip(\beta, M)) \subset Lip(\beta, Me^{-(a+b+2)t\beta})$.

Proof. (i) is a consequence of (9) and Corollary 2.3, while (ii) follows from (9) and (5). Remark that the Hölder constant diminishes under $T(t)$ and vanishes when $t \to \infty$. □

If $f \in D(A)$, the unique solution of the problem

$$w_t(t,x) = Aw(t,x) \quad 0 < x < 1, \quad t \geq 0,$$

$$w(0,x) = f(x) \qquad 0 \leq x \leq 1 \tag{11}$$

is given by $w(t,x) = T(t)f(x)$, $t \geq 0$, $0 \leq x \leq 1$. So, if f is convex of order m, then $w(t,.)$ is convex of order m for all $t \geq 0$; if f is in $Lip(\beta, M)$, then $w(t;.)$ is in $Lip(\beta, Me^{-(a+b+2)t\beta})$ for all $t \geq 0$.

On the other hand, consider (11) for $f = e_m$; it is not difficult to prove that the corresponding solution has the form

$$T(t)e_m = w(t,.) = \sum_{i=0}^{m} c_i(t)e_i$$

where $\lim_{t\to\infty} c_i(t) = 0$, $i = 1, \ldots, m$ and $\lim_{t\to\infty} c_0(t)$ is finite.

This means that for every polynomial p there exists a constant function Kp such that $\lim_{t\to\infty} T(t)p = Kp$. As a consequence, the semigroup has the following important property (which is present in a suitable form, see [2], Theorem 6.2.6 also in the general Altomare theory):

Corollary 3.3 *For each $f \in C[0,1]$ there exists a constant function Kf such that $\lim_{t\to\infty} T(t)f = Kf$. So, if in (11) $f \in D(A)$, then $\lim_{t\to\infty} w(t,.) = Kf$.*

At this point we state

Conjecture 3.4 $K = P_0$, *that is, the constant value of the function Kf is given by*

$$A_0(f) = \left(\int_0^1 t^a(1-t)^b f(t)\,dt\right) / \int_0^1 t^a(1-t)^b\,dt.$$

Still in the case $\alpha = 0$, suppose that $-1 < a, b < 0$.
Let the operator A be defined by (7), but this time

$$D(A) := \{f \in C[0,1] \cap C^2(0,1) \mid \lim_{x\to 0,1}(u(x)f''(x) + v(x)f'(x)) = 0\}. \qquad (12)$$

Consider the function

$$W(x) = \exp(-\int_{\frac{1}{2}}^{x} \frac{v(t)}{u(t)}\,dt) = 2^{-a-b-2}x^{-a-1}(1-x)^{-b-1}.$$

Since $a < 0$ and $b < 0$, we have $W \in L^1(0, \frac{1}{2})$ and $W \in L^1(\frac{1}{2}, 1)$. By the results of [14] it follows that $(A, D(A))$ generates a semigroup $(T(t))$ of positive contractions on $C[0,1]$. Corollary 3.3 *and the Conjecture* 3.4 *have the same signification as before; maybe* (9) *is true also in this case.*

4 The Feller semigroup in the case $\alpha > 0$

Now let

$$Af(x) = \frac{x(1-x)}{2}f''(x), \qquad 0 < x < 1, \qquad f \in D(A) \qquad (13)$$

where

$$D(A) := \{f \in C[0,1] \cap C^2(0,1) \mid \lim_{x\to 0,1} \frac{x(1-x)}{2}f''(x) = 0\}. \qquad (14)$$

It is well known (see [2], Theorem 6.3.5 or [18], Sect. III.5.7) that $(A, D(A))$ generates a semigroup $(T(t))$ of positive contractions on $C[0,1]$ such that

$$T(t) = \lim_{n\to\infty} B_n^{[nt]} \qquad \text{strongly on} \qquad C[0,1], \qquad (15)$$

where B_n are the classical Bernstein operators on $C[0,1]$.

Since $C^2[0,1]$ is a core for A, Theorem 2.1 and Chernoff's product formula yield an alternative representation of this semigroup:

$$T(t) = \lim_{n\to\infty} P_n^{[nt]} \qquad \text{strongly on} \qquad C[0,1]. \qquad (16)$$

By using the preservation properties of Bernstein operators B_n or those of Mache operators P_n, we find that $T(t)$ preserves the m-convexity and $T(t)(Lip(\beta, M)) \subset Lip(\beta, M)$.

5 A qualitative Chernoff-type approach

In the sequel we consider only the case $\alpha = 0$, $a, b \geq 0$. Let $q := \max\{a+1, b+1\}$ and $\frac{1}{p_0} = 2\max\{2, q\}$. For $0 \leq p \leq 1$, $x \in [0,1]$ and $f \in C[0,1]$ let

$$T(p)f(x) := (1-x)f((1-\sqrt{p})x) + xf(\sqrt{p} + (1-\sqrt{p})x). \tag{17}$$

For $0 \leq p \leq p_0$ we may consider the linear positive contractions $E(p) : C[0,1] \to C[0,1]$,

$$E(p)f(x) := \frac{1}{2}F(4p)f(x) + \frac{1}{2}f(x + 2pv(x)). \tag{18}$$

It is not difficult to prove that for $f \in C^2[0,1]$,

$$\lim_{p\to 0} \frac{1}{p}(E(p)f - f) = Af \tag{19}$$

where A is described by (7).

Let $(T(t))$ be the semigroup from (9). An application of Chernoff's product formula ([18], Theorem III.5.2) yields an alternative description of the semigroup, namely:

Theorem 5.1 *For $t \geq 0$ we have*

$$T(t) = \lim_{n\to\infty} E(\frac{t}{n})^n \qquad \textit{strongly on} \qquad C[0,1]. \tag{20}$$

The approximations of $T(t)$ furnished by (9) and (20) can be compared in the spirit of [25], Ex.5.3.

6 A quantitative approach

Consider the Altomare projection T on $C[0,1]$, $Tf(x) = (1-x)f(0) + xf(1)$. With notation as in the preceding section, (5.18) of [25] yields for $0 < p \leq p_0$ and $f \in C^3[0,1]$:

$$\left\| \frac{1}{2p}(F(4p)f - f) - uf'' \right\| \leq \frac{\sqrt{p}}{12}\|f^{(3)}\|. \tag{21}$$

This means that

$$\left| \frac{1}{p}(E(p)f(x) - f(x)) - Af(x) \right| \leq \frac{\sqrt{p}}{12}\|f^{(3)}\| + |v(x)|\,|f'(y) - f'(x)|$$

for some y between x and $x + 2pv(x)$. Since $\|v\| = q$, we obtain for $f \in C^3[0,1]$:

$$\left\| \frac{1}{p}(E(p)f - f) - Af \right\| \leq \frac{\sqrt{p}}{12}\|f^{(3)}\| + 2pq^2\|f''\|. \tag{22}$$

On the other hand, for $f \in C^4[0,1]$ we have (see, e.g., [24], p. 7):

$$\left\| \frac{1}{p}(T(p)f - f) - Af \right\| \leq \frac{p}{2}\|A^2 f\|. \tag{23}$$

So, for $f \in C^4[0,1]$ we get

$$\|E(p)f - T(p)f\| \leq p^2\left(\frac{1}{2}\|A^2 f\| + 2q^2\|f''\|\right) + \frac{p\sqrt{p}}{12}\|f^{(3)}\|. \tag{24}$$

Let U be the space of those $f \in C^4[0,1]$ for which $T(t)f \in C^4[0,1]$, $t \geq 0$, and $\sup_{t\geq 0} \|(T(t)f)^{(j)}\| < \infty$, $j = 2, 3$. For $f \in U$ set

$$|f|_U := \max\{\frac{1}{2}\|A^2 f\| + 2q^2 \sup_{t\geq 0}\|(T(t)f)''\|, \frac{1}{12}\sup_{t\geq 0}\|(T(t)f)^{(3)}\|\}.$$

Let $f \in U$. Then

$$\|A^2 T(t)f\| = \|T(t)A^2 f\| \leq \|A^2 f\|,$$

hence (24) implies

$$\|(E(p) - T(p))T(t)f\| \leq (p^2 + p\sqrt{p})\,|f|_U. \tag{25}$$

This is the *consistency condition* (2.3) from [17], with $C = 1$, $\omega = 0$, $\varphi(p) = p + \sqrt{p}$. Now Theorem 1 of [17] yields the following quantitative result related to the Theorem 5.1:

Theorem 6.1 *For $f \in U$, $n \geq 1$ and $t \geq 0$ we have*

$$\left\| E(\frac{t}{n})^n f - T(t)f \right\| \leq t\left(\frac{t}{n} + \sqrt{\frac{t}{n}}\right)|f|_U. \tag{26}$$

The subspace U is dense in $C[0,1]$. The proof is the same as in [25], Remark 5.5.b, and uses Corollary 3.3 as well as the fact that each P_n transforms the polynomials of degree $\leq m$ into polynomials of degree $\leq m$.

Appendix. Let $f \in D(A)$; we know that the unique solution of (11) is $w(t,x) = T(t)f(x) \sim E(\frac{t}{n})^n f(x)$. We shall devise an algorithm for computing $E(p)^n f(x)$, given $f \in C[0,1]$, $x \in [0,1]$, $n \geq 1$ and $0 \leq p \leq p_0$.

With obvious notations (see (18)) we have

$$E(p)f(x) = s_1(x)f(E_1 x) + s_2(x)f(E_2 x) + s_3(x)f(E_3 x). \tag{27}$$

Consider the 3^n numbers

$$y_{j_1\ldots j_n} = f(E_{j_n}\ldots E_{j_1}x),$$

where $j_1, \ldots, j_n \in \{1,2,3\}$.

For $k \in \{n-1, \ldots, 1\}$ define successively

$$y_{j_1 \ldots j_k} = s_1(E_{j_k} \ldots E_{j_1} x) y_{j_1 \ldots j_k 1} + s_2(E_{j_k} \ldots E_{j_1} x) y_{j_1 \ldots j_k 2} + s_3(E_{j_k} \ldots E_{j_1} x) y_{j_1 \ldots j_k 3}.$$

By induction it is easy to prove that

$$y_{j_1 \ldots j_k} = E(p)^{n-k} f(E_{j_k} \ldots E_{j_1} x), \qquad k \in \{n-1, \ldots, 1\}.$$

In particular,

$$y_j = E(p)^{n-1} f(E_j x), \qquad j = 1, 2, 3;$$

due to (27), this implies

$$E(p)^n f(x) = s_1(x) y_1 + s_2(x) y_2 + s_3(x) y_3.$$

References

[1] Altomare, F., *Limit semigroups of Bernstein-Schnabl operators associated with positive projections,* Ann. Sc. Norm. Sup. Pisa, cl. Sci. IV, 16 (1989), 259–279.

[2] Altomare, F. and Campiti, M., *Korovkin-type Approximation Theory and its Applications,* W. de Gruyter, Berlin-New York, 1994.

[3] Altomare, F. and Carbone, I., *On some degenerate differential operators on weighted function spaces,* J. Math. Anal. Appl. 213 (1997), 308–333.

[4] Altomare, F. and Attalienti, A., *Forward diffusion equations and positive operators,* Math. Z. 225 (1997), 211–229.

[5] Altomare, F. and Attalienti, A., *Degenerate evolution problems and related Markov processes arising from Mathematical Finance, Part I and II,* preprints, Bari University (2001).

[6] Altomare, F. and Mangino, E.M., *On a class of elliptic-parabolic equations on unbounded intervals*, Positivity, 2000.

[7] Altomare, F. and Rasa, I., *Towards a characterization of a class of differential operators associated with positive projections,* Atti Sem. Mat. Fis. Univ. Modena Suppl. 46 (1998), 3–38.

[8] Altomare, F. and Rasa, I., *Feller semigroups, Bernstein type operators and generalized convexity associated with positive projections,* in: New Developments in Approximations Theory (eds.) M.D. Buhmann, M. Felten, D.H. Mache und M.W. Müller), Int. Series of Numerical Mathematics Vol. 132, Birkhäuser Verlag, Basel (1999), 9–32.

[9] Attalienti, A., *Generalized Bernstein-Durrmeyer operators and the associated limit semigroup,* J. Approx. Theory 99 (1999), 289–309.

[10] Adell, J.A. and de la Cal, J., *Bernstein-Durrmeyer operators,* Comp. Math. Appl. 30 (1995), 1–14.

[11] Campiti, M., *Binomial-type coefficients and classical approximation processes,* in: Handbook of analytic-computational methods in applied mathematics, (ed.) G. Anastassiou, Boca Raton, FL: Chapman & Hall/CRC, (2000), 947–996.

[12] Campiti, M. and Metafune, G., *Evolution equations associated with recursively defined Bernstein-type operators,* J. Approx. Theory 87 (1996), 270–290.

[13] Campiti, M., Metafune, G. and Pallara, D., *Generalized Voronovskaja formula and solutions of second-order degenerate differential equations,* Rev. Roum. Math. Pures Appl. 44 (1999), 755–766.

[14] Clement, Ph. and Timmermans, C.A., *On C_0-semigroups generated by differential operators satisfying Ventcel's boundary conditions,* Indag. Math. 89 (1986), 379–387.

[15] Durrmeyer, J.J., *Une formule d'inversion de la transformée de Laplace: Applications à la théorie des moments,* Thèse de 3ième cycle, Univ. Paris, 1967.

[16] Derriennic, M.M., *Sur l'approximation des fonctions d'une ou plusieurs variables par des polynômes de Bernstein modifiés et application au problème des moments,* These de 3ième cycle, Univ. de Rennes, 1978.

[17] Dickmeis, W. and Nessel, R.J., *Classical approximation processes in connection with Lax equivalence theorems with orders,* Acta Sci. Math. 40 (1978), 33–48.

[18] Engel, K.J. and Nagel, R., *One-parameter semigroups for linear evolution equations,* Springer-Verlag (2000).

[19] Gavrea, I., Gonska, H.H. and Kacso, D.P., *On the variation-diminishing property,* Resultate Math. 33 (1998), 96–105.

[20] Lupas, A., *Die Folge der Beta Operatoren,* Dissertation. Stuttgart (1972).

[21] Mache, D.H., *Gewichtete Simultanapproximation in der L_p-Metrik durch das Verfahren der Kantorovič Operatoren,* Dissertation, Univ. Dortmund (1991).

[22] Mache, D.H., *A Link between Bernstein Polynomials and Durrmeyer Polynomials with Jacobi Weights,* in: Approximation Theory VIII, Vol. 1, Approximation and Interpolation, Charles K. Chui and Larry L. Schumaker (eds.), World Scientific Publishing Co., (1995), 403–410.

[23] Mache, D.H. and Zhou, D.X., *Characterization Theorems for the Approximation by a Family of Operators,* J. Approx. Theory 84 (1996), 145–161.

[24] Pazy, A., *Semigroups of linear operators and applications to partial differential equations,* Springer-Verlag, (1983).

[25] Rasa, I., *Feller semigroups, elliptic operators and Altomare projections,* to appear in Proc. 4th FAAT Maratea (2000).

[26] Rasa, I. and Vladislav, T., *A Voronovskaja-type formula and preservation properties of a class of operators,* Proc. Roger 2000, Univ. Duisburg, Schriftenreihe Fach. Math. SM-DU-485 (2000), 120–123.

[27] Vladislav, T. and Rasa, I., *Analiza Numerica, Aproximare, problema lui Cauchy abstracta, proiectori Altomare,* Editura Tehnica, Bucuresti (1999).

Institute of Mathematics
Technical University Cluj-Napoca
RO-3400 Cluj-Napoca, Romania
Email address: Ioan.Rasa@math.utcluj.ro

Advanced Problems in Constructive Approximation
(Eds.) M.D. Buhmann and D.H. Mache
International Series of Numerical Mathematics Vol. 142, 153–163

A Survey on Lagrange Interpolation Based on Equally Spaced Nodes

Michael Revers

Abstract

Lagrange interpolation polynomials based on the equidistant node system have not been a popular subject in approximation theory. This is due to some famous examples published by C. Runge in 1901 and later by S.N. Bernstein in 1918 which discouraged mathematicians from considering this method of interpolation. This paper provides a brief survey of Lagrange interpolation polynomials which are based on equidistant nodes including recent results on pointwise divergence properties and certain limit relations.

1 Introduction

We begin with some definitions and notation. Let $C = C(I)$ denote the Banach space of continuous functions on the interval $I := [-1, 1]$ equipped with the usual uniform norm $\|\cdot\|$. Further denote by Π_n the set of algebraic polynomials of degree at most n. Let X be an **interpolation array**, i.e.,

$$X = \{x_{j,n} : j = 0, 1, \ldots, n; n = 0, 1, 2, \ldots\},$$

with

$$-1 \leq x_{0,n} < x_{1,n} < \cdots < x_{n,n} \leq 1, \quad (n = 0, 1, 2, \ldots),$$

and consider the corresponding **Lagrange interpolation polynomial**

$$L_n(f, X, x) := \sum_{j=0}^{n} f(x_{j,n})\, l_{j,n}(X, x), \quad n \in \mathbf{N}_0.$$

Here, for $n \in \mathbf{N}_0$,

$$l_{j,n}(X,x) := \frac{w_n(X,x)}{w_n'(X,x_{j,n})(x-x_{j,n})}, \quad 0 \le j \le n,$$

with

$$w_n(X,x) = \prod_{j=0}^{n}(x-x_{j,n}),$$

are polynomials of exact degree n. They are called the **fundamental polynomials** associated with the **nodes** $\{x_{j,n} : j = 0,1,\ldots,n\}$. In other words, to each $f \in C(I)$ there corresponds a unique interpolating polynomial $L_n(f,X,.)$ of degree at most n coinciding with f at the nodes $\{x_{j,n} : 0 \le j \le n\}$.

The main question is, of course, the convergence, i.e., to understand for what choices of the interpolation array X we can expect that

$$L_n(f,X) \to f, \quad \text{as } n \to \infty.$$

Since, by the Chebyshev alternation theorem, the best uniform approximation $p_n^*(f)$ to $f \in C(I)$ from Π_n interpolates in at least $n+1$ points, there exists, for each $f \in C(I)$, an interpolation array Y (unfortunately depending on f) for which

$$\|L_n(f,Y) - f\| = E_n(f) := \min_{p\in\Pi_n} \|f-p\|$$

tends to 0 as $n \to \infty$. However, for the whole class $C(I)$, the situation is much less favorable, since there is no "universally effective" set of nodes.

2 Divergence results

The main result, which can be considered as the starting point of the divergence theory of Lagrange interpolation, is due to G. Faber [10] in 1914. In order, to formulate the mentioned negative result of Faber, we quote some estimates and introduce some further definitions. By the classical Lebesgue estimate

$$\begin{aligned} |L_n(f,X,x) - f(x)| &\le |L_n(f,X,x) - p_n^*(f,x)| + |p_n^*(f,x) - f(x)| \\ &= |L_n(f-p_n^*,X,x)| + E_n(f) \\ &\le \left(\sum_{j=0}^{n}|l_{j,n}(X,x)| + 1\right) E_n(f), \end{aligned}$$

therefore, with the notations

$$\begin{aligned} \lambda_n(X,x) &= \sum_{j=0}^{n}|l_{j,n}(X,x)|, \quad n \in \mathbf{N}_0, \\ \Lambda_n(X) &= \|\lambda_n(X,.)\|, \quad n \in \mathbf{N}_0, \end{aligned}$$

(**Lebesgue function** and **Lebesgue constant** of Lagrange interpolation, respectively), we have for $n \in \mathbf{N}_0$

$$|L_n(f, X, x) - f(x)| \leq (\lambda_n(X, x) + 1) E_n(f)$$

and

$$\|L_n(f, X) - f\| \leq (\Lambda_n(X) + 1) E_n(f).$$

It is not hard to see that the Lebesgue constant equals the operator norm of $L_n(., X)$ with

$$\begin{aligned} L_n(., X) : C[-1, 1] &\rightarrow C[-1, 1] \\ f &\longmapsto L_n(f, X). \end{aligned}$$

G. Faber [10] proved the then rather surprising lower bound

$$\Lambda_n(X) \geq \frac{1}{8\sqrt{\pi}} \log(n+1), \quad n \in \mathbf{N}_0, \tag{1}$$

for any interpolation array X. Based on this result he obtained

Theorem 2.1 *(Faber, 1914) For any interpolation array X there exists a function $f_1 \in C[-1, 1]$ such that*

$$\limsup_{n \to \infty} \|L_n(f_1, X) - f_1\| = \infty.$$

But, of course, this result does not exclude a pointwise convergence result at least at a single point. This question was (negatively) answered by S.N. Bernstein [3] in 1931:

Theorem 2.2 *(Bernstein, 1931) For any interpolation array X there exists a point $x_0 \in [-1, 1]$ and an $f_2 \in C[-1, 1]$ such that*

$$\limsup_{n \to \infty} |L_n(f_2, X, x_0)| = \infty.$$

The next natural question is: are there divergence results on a set of positive measure? As has been observed independently by Ch. Méray [15] in 1884 and some years later by C. Runge [22] in 1901, Lagrange interpolation polynomials which are based on an equidistant interpolation array need not provide a good approximation method. Runge considered the interpolation array

$$E = \left\{-1 + \frac{2j}{n} : j = 0, 1, \ldots, n; n \in \mathbf{N}\right\}.$$

We shortly call this array the **equidistant node system**. Then we have the following divergence result which is due to Runge [22]:

Theorem 2.3 *(Runge, 1901) Let* $f(x) = (1 + 25x^2)^{-1}, x \in [-1,1]$. *Then, for* $0.72 \ldots < |x_0| < 1$,

$$\limsup_{n\to\infty} |L_n(f, E, x_0)| = \infty,$$

whereas, for $0 \le |x_0| < 0.72\ldots$,

$$\limsup_{n\to\infty} L_n(f, E, x_0) = f(x_0).$$

However, S.N. Bernstein [2] produced an example which is even more dramatic.

Theorem 2.4 *(Bernstein, 1918) Let* $f(x) = |x|, x \in [-1,1]$. *Then*

$$\limsup_{n\to\infty} |L_n(f, E, x_0)| = \infty, \quad \forall x_0 \in [-1,1], x_0 \ne -1, 0, +1.$$

In other words, the Bernstein example exhibits a particularly simple function for which the interpolating polynomials on E diverge throughout $[-1,1]$, except at a few points.

The next result states a similar theorem concerning the **Chebyshev interpolation array**

$$T = \left\{ -\cos\left(\frac{2j+1}{2n+2}\pi\right) : j = 0, 1, \ldots, n; n \in \mathbf{N}_0 \right\}.$$

G. Grünwald [12] and J. Marcinkiewicz [14] (independently) obtained

Theorem 2.5 *(Grünwald, Marcinkiewicz, 1936/37) There exists a function* $f_3 \in C[-1,1]$, *for which*

$$\limsup_{n\to\infty} |L_n(f_3, T, x_0)| = \infty, \quad \forall x_0 \in [-1,1]. \tag{2}$$

In a joint paper, [13], P. Erdős and G. Grünwald sharpened this result. They constructed a function $f \in C[-1,1]$ satisfying (2), where, at the same time, the even function $f(\cos\vartheta)$ has a uniformly convergent Fourier series on $[0,\pi]$.

In 1980, P. Erdős and P. Vértesi [9] settled the case for general interpolation arrays. They proved the following remarkable result:

Theorem 2.6 *(Erdős, Vértesi, 1980) For any interpolation array* X *one can find a function* $f_4 \in C[-1,1]$ *such that*

$$\limsup_{n\to\infty} |L_n(f_4, X, x_0)| = \infty, \quad \textit{for almost all } x_0 \in [-1,1].$$

Moreover, the divergence set is of second category on $[-1,1]$.

It is not hard to see that the word "almost" in the above statement is best possible. It is fairly easy to construct an interpolation array for which $\lim_{n\to\infty} L_n(f,X,x) = f(x), f \in C[-1,1]$, on a dense subset (again independent of f) of $[-1,1]$. For example, we can take

$$X = \begin{pmatrix} x_0 & & \\ x_0 & x_1 & \\ x_0 & x_1 & x_2 \\ . & . & . \end{pmatrix},$$

where $\{x_k\}_{k=0}^{\infty}$ is a dense subset of $[-1,1]$.

As a final cautionary note about divergence results which are based on general interpolation arrays we mention two papers (I. Muntean [16] and S. Cobzas and I. Muntean [8]).

Theorem 2.7 *(Muntean, 1976) Given an arbitrary interpolation array X and denote by*

$$U_X = \left\{ f \in C[-1,1] : \limsup_{n\to\infty} \|L_n(f,X)\| = \infty \right\}.$$

Then U_X is an uncountable, G_δ, dense subset of $C[-1,1]$.

The examples of Méray, Runge, Bernstein, Grünwald, Marcinkiewicz were not isolated examples of "bad" functions. The result of Muntean, loosely speaking, states that "bad" functions are everywhere.

3 Equidistant interpolation

We have seen in the previous chapter that the efficiency of approximation by $L_n(.,X)$ is governed by

$$\|L_n(f,X) - f\| \leq (1 + \Lambda_n(X)) E_n(f).$$

For certain sets of interpolation nodes the behavior of the Lebesgue function has been well investigated. An excellent survey is given in Brutman [5]. Among the many results available up to now, we collect some facts which are placed within the relevance of this survey. For the Chebyshev nodes T the behavior of $\lambda_n(T,.)$ has been investigated for the first time by Bernstein [2] who established an asymptotic value for $\Lambda_n(T)$ by

$$\Lambda_n(T) \sim \frac{2}{\pi}\log(n+1), \quad \text{as } n \to \infty. \tag{3}$$

In view of Faber's result (1) it becomes clear that the Chebyshev nodes T are a particularly good choice for interpolation if good uniform approximation is desired. However, T is not an optimal node system, that is, there exists some set of nodes X such that

$$\Lambda_n(X) < \Lambda_n(T), \quad n \geq 1.$$

Now, turning to the case of equidistant nodes, Schönhage [24] published in 1961 the asymptotic expression

$$\Lambda_n(E) \sim \frac{2^{n+1}}{en(\log n + \gamma)}, \quad \text{as } n \to \infty,$$

where $\gamma = 0.577\ldots$ is Euler's constant. The above mentioned divergence examples of Runge and Bernstein together with Schönhage's result made the study of interpolation polynomials which are based on equidistant nodes very unpopular.

For example, many papers have been written about Chebyshev interpolation. Also, much attention has been devoted to the case $X^{(\alpha,\beta)}$ consisting of the zeroes of the Jacobi polynomial $P_n^{(\alpha,\beta)}$ (see, for example, Szegő [25]). But calculating the zeroes of Jacobi polynomials is (numerically) not easy. On the other hand, if we are interested in equidistant interpolation then splines are more popular than polynomials.

However, despite the negative results of Runge and Bernstein and the undisputed success of spline interpolation, the study of Lagrange interpolation based on equidistant nodes offers many interesting unsolved problems. Here we shall treat some problems which are related to equidistant Lagrange interpolation.

4 New results

Let us recall the divergence result of S.N. Bernstein [2]: Let $f(x) = |x|, x \in [-1,1]$. Then

$$\limsup_{n\to\infty} |L_n(f, E, x_0)| = \infty, \quad \forall x_0 \in [-1,1], x_0 \neq -1, 0, +1.$$

Bernstein's proof is short but rather technically. It is based on the Newton representation of the interpolating polynomials. Another source for the proof is the book of I.P. Natanson ([17], pp. 30–35). Since the points ± 1 are interpolation points for each Lagrange polynomial $L_n(f, E, .), n \in \mathbf{N}$ we simply have $L_n(|x|, E, -1) = L_n(|x|, E, 1) = 1$ for all indices. Hence divergence can never occur at the endpoints (for any f). However, at the point $x_0 = 0$ the (convergence) behavior of the interpolating polynomials to the function $|x|$ is not so clear. We mention that Bernstein has not investigated the case $x_0 = 0$, probably because 0 is a node for all even integers n. It was first noted in the student term paper of D.L. Berman in 1939 that

$$\lim_{n\to\infty} L_n(|x|, E, 0) = 0,$$

and S.M. Lozinskii showed more exactly that $|L_n(|x|, E, 0)| \leq C/n$. A short survey on this topic is given, for example, in ([23], p. 285). S.M. Lozinskii again used the Newton representation for the upper estimate of the order of convergence. In [19] Bernstein's result is extended to the following

Theorem 4.1 *Let $\alpha \in (0,1]$. Then*

$$\limsup_{n\to\infty} |L_n(|x|^\alpha, E, x_0)| = \infty, \quad \forall x_0 \in [-1,1], x_0 \neq -1, 0, +1. \tag{4}$$

The proof of this fact is completely different (even for the case $\alpha = 1$) from Bernstein's idea since it is based on the Lagrange representation formula. This formula has not become very popular in proving divergence result (see also the recent work of L. Brutman and E. Passow [6] which deals with divergence properties for $|x|$ at a broad family of interpolation nodes (including the so-called Newman nodes)). Motivated by some numerical computations I conjectured that the everywhere divergence (apart from $-1, 0, +1$) of the sequence $(L_n(|x|^\alpha, E, x_0))_{n\geq 1}$ takes place for all $\alpha > 0$, $(\alpha \neq 2, 4, 6, \ldots)$.

It is worth mentioning that Bernstein's divergence result has been sharpened in 1990 by G. Byrne, T.M. Mills and S. Smith [7] to the following

Theorem 4.2 *Let $0 < |x_0| < 1$. Then*

$$\limsup_{n\to\infty} \frac{1}{n+1} |L_n(|x|, E, x_0) - |x_0||$$

$$= \frac{1}{2}\left[(1+x_0)\log(1+x_0) + (1-x_0)\log(1-x_0)\right]. \tag{5}$$

In other words, the rate of divergence of the interpolation polynomials is geometrically fast, but depends on the location of x_0 in $[-1,1]$. It is somewhat surprising that the nth root asymptotics (5) also holds for the "smoother" function $|x|^3$. In [20] the following nth root asymptotics is proved:

Theorem 4.3 *Let $0 < |x_0| < 1$. Then*

$$\limsup_{n\to\infty} \frac{1}{n+1} \left| L_n\left(|x|^3, E, x_0\right) - |x_0|^3 \right|$$

$$= \frac{1}{2}\left[(1+x_0)\log(1+x_0) + (1-x_0)\log(1-x_0)\right]. \tag{6}$$

It is somewhat surprising to see that the rate of divergence is not influenced by the parameter α (at least for the cases $\alpha = 1, 3$). Now, looking to the convergence behavior at the point $x_0 = 0$ we have (see [20]):

Theorem 4.4 *Let $n = 2m-1, m \in \mathbf{N}, m \geq 2$. Then*

$$\frac{2}{\pi}\frac{1}{n^3}\left[1+\frac{2}{n-2}\right] \leq \left| L_n\left(|x|^3, E, 0\right)\right| \leq \frac{2}{\pi}\frac{1}{n^3}\left[1+\frac{2}{n-2}\right]\left[1+\frac{2}{n-1}\right].$$

This result extends the well-known result of D.L. Berman and S.M. Lozinskii (which handles the case $|x|$) and it shows that the **exact** rate of convergence at $x_0 = 0$ is $O(n^{-3})$. A careful investigation into the proof of (4) and (6) strongly motivates the following question: Let $\alpha \in \mathbf{R}_+ \backslash 2\mathbf{N}$ and $0 < |x_0| < 1$. Is it true that

$$\limsup_{n\to\infty} \frac{1}{n+1} \log |L_n(|x|^\alpha, x_0) - |x_0|^\alpha| = \frac{1}{2}[(1+x_0)\log(1+x_0) + (1-x_0)\log(1-x_0)]\ ?$$

Very recently M. Ganzburg [11] established the following strong asymptotics for the rate of divergence of $||x_0|^\alpha - L_n(|x|^\alpha, E, x_0)|$ for $0 < |x_0| < 1$:

Theorem 4.5 *For $x \in [-1,1]$ and $\alpha > 0$ let (we denote the gamma function by $\Gamma(.)$)*

$$\begin{aligned}
\varphi_N(x) &:= \sqrt{1-x^2}\left((1+x)^{1+x}(1-x)^{1-x}\right)^{N/2}, \\
C_1(\alpha) &:= \int_0^\infty \frac{y^{\alpha-1}}{e^y + e^{-y}} dy = \Gamma(\alpha) \sum_{k=0}^\infty (-1)^k (2k+1)^{-\alpha}, \\
C_2(\alpha) &:= \int_0^\infty \frac{y^\alpha}{e^y - e^{-y}} dy = \Gamma(\alpha+1) \sum_{k=0}^\infty (2k+1)^{-(\alpha+1)}, \\
c(x) &:= \begin{cases} \cos\frac{\pi}{2m}, & x = p/m, (p,m) = 1, m \text{ odd}, |p| \in \mathbf{N}, \\ 1, & \text{otherwise}, \end{cases} \\
s(x) &:= \begin{cases} \cos\frac{\pi}{2m}, & x = p/m, (p,m) = 1, p \text{ odd}, |m| \in \mathbf{N}, \\ 1, & \text{otherwise}. \end{cases}
\end{aligned}$$

Further let $x_0 \in (-1,1), x_0 \neq 0$ be a fixed point.

(i)

$$\limsup_{N=2n-1\to\infty} \left((\pi N/2)^{\alpha+2} / \varphi_N(x_0)\right) ||x_0|^\alpha - L_N(|x|^\alpha, E, x_0)| = \frac{4}{\pi}\left|\sin\frac{\pi\alpha}{2}\right| C_1(\alpha+2)\, x_0^{-2} c(x_0) \quad \text{for } \alpha > -2, \tag{7}$$

(ii)

$$\limsup_{N=2n\to\infty} \left((\pi N/2)^{\alpha+1} / \varphi_N(x_0)\right) ||x_0|^\alpha - L_N(|x|^\alpha, E, x_0)| = \frac{4}{\pi}\left|\sin\frac{\pi\alpha}{2}\right| C_2(\alpha)\, x_0^{-1} s(x_0) \quad \text{for } \alpha > 0. \tag{8}$$

As immediate consequences of (7) and (8), Ganzburg obtains the nth root asymptotic relation for $||x_0|^\alpha - L_n(|x|^\alpha, E, x_0)|$ and extends (5) and (6) to all $\alpha > 0, (\alpha \neq 2, 4, 6, \ldots)$.

Now, what can be said about the convergence behavior of $L_n(|x|^\alpha, E, 0), \alpha > 0$? In [21] the following result is proved:

Theorem 4.6 *Let $\alpha \in (0,1]$. Then*

$$C(\alpha) := \lim_{N=2n-1\to\infty} N^\alpha L_N(|x|^\alpha, E, 0) = \frac{4}{\pi}\left(\frac{2}{\pi}\right)^\alpha \sin\frac{\pi\alpha}{2}\int_0^\infty \frac{t^{\alpha-1}}{e^t+e^{-t}}dt. \quad (9)$$

Since the proof of (9) is based on a certain representation of arbitrary powers of n by the gamma function – which only holds for $\alpha \in (0,1]$ – it seems that the restriction for $\alpha \in (0,1]$ is rather technical than essential for obtaining the result (9). Thus I conjectured that (9) also holds for **all** relevant $\alpha > 0$. Actually, in the previous mentioned paper [11], M. Ganzburg was able to establish relation (9) to all $\alpha > 0$. The result is surprising because it indicates a (possible) connection to the most prominent constant in polynomial uniform approximation for $|x|^\alpha$ – the so-called Bernstein constant (in approximation theory). We give a short exposition of this topic:

In 1913 and later (for the general case in 1938), S.N. Bernstein ([1], [4]) investigated into the behavior of the **best uniform** approximation polynomials to the function $|x|$ (and $|x|^\alpha$, respectively). Recall the definition of Π_n to be the space of (real) algebraic polynomials with degree at most n and the minimal approximation error $E_n(|x|^\alpha, [-1,1])$ to be defined by

$$E_n = \min_{p\in\Pi_n} \| |x|^\alpha - p\| .$$

From Jackson's and Bernstein's theorems about the dependence of approximation speed and smoothness of the function to be approximated we know that in case of $\alpha \in \mathbf{R}_+ \setminus 2\mathbf{N}$ the minimal error $E_n(|x|^\alpha, [-1,1])$ behaves like $O(n^{-\alpha})$ as $n\to\infty$. In [4] S.N. Bernstein proved the following deep result. He showed that the limit

$$\lim_{n\to\infty} n^\alpha E_n(|x|^\alpha, [-1,1]) =: \beta(\alpha)$$

exists for each $\alpha > 0$. Unfortunately, an explicit expression for the constant $\beta(\alpha), \alpha > 0$ is still unknown. In case $\alpha = 1$ the number $\beta(1) = 0.28016\ldots$ is known as Bernstein's constant (see also [26]). For large values of α, Bernstein was able to establish an asymptotic expression. He showed (see [4], p. 190) that

$$\beta(\alpha) \sim \frac{1}{\pi}\left|\sin\frac{\pi\alpha}{2}\right|\Gamma(\alpha), \text{as } \alpha\to\infty,$$

and moreover, he obtained a both-sided estimate (see also [18], p. 505) for the approximation error $E_n(|x|^\alpha, [-1,1])$ ($\alpha > 0, n$ sufficiently large) from which we deduce that

$$\eta_\alpha\left(\frac{\pi}{\pi+4}\right)^\alpha \frac{1}{2\sqrt{2}} \le \beta(\alpha) \le \eta_\alpha, \quad \text{with} \quad \eta_\alpha = \frac{4}{\pi}\left|\sin\frac{\pi\alpha}{2}\right| \int_0^\infty \frac{t^{\alpha-1}}{e^t+e^{-t}}dt.$$

It is an interesting open problem to reveal whether there is a "simple" connection between the both approximation constants $\beta(\alpha)$ and $C(\alpha)$ or not?

References

[1] S.N. Bernstein, Sur la meilleure approximation de $|x|$ par des polynômes de degrés donnés, *Acta Math.* **37** (1913), 1–57.

[2] S.N. Bernstein, Quelques remarques sur l'interpolation, *Math. Ann.* **79** (1918), 1–12.

[3] S.N. Bernstein, Sur la limitation des valeurs d'un pôlynome $P_n(x)$ de degré n sur tout un segment par ses valeurs en $n+1$ points du segment, *Bull. Acad. Sci. de l'USSR*, **8** (1931), 1025–1050.

[4] S.N. Bernstein, Sur la meilleure approximation de $|x|^p$ par des polynômes de degrés très élevés, *Bull. Acad. Sci. USSR Sér. Math.* **2** (1938), 181–190.

[5] L. Brutman, Lebesgue functions for polynomial interpolation – a survey, *Ann. Numer. Math.* **4** (1997), 111–127.

[6] L. Brutman, E. Passow, On the Divergence of Lagrange Interpolation to $|x|$, *J. Approx. Theory* **81** (1995), 127–135.

[7] G. Byrne, T.M. Mills, S. Smith, On Lagrange interpolation with equidistant nodes, *Bull. Austral. Math. Soc.* **42** (1990), 81–89.

[8] S. Cobzas, I. Muntean, Condensation of singularities and divergence results in approximation theory, *J. Approx. Theory* **31** (1981), 138–153.

[9] P. Erdős, P. Vértesi, On the almost everywhere divergence of Lagrange interpolating polynomials on arbitrary system of nodes, *Acta Math. Acad. Sci. Hungar.* **36** (1980), 71–89.

[10] G. Faber, Über die interpolatorische Darstellung stetiger Funktionen, *Jahresber. Deutsch. Math.-Verein.* **23** (1914), 190–210.

[11] M. Ganzburg, Strong Asymptotics in Lagrange Interpolation with Equidistant Nodes, *Preprint* 2002.

[12] G. Grünwald, Über Divergenzerscheinungen der Lagrangeschen Interpolationspolynome stetiger Funktionen, *Annals of Mathematics* **37** (1936), 908–918.

[13] G. Grünwald, P. Erdős, Über einen Faber'schen Satz, *Annals of Mathematics* **39** (1938), 257–261.

[14] J. Marcinkiewicz, Sur la divergence des polynômes d'interpolation, *Acta Sci. Math. (Szeged)* **8** (1937), 131–135.

[15] Ch. Méray, Observations sur la légitimité de l'interpolation, *Asens* **1** (1884), 165–176.

[16] I. Muntean, The Lagrange interpolation operators are densely divergent, *Studia Univ. Babes-Bolyai Mat.* **21** (1976), 28–30.

[17] I.P. Natanson, *Constructive function theory*, Vol. III (1965), Frederick Ungar, New York.

[18] M. Revers, On the approximation of certain functions by interpolating polynomials, *Bull. Austral. Math. Soc.* **58** (1998), 505–512.

[19] M. Revers, The divergence of Lagrange interpolation for $|x|^\alpha$ at equidistant nodes, *J. Approx. Theory* **103** (2000), 269–280.

[20] M. Revers, On Lagrange interpolation with equally spaced nodes, *Bull. Austral. Math. Soc.* **62** (2000), 357–368.

[21] M. Revers, Approximation constants in equidistant Lagrange interpolation, *Periodica Mathematica Hungarica* **40**(2) (2000), 167–175.

[22] C. Runge, Über empirische Funktionen und die Interpolation zwischen äquidistanten Ordinaten, *Zeitschrift für Mathematik und Physik* **46** (1901), 224–243.

[23] B.L. Sendov, A. Andreev, Interpolation and Approximation, in *Handbook of Numerical Analysis*, Vol. III, P.G. Ciarlet, J.L. Lions, Eds., (1994) North Holland.

[24] A. Schönhage, Fehlerfortpflanzung bei Interpolation, *Numerische Mathematik* **3** (1961), 62–71.

[25] G. Szegő, *Orthogonal polynomials*, 4th edn. Colloq. Pub. **23**, Am. Math. Soc., (1975) Providence, RI.

[26] R.S. Varga, A.J. Carpenter, On the Bernstein Conjecture in approximation theory, *Const. Approx.* **1** (1985), 333–348.

Department of Mathematics
University of Salzburg
Hellbrunnerstrasse 34
A-5020 Salzburg, Austria
Email address: michael.revers@sbg.ac.at

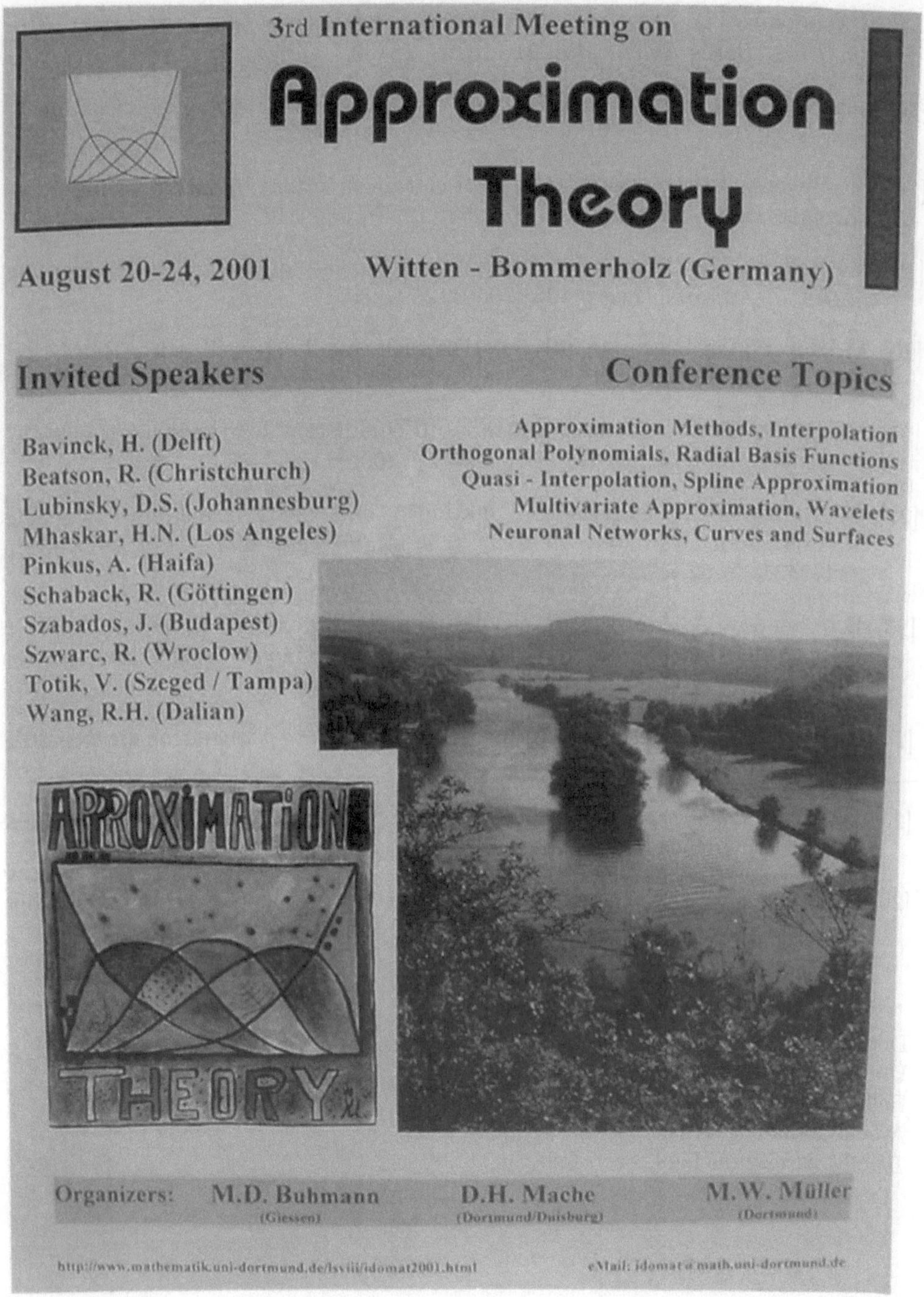
3rd International Meeting on
Approximation Theory
August 20-24, 2001
Witten - Bommerholz (Germany)
Invited Speakers
Conference Topics
Approximation Methods, Interpolation
Orthogonal Polynomials, Radial Basis Functions
Quasi - Interpolation, Spline Approximation
Multivariate Approximation, Wavelets
Neuronal Networks, Curves and Surfaces
Bavinck, H. (Delft)
Beatson, R. (Christchurch)
Lubinsky, D.S. (Johannesburg)
Mhaskar, H.N. (Los Angeles)
Pinkus, A. (Haifa)
Schaback, R. (Göttingen)
Szabados, J. (Budapest)
Szwarc, R. (Wroclow)
Totik, V. (Szeged / Tampa)
Wang, R.H. (Dalian)
APPROXIMATION
THEORY
Organizers:
M.D. Buhmann (Giessen)
D.H. Mache (Dortmund/Duisburg)
M.W. Müller (Dortmund)
http://www.mathematik.uni-dortmund.de/lsviii/idomat2001.html
eMail: idomat@math.uni-dortmund.de

Advanced Problems in Constructive Approximation
(Eds.) M.D. Buhmann and D.H. Mache
International Series of Numerical Mathematics Vol. 142, 165–186

Multiresolution Analysis with Pulses

Carl H. Rohwer

Abstract

Multiresolution analysis has recently received considerable attention in relation to wavelets. The word "multiresolution" is appropriate in so far as wavelets are local in some sense, and therefore have exponentially decaying impulse response. In image processing it is clear that edges and impulses yield undesirable synthetic features in partially reconstructed images from linear multiscale decompositions. Median decompositions are regarded as better in practice, but computational complexity and lack of theory are problems. An alternative, from mathematical morphology is possible, yielding results demonstrably similar to the median decomposition, but computationally simpler, and having a strong theory for deriving qualitative and quantitative properties.

1 Introduction

The fast Fourier transform (FFT) has often been called the most important mathematical tool in modern technology. In a similar way the fast wavelet transform (FWT) will have an impact, for instance, in image processing and transmission. Recently a new idea emerged, which some consider to be even better in the "multiresolution analysis" of digital images. In the book, image processing and data analysis, Stark Murtagh and Bijaoui (1) follow a discussion of the wavelet transform with a section on multiresolution analysis (MRA) based on the median transform. Roughly speaking, the averaging filter that maps a function onto a function in a space of lower dimension in the wavelet transform is replaced by a median smoother. As usual this can heuristically be motivated by yielding a more robust estimator of an average and therefore outliers have a lesser damage on the "smoother component" in the mapping. Iterative and non-iterative algorithms for median transforms are presented. The claim is made that this MRA is well suited when image reconstruction is done from a subset of the (additive-) decomposition for purposes of restoration, compression and partial reconstruction. The reconstructed image is often found to have fewer artifacts than in the case of wavelet decomposition, where these artifacts are often in the form of the specific wavelets chosen. An example is the negative ring surrounding bright point sources. Shapes are found to be closer to those of the input image. Computational requirements are listed as high, although there is a saving in that these transforms can be performed in integer values only and decimation can yield a considerable economisation. Other morphological tools, specifically N erosions followed by N dilations are mentioned, with the observation that results were found to be better with the median. Morphological transform are presented, with mention of the good estimate of the image background that is obtained, especially for images with small structures, as in astronomy.

In one-dimensional signal analysis the origin of many artifacts (irritating significant distortions) in signals partially reconstructed from wavelet decompositions, can often be understood in the context of approximation theory. The linear projections used in wavelet decompositions perform relatively well in audio signals. The explanation may ultimately be physiological, but for the purpose at hand, it is sufficient to observe that an audio signal is often well approximated locally by trigonometric polynomials, so that the FFT can be truncated for compression. Such local approximation does have some damage (Gibb's phenomenon), but "softer" windows can be employed to lessen this. Depending upon the support of a specific wavelet, and the norm of an associated mapping onto a subset of lesser dimension, the Gibb's-Phenomenon may be acceptable for the partial reconstruction.

For visual signals, or images, sharp edges and constant regions play a significant part in the acceptability or recognisability of an image. Such data, approximated by smooth functions, or sequences of sampled smooth functions, bring Gibbs into play, with possible "overshoot" and "undershoot" near edges. An edge can be interpreted as having an impulse in the derivative, or an impulse in the sequence of differences of samples of such an image. Impulses are not handled well by linear filters, or smooth functions in approximation theory. Convoluting a wavelet with an impulse yields precisely the wavelets concerned so that impulsive noise on a sequence can be expected to yield spurious features, often close to the shape of the wavelets, or sums of these, at all resolution levels.

In the extreme case of the non-local Fourier transform, the impulse results in a constant function in the frequency domain. This contamination is difficult to remove afterwards. A linear filter preserves the "energy" in an impulse. Even in the case of the simplest Haar-wavelet decomposition, where the projection onto the lower frequency sequence has minimal norm, an impulse is merely spread onto all levels of the wavelet decomposition. The general rule of thumb is to handle outliers (impulses) before any linear mapping is done. This was the original motivation of Tukey and others to use running medians ("median smoothers"), or other selectors, before any linear mapping, or at least as soon as possible. It is also in this light that the median smoothers were considered the "basic" smoothers. Despite the almost universal lament at the lack of an underlying theory (11), useful ideas emerged. A sensible set of axioms emerged, as well as a characterization of the so-called "roots" of median smoothers in terms of edges, constant regions and monotone sections. As it turns out a simple, illuminating characterization of the eigensequence w.r.t. the eigenvalue 1 ("root"), turns out to be that of local monotonicity, provided the sequences considered are square summable. This is always the case in practice, as the only spurious roots are periodic, a fact that was proved recently (12). Furthermore, a slight relaxation of one of the axioms of Mallows allows an even simpler set of selectors, which can essentially be reduced further to composition of minimum- and maximum operators, and which allow a framework for comparison and analysis. This leads to the so-called $LULU$-theory, developed for the purpose of removing impulsive noise from one-dimensional data (5), (6), (7). This theory was later shown to overlap particular cases of the theory of mathematical morphology, as developed by Serra (10) and others. The $LULU$-theory has the special advantages of demonstrating the concept of local-monotonicity ("trend") that is complementary to a concept of impulsive noise in a strict sense. It yields a collection of nonlinear smoothers that are well structured in an ordering on such operators, and some illuminating theory. Computational and other advantages can be demonstrated, leading to a theory for selecting, constructing and comparing morphological filters, median filters and others that can lead to a multiresolution analysis with the advantages observed as well as computational efficiency comparable to the FWT.

For the purpose at hand a sequence can conveniently be identified with an interpolating spline function of order 2 or 1 (degree 0 or 1), as required, since there is a one-to-one correspondence from the B-spline basis. This idea leads to a useful generalisation, since a spline function of arbitrary degree can be decomposed into a sequence of coefficients representing pulses of B-spline shape. Since B-splines have multiscale relations it is clear that there exist an infinite number of possibilities to decompose a polynomial spline into linear combinations of B-splines at all resolution levels. A diadic wavelet decomposition is one such decomposition, of a specific type, where resolution is halved at each stage and a specific choice is made to have pulses at all levels exhibiting some local periodicity.

A beautiful, comprehensive theory of spline wavelets exists (2), but often we pretend to want what the theory delivers because the theory does not quite deliver what we want. Rethinking the requirements of a multiresolution analysis can be shown to be useful. It could start with a clear definition of the concept of resolution, along the lines of the practical heuristic uses of the term in measurements of physics or chemistry. In chemistry, for example, a spectroscopy apparatus can be said to have a resolution sufficient to distinguish between spectral lines of two chemical compounds. These may be of Gaussian-shape and the instrument must be able to separate two definite maxima, even in the presence of some reasonable expected noise. If the instrument has a higher resolution it can separate a spectrum of these compounds into several pulses from the presence of a few ionisation species in the compounds. When the instrument is even better it could even break up these pulses into pulses representing different isotopes in the species. Thus resolution is associated with ability to distinguish or separate different "pulses."

We should like a more rigorous concept of "resolution" along these lines and an associated multiresolution analysis. If at all possible we would like to have a decomposition that is translation invariant in both axes, so as to remove the problem of the phase ambiguity (the dependence of the decomposition on the choice of index). A further useful property would be a linear decrease in resolution instead of a geometric one, as is the case with classic Fourier analysis.

It will be demonstrated that all these aims can be realised to a remarkable extent to provide an alternative system of multiresolution analysis.

2 Multiresolution analysis

For the purpose at hand it is convenient to briefly sketch the ideas of a simple wavelet decomposition.

Given a function f, a subset of a function space is chosen so that it is spanned by translations of a so-called scaling function φ, which is itself a linear combination of the type

$$\phi(t) = \sum_{i=-\infty}^{\infty} \alpha_i \phi(2t - i).$$

It is sufficient for the purpose at hand to consider only the simple Haar-wavelet decomposition. The "scaling function" ϕ in this case is simply the characteristic function of the interval $[0, 1)$, thus a B spline of order 1 (degree 0). It is clear that $\phi(t)$ is a linear combination of $\phi(2t)$ and $\phi(2t-1)$, and that, in this case, it is simply the sum.

Thus $\phi(t) = \phi(2t) + \phi(2t-1)$, and by induction

$$\phi(t) = \sum_{i=0}^{2^k} \phi(2^k t - i).$$

Consider the span of these functions, and $x = \{x_i : i = 0, \ldots, N-1\}$, which is a sampling of a function of f at the values $t_i = \dfrac{i+\delta}{2^k}$, where $0 \leq \delta < 1$.

Clearly, in this case, the sequence can be identified with the sequence of coefficients w.r.t. the basis $\{\phi_i; \phi_i(k) = \phi(2^k t - i)\}$. Letting

$$B_j = \operatorname{span}\{\phi_i; \phi_i(t) = \phi(2^j t - i)\},$$

a best least squares estimate Px from B_{k-1} to a function

$$x = \sum_{i=0}^{N-1} \alpha_i \phi(2^k t - i)$$

from B_k is easily obtained as

$$Px(t) = \sum_{j=0}^{\frac{N}{2}} \frac{1}{2}(\alpha_{2j} + \alpha_{2j+1})\phi(2^{k-1}t - j)$$

and

$$(x - Px)(t) = \sum_{j=0}^{\frac{N}{2}} \frac{1}{2}(\alpha_{2j+1} - \alpha_{2j})\psi(2^{k-1}t - j),$$

where $\psi(t) = \varphi(2t-1) - \varphi(2t)$, is the Haar-wavelet.

The basis is orthonormal w.r.t. the inner product

$$(x, y) = \int_{-\infty}^{\infty} x(t)y(t)dt,$$

and the set of wavelets ψ_i span the orthogonal complement of B_{k-1}, if $\psi_i(t) = \psi(2^{k-1}t - i)$.

For later comparison it is sufficient to note that there is a preservation of "energy" in the sequences in that

$$||\alpha||_2^2 = \sum_{i=-\infty}^{\infty} \alpha_i^2 = \sum_{i=-\infty}^{\infty} 2\left(\frac{\alpha_{2j} + \alpha_{2j+1}}{2}\right)^2 + \sum_{i=-\infty}^{\infty} 2\left(\frac{\alpha_{2j+1} - \alpha_{2j}}{2}\right)^2$$

Thus the sequence x is decomposed into a "smoother" sequence Px, which is pairwise constant, and a "rougher" sequence $x - Px$ which has pairwise elements equal in absolute value, but differing in signs. Significant, for later argument, is that Px has every three consecutive elements monotone.

Since the wavelet ψ has a definable frequency, it is natural to view the decomposition as a "smoother" spline Px and a wavelet component, which is a sequence with an associated frequency locally. Repeating such a decomposition the original sequence x can eventually be decomposed into a constant sequence and several "layers" of wavelets at frequencies that are successively an octave lower than the previous.

Schematically the wavelet decomposition can be viewed in the following diagram.

$$\begin{array}{ccccccc} f & \longrightarrow & P_1 f & \longrightarrow & P_2P_1 f & \longrightarrow & P_3P_2P_1 f & \longrightarrow \\ \downarrow & & \downarrow & & \downarrow & & \downarrow & \\ W_1 f = (I - P_1)f & & W_2 f = (I - P_2)P_1 f & & W_3 f = (I - P_3)P_2P_1 f & & & \end{array}$$

From the theory of wavelets the projections P_i have the following properties;

(i) $P_i f \perp \{f, P_1 f, \ldots, P_{i-1} f\}$

(ii) P_i is idempotent, linear and eigenvalues are only 0 and 1.

Reconstruction can be achieved by adding the different "layers", and partial reconstruction by a subset of the layers, or a set of subsets of each layer. For data compression, for instance, the coefficients at each wavelet level can be "quantized" and only the nonzero quantized values stored or transmitted. Thus it is reasonable to speak of a "local frequency content" by considering the size of coefficients of wavelets with support at a chosen location.

What seems clear is that a signal with sections of almost constant value will have small frequency content at all frequencies there. Only in regions where the sequence varies significantly will there be a need for high frequency information being transmitted. For purposes of automatic analysis, significant changes in the sequences are identified with "wavelet activity" nearby. The effect seems local due to the small support of the scaling function φ and the wavelet ψ. In a sense therefore it is reasonable to say that global shape is determined by the low frequency content and higher resolution features reside in the wavelet coefficients. Hence the name multiresolution analysis (MRA). But should this name not be reserved for a stricter interpretation?

A linear transform can be characterised by its response to an "impulse". Letting d_i be the sequence $\{\delta_{ij} : j = 0, \ldots, N-1\}$, where δ_{ij} is the Kronecker-delta it is easy to see that, in spite of its minimal support, the wavelet decomposition will have exponential decaying amplitudes in exponentially growing support intervals. A partial reconstruction will therefore inevitably have deviation from the original sequence in an arbitrarily large region. This is the essential problem of the response of a linear mapping to "impulsive noise", if impulsive noise is precisely defined as an arbitrary multiple of a Kronecker-delta sequence. The "energy" in such an impulse is "spread" or "smeared", and, depending on the amplitude of the impulse, can completely swamp the essential signal in an arbitrarily large region. If impulses like these can be expected in a measuring (or transmitting-) device, there will have to be some precaution taken, preferably before any linear transformation is performed. This led to the widespread use of pre-smoothing with running medians. The problem is essentially similar in all wavelet decompositions, and a progressively more local feature has progressively wider frequency content in general; the time-frequency window has an area exceeding a fixed positive quantity. Moreover, the damage done is not restricted to impulses, but to impulses in differences of the sequences too. Thus a sampling of a simple step function will generally have significant distortions spreading into all frequency levels. This behaviour is also phase dependent.

It is illuminating to consider the above observations in the simple case of a sequence of samplings of a smooth function, with a simple unit step function (Heaviside function), uniformly distributed random noise and two isolated impulses (Kronecker-delta sequences) added.

It is clear that the synthetic wavelet activity due to the edge and the impulses is amplified by the factor α if the original impulses and jump discontinuity are, resulting in arbitrary synthetic features arbitrarily far in the decomposition.

In image processes edges are significant for picture quality. Linear mappings generally do not preserve monotonicity in a sequence as median smoothers do, and do so in a precise local sense, as do all rank order selectors. Can this be exploited? Clearly experience suggest that multiresolution analysis with medians works well in image processing (1). Are there computationally efficient alternatives, and can an underlying theory provide reassurance of predictable, comparable performance? And if a "feature" or a deviation from a "smooth" surrounding trend is at a precisely defined "resolution", in that it is sufficiently local, can it be separated without excessive contamination?

3 An alternative multiresolution analysis

A first possible development can start with a precise definition of impulsive noise, and the so-called *LULU*-operators (5). Operators are constructed that remove, from constant sequences, the impulses that create problems in linear decompositions.

Definition 3.1 *A pure n-impulse is a sequence p with $p_i = \alpha$, for $i \in [j, j+1, \ldots, j+n-i]$ and 0 elsewhere. (α is any nonzero scalar amplitude.) A n-impulse is any sequence which is between $-p$ and p, for some pure n-impulse.*

For each integer $n \geq 0$, the following operators are defined.

Definition 3.2 *Let x be a sequence in X, then;*

$$\begin{aligned}
(U_n x)_i &= \min\{\max\{x_{i-n}, \ldots, x_i\}, \ldots, \max\{x_i, \ldots, x_{i+n}\}\} \\
(L_n x)_i &= \max\{\min\{x_{i-n}, \ldots, x_i\}, \ldots, \min\{x_i, \ldots, x_{i+n}\}\} \\
(M_n x)_i &= \text{median}\{\mathrm{x_{i-n}}, \ldots, \mathrm{x_i}, \ldots, \mathrm{x_{i+n}}\}.
\end{aligned}$$

These operators are easily seen to remove n-impulses from constant sequences. Analysis shows that U_n, L_n, U_nL_n and L_nU_n are all idempotent. U_n and L_n are morphological filters (10) in one dimension and are just compositions of "erosions" and "dilations". Furthermore, they are each others "duals", since $-U_n x = L_n(-x)$, where $-x = Nx = \{y_i;\ y_i = -x_i\}$. Furthermore it should be clear that U_n annihilates any sequence x that is a nonnegative n-impulse (and also that L_n annihilate any nonpositive n-impulse) since the maxima of $n+1$ consecutive elements of a non-negative impulse must contain a zero. Careful analysis (6),(7) shows that the compositions L_nU_n and U_nL_n annihilate any n-impulse, are invariant w.r.t. the choice of axes for the data under consideration, since they are translation invariant and if x is any sequence and c a constant sequence, $U_n(x+c) = U_nx+U_nc = U_nx+c$, and $L_n(x+c) = L_nx + c$. They are also scale invariant in that $L_n(\gamma x) = \gamma L_n x$ and $U_n(\gamma x) = \gamma U_n x$, for any scalar $\gamma \geq 0$. Clearly the operators are non-linear, as the sum of n-impulses need not be in the nulset of U_n or L_n. Defining the usual (partial-) order relation on X, and on the operators, also shows (5) the operators

U_n, L_n and M_n to be syntone and $U_nL_n \leq M_n \leq L_nU_n$. Clearly U_nL_n, M_n, L_nU_n are therefore equivalent in the power of removing impulses, at least from constant sequences. Due to the nonlinearity of the operators it is unclear in how far they remove an impulse superimposed on a general sequence ("signal"), but is so in some precise sense, in that the interval of sequences can be considered as an "interval of fundamental ambiguity" associated with the concept of "impulsive noise" (6). Fundamental is that the common range $\mathcal{M}$ of U_nL_n, L_nU_n consists precisely of those sequences that are n-monotone.

Definition 3.3 *A sequence x is n-monotone if each set of $n+2$ consecutive elements $\{x_i, x_{i+1}, \ldots, x_{i+n+1}\}$ are monotone increasing (-nondecreasing) or monotone decreasing (-nonincreasing).*
$\mathcal{M}_n$ is the subset of all n-monotone sequence in ℓ_1.

The concept of local monotonicity is a more compact characterisation of the so-called "roots" of median smoothers. The behaviour of M_n itself is enigmatic, and this can be associated with the existence of sections of "spurious" roots in a sequence. The "spurious" roots have recently been shown to be precisely the periodic ones (12) that are not in ℓ_p. In the case of M_1 it is essentially only one sequence, namely $x_i = (-1)^i$, and its multiples.

It is easy to see that if $x \in X$ then x is 0-monotone and therefore $\mathcal{M}_0 = X$. Furthermore $\mathcal{M}_0 \supset \mathcal{M}_1 \supset \cdots \supset \mathcal{M}_n \supset \cdots$ form a sequence of nested subsets. (A Haar-decomposition projects into $\mathcal{M}_1$ with the first decomposition onto $\mathcal{M}_3$ with the second and so forth, and this is dependent on the phase, or therefore of the choice of the nodes of the splines involved.) If an alternative decomposition is to be constructed it can be considered prudent to aim for an elementary separator P in such a way that the following criteria are not compromised too much.

Effectiveness : The output P must be a sequence without higher resolution detail.

Efficiency : The computations must be local and economical in terms of basic digital operations like logical comparisons, additions, multiplications, divisions etc.

Consistency : Mapping the output again should preserve it, or confirm it as good.

Stability : Input perturbations should not result in excessive output perturbations.

Considering the simple Haar-wavelet, the projection operator P, is effective since it is a projection onto a spline-subspace of half the original dimension of the (order 1-) spline space at the sampling resolution. It is achieved efficiently by a simple averaging filter on the sequence. Since a projection operator is idempotent it preserves its own output. (It is noteworthy that this necessitates a filter that is not translation invariant (and therefore phase dependent as it depends on the nodes of the spline subspace of lower dimension onto which is mapped.) Thus it does not meet the requirements of one of the axioms of a smoother, as introduced by Mallows. Stability has been argued to be suspect in the case of features that are "brief" impulses. Excessive output perturbations can result.

Choosing, as the first separator in an alternative multiresolution analysis the operator $P_1 = L_1U_1$, it is clear that the output is 1-monotone. (U_1L_1 would be another choice leading to a similar scheme.) A general sequence $x \in X$ is then effectively mapped onto a 1-monotone sequence P_1x in $\mathcal{M}_1$. This operator is efficient, requiring only $0(N)$ comparisons, as will be shown later. Since L_1U_1 is idempotent it is consistent, in preserving its output, but, since the operator is nonlinear, consistency demands somewhat more. Since P_1 is not a projection, the component $(I - P_2)x$ that is removed must also be consistently removed, in that $(I - P_1)(I - P_1)x = (I - P_1)x$.

This means that $I - P_1$ must also be idempotent. This turns out to be so for all the operators L_nU_n (and U_nL_n) and this "co-idempotence" of $P_n = L_nU_n$ is equivalent to having $(I - P_n)x$ being a null-sequence of the operator P_n, for each x (7). The separation can thus be considered to be consistent. Stability is good since the operator has a Lipschitz constant, so that small amplitude perturbations cannot be amplified (8). Furthermore a single (large) impulse has an influence restricted to amplitudes of neighbours, and the influence is local. Under the heading of effectiveness, a further consideration arises when P is not a projection. A projection onto a subspace S is automatically a good approximation in the appropriate norm, and since the Lebesgue inequality is applicable, also if the norm of P is finite, in any other norm. If P_n is merely a separator, (a idempotent and co-idempotent mapping), it is important to consider whether the image P_nx is a good approximation from $\mathcal{M}_n$ to the sequence x. This can again be shown to be so in the cases of L_nU_n and U_nL_n (8).

Thus the separator L_1U_1 effectively separates a sequence x into a good approximation L_1U_1x in $\mathcal{M}_1$ and a (high resolution) sequence $(I - L_1U_1)x$, which is a null-sequence of L_1U_1, and thus can be considered to consist of sufficiently local impulses, sufficiently separated not to yield a lower resolution non-zero output when mapped by L_1U_1. It can be considered as a sum of "noiselets". It must be stressed that the sum of such null-sequences is not automatically a null-sequence again! Furthermore, the operator U_nL_n, although considering L_nU_nx as a "signal", since it is in $\mathcal{M}_n$, does not necessarily consider (confirm) $(I - L_nU_n)x$ as being noise. This is because $U_nL_n(I - L_nU_n)$ is not the zero operator, although $L_nU_n(I - L_nU_n)$ is. L_nU_n and U_nL_n have a common range but $I - L_nU_n$ and $I - U_nL_n$ not. These observations are associated with a fundamental "uncertainty principle" with respect to the concept of impulse (6), and will result in a similar uncertainty in the concept of resolution, if made strict. The two decompositions, with LU and UL can be effectively done in parallel, but separately.

It is instructive to consider a sequence with two large impulses, as well as a "jump-discontinuity", sufficiently large so that the "smoother" parts L_1U_1x and U_1L_1x will not be affected at all if the impulses are multiplied by an arbitrarily larger number. All the extra amplitude will be restricted to the noise-components $x - L_1U_1x$ and $x - U_1L_1x$. No change will result in lower layers of decompositions. Only when the amplitude decreases to the level of the local variation of the uncontaminated signal will there be a minor change in L_1U_1x and U_1L_1x, (minor meaning; of the order of the difference between the two). Since the Haar-decomposition is based on linear operators the multiplication of the impulses by an arbitrary α will result in a proportional distortion by the same factor in the wavelet component resulting in proportional distortion with exponentially increasing width. With sufficiently large amplitude impulses this can swamp all significant features of the original sequence.

In the Haar-decomposition the large scale level change smear this significant feature to exponentially growing large sections of the successively smoother decompositions. The synthetic wavelet activity at the jump will have a correspondingly large amplitude in all subsequent wavelet layers. The well-known idea of thresholding the wavelet coefficient sequences to handle impulsive noise is clearly limited in its effectiveness, quite apart from the difficulty of choosing an appropriate threshold. The observations above are demonstrated already in the simple example of a constant signal and a single impulse of width 1. The first two Haar-decompositions demonstrate the exponentially growing width in both components of the decompositions, and because the operators involved are linear this behaviour is scale independent. Omitting one or more levels of wavelet components will result in a large distortion in the reconstructed signal.

The $LULU$-decomposition of the same signal and impulse has the full energy of the impulse in its first noise-component and omitting this (highest resolution) component and any other will leave a perfectly smoothed original constant signal.

Since the Haar decomposition is linear, this behaviour will result in a similar distortion when added to any signal. Since the $LULU$-decompositions are not linear, the superposition of this impulse on a given signal x will not necessarily result in an undistorted removal of the impulse from x. But all the distortions will not exceed the magnitude of the local variation of the signal at the position of the impulse. Strictly speaking L_1U_1x and U_1L_1x cannot be distorted by more than the factor $\max\{|x_{i+1}-x_i|, |x_i-x_{i-1}|\}$ at i, if the arbitrarily large impulse is added at i, since both have either the value x_{i-1} or x_{i+1}. No distortion larger than this can result in any lower level of decompositon, since the norms of all the operators L_nU_n and U_nL_n are 1. A further distortion occurs at the two neighbouring points $i-1$ and $i+1$, but this distortion cannot exceed $|x_{i-2}-x_{i-1}|$ and $|x_{i+2}-x_{i+1}|$ respectively. Clearly the contamination can spread, but this cannot exceed the maximum amplitude of the signal in the corresponding region. Furthermore, it does not spread far unless there is some Nyquist frequency present. A precise analysis is not intended here, and may become exceedingly difficult. Experience suggest very limited growth in the contamination support.

A practical typical comparison that is illustrative of the specific advantage of the $LULU$-decomposition argued above is where a broad n-impulse, with random noise added, is decomposed by $LULU$-decomposition and Haar decomposition. Except for the, progressively more unlikely event, as n increases, of the n-pulse starting and ending at a node of the lower-dimensional spline subspaces, there will be wavelet activity at arbitrary levels due to the sharp edge.

Having suggested and heuristically argued many "advantages" the $LULU$-decompositions have, it would be appropriate to derive some theorems that can support some of these known, observed or believed properties of the $LULU$-decomposition.

In linear orthogonal decompositions the squares of 2-norm are preserved. A pure n-pulse has the same "energy" as n 1-pulses of the same amplitude, but, depending upon how these 1-pulses are distributed, can have a total variation between 2α and $2n\alpha$. The total variation is therefore a measure of the "resolution" of features. Pursuing this idea leads to the following substantial results.

4 Decomposition, smoothing and variation reduction

The operators involved were originally intended for nonlinear smoothing, or presmoothing for the removal of impulsive noise. On attempting to clarify and quantify some experimental observations and case studies, it is natural to choose total variation of a sequence as a measure of smoothness. For each sequence x, the total variation $T(x)$ is defined by;

Definition 4.1

$$T(x) = \lim_{n \to \infty} \sum_{-N}^{N} |x_{i+1} - x_i|.$$

As is well known, and easy to prove, $T(x)$ is a semi-norm. If $x \in \ell_p$, the vectorspace of sequences such that the p-norm,

$$||x||_p^p = \lim_{N \to \infty} \sum_{i=-N}^{N} |x_i|^p,$$

is finite, then $T(x)$ is also a norm. This is because

$$\lim_{i \to -\infty} |x_i| = \lim_{i \to \infty} |x_i| = 0,$$

and thus the only sequence with zero variation is the null-sequence.

A wavelet decomposition of a sequence $x \in \ell_p$ with order 1 or 2 spline wavelets will eventually lead to a constant sequence, which has to be the null-sequence. Thus all the "energy" has been peeled off into the (wavelet-) frequency layers, since the squares of the 2-norms of Px and $x - Px$ add up to the square of the 2-norm of x, so that

$$||x - P_1x||_2^2 + ||P_1x - P_2P_1x||_2^2 + \cdots = ||x||_2^2.$$

The behaviour of the total variation of x and all the components of the decomposition at the intermediate stages is of interest, if smoothing is desired for the purpose of exposing significant features in the sequence x without too much damage. In a wavelet-decomposition, like in a Fourier decomposition, it is natural to stop the decomposition process when the frequency layers do not contain significant energy any more. Similarly it would be convenient if the reduction in variation has some natural measure giving precise indication of what fraction of the total variation has been removed. Assuming that the features to be exposed by smoothing are given by a sequence e to which higher resolution noise, given by a sequence r, has been added to produce a significant increase in variation. Since $T(x) = T(e+r) \leq T(e) + T(r)$ in general we are therefore assuming that $T(x) \approx T(e) + T(r)$. We should like to stop smoothing when the total variation goes significantly below that of $T(e)$, which is not known but perhaps estimated. If the noise r is of much higher "resolution" than the desired features of the sequence, the successive peeling off of resolution layers could be expected to decrease the variation steadily until most of the noisy features are removed. Reaching the expected resolution level of e should result in a further strong reduction, perhaps identifying that the unknown resolution level of e has been reached, and further "smoothing" would partially erase these. This heuristic motivation could be experimentally supported, but clearly the underlying assumption of a "proportional" allocation of the total variation to each resolution layer, is crucial in the appropriateness.

The *LULU*-decompositions have such a remarkable property. It can be proved in the following way.

Lemma 4.2 *Let x be $(n-1)$-monotone. If j is a point where $U_n x_j \neq x_j$ and $x_{j-1} \neq x_j$, then $U_n x_j = \min\{x_{j-1}, x_{j+n}\}$ and $U_n x_{j-1} = x_{j-1}$.*

Proof. If $U_n x_j$ differs from x_j it must be larger, since $U_n x \geq x$.
$U_n x_j = \min\{\max\{x_{j-n}, \ldots, x_j\}, \ldots, \max\{x_j, \ldots, x_{j+n}\}\} > x_j$, implies that each of the maxima is larger that x_j, so that there are at least two values $x_e \in \{x_{j-n}, \ldots, x_j\}$ and $x_r \in \{x_j, \ldots, x_{j+n}\}$, such that $x_e, x_r > x_j$.

Each set of $n+1$ successive elements of x are monotone. Then $x_{j-n} \geq \cdots \geq x_e \geq \cdots \geq x_{j-1} \geq x_j \leq x_{j+1} \leq \cdots \leq x_r \leq \cdots \leq x_{j+n}$. Noting $x_{j-1} \neq x_j$, from the assumption of the lemma, it must follow that $x_{j-1} > x_j$. From this it follows that there must be a constant section of equal values, since $x_{j-1} > x_j \geq \cdots \geq x_{j+n-1}$ and $x_j \leq \cdots \leq x_{j+n-1} \leq x_{j+n}$, imply that $x_j = x_{j+1} = \cdots = x_{j+n-1}$. At least one of the points $x_{j+1}, \ldots, x_{j+n}$ must be strictly larger than x_j, so that this value must be x_{j+n}.

Therefore also $U_n x_{j+n} = x_{j+n}$. Thus $U_n x_j = \min\{\max\{x_{j-1}, \ldots, x_{j+n-1}\}, x_{j+n}\} = \min\{x_{j-1}, x_{j+n}\}$.
Furthermore, clearly $U_n x_{j-1} = x_{j-1}$, since $\max\{x_{j-1}, x_j, \ldots, x_{j-1}\} = x_{j-1}$. □

Theorem 4.3 *For $x \in M_{n-1}$, $n \geq 1$,*

$$T(x) = T(\bigvee^{n} x) + T(U_n x - x) = T(\bigwedge^{n} x) + T(L_n x - x)$$

Proof. Consider the sequence $t = \{t_j\}$ of integers where $U_n x_{t_j} > x_{t_j}$ and $x_{t_j+1} \neq x_{t_j}$. By the lemma, and noting that $\{x_{t_j-n}, \ldots, x_{t_j}\}$ are $(n-1)$-monotone, $x_{t_j-1} > x_{t_j}$, it follows that. The sequence t cannot contain two consecutive integers, so that

$$T(x) = \sum_{j=-\infty}^{\infty} \tau_j, \quad \text{with} \quad \tau_j = \sum_{i=t_j-n}^{t_{j+1}-n-1} |x_{i+1} - x_i|$$

Consider a specific j, with $k = t_j$ and $m = t_{j+1}$ for notational convenience, let $c_{j+1} = w$ be the first index after k such that $x_w > x_{w+1}$. Clearly w exists in $[k+1, m-1]$. Since x is $(n-1)$-monotone, and therefore $x_{k-1} > x_k \geq x_{k+1} \geq \cdots \geq x_{k+n-1}$ and $x_k \leq x_{k+1} \leq \cdots \leq x_{k+n}$, it follows that $x_k = x_{k+1} = \cdots = x_{k+n-1}$. A similar argument yields $x_{w-n} = \cdots = x_{w-1} = x_w$. By the previous lemma $Ux_k = \min\{x_{k-1}, x_{k+n}\}$ implies that x_{k-1}, x_{k+n} and w must be larger than x_k.

Since t_{j+1} is the first integer after t_j where the sequence changes from a monotone decreasing to monotone increasing set of values, $\{x_{k+n}, \ldots, x_{m-n-1}\}$ has the following structure; $\{x_{k-n}, \ldots, x_{k-1}\}$ is a monotone decreasing section, $x_{k-1} > x_k = x_{k+1} = \cdots = x_{k+n-1}$, $\{x_k, \ldots, x_w\}$ is monotone increasing and $x_w > x_{w+1} \geq \cdots \geq x_m$. Let:

$$\begin{aligned}
\mu_j &= \sum_{i=k-n}^{m-n-1} |\bigvee^{n} x_{i+1} - \bigvee^{n} x_i| \\
&= \sum_{i=k-n}^{k-2} |\bigvee^{n} x_{i+1} - \bigvee^{n} x_i| + \sum_{i=k-1}^{k-1} |\bigvee^{n} x_{i+1} - \bigvee^{n} x_i| \\
&\quad + \sum_{i=k}^{m-n-1} |\bigvee^{n} x_{i+1} - \bigvee^{n} x_i|
\end{aligned}$$

The first of these three sums is equal to $\sum_{i=k-n}^{k-2} |x_{i+1} - x_i|$, since the set $\{x_{k-n}, \ldots, x_{k+n-1}\}$ is monotone decreasing, and the next sum is

$$|\bigvee^{n} x_k - \bigvee^{n} x_{k-1}| = |x_{k+n} - x_{k-1}|.$$

This sum has two cases;

(i) Suppose $w > m - n - 1$. Then

$$\sum_{i=k}^{m-n-1} |\bigvee^{n} x_{i+1} - \bigvee^{n} x_i| = |x_{k+n+1} - x_{k+n}| + \cdots |x_{w-n} + x_{w-n-1}|,$$

since $m - n - 1 > w - n - 1$ if $w < m$ and $x_{w-n} = x_{w-n+1} = \cdots = x_{w-1} = x_w$. This is the variation on a monotone increasing section and yields $|x_w - x_{k+n}|$.

Thus

$$\begin{aligned}\mu_j &= \sum_{i=k-n}^{k-2} |x_{i+1} - x_i| + |x_{k+n} - x_{k-1}| + x_w - x_{k+n} \\ &= \sum_{i=k-n}^{k-1} |x_{i+1} - x_i| - |x_k - x_{k-1}| + |x_{k+n} - x_{k-1}|\end{aligned}$$

But

$$\begin{aligned}x_w - x_{k+n} &= x_w - x_{k+n-1} + x_{k+n-1} - x_k \\ &= \sum_{i=k}^{m-n-1} |x_{i+1} - x_i| + x_{k+n-1} - x_k. \\ &= \tau_j + (x_k - x_{k-1}) + |x_{k+n} - x_{k-1}| + x_k - x_{k+n}\end{aligned}$$

Therefore

$$\begin{aligned}\mu_j &= \tau_j + 2x_k - 2\min\{x_{k+n}, x_{k-1}\} \\ &= \tau_j - 2(U_n x_k - x_k),\end{aligned}$$

from the lemma. But

$$2(U_n x_k - x_k) = \frac{2}{m} \sum_{i=k-n}^{m-n-1} |U_n x_i - x_i| = \sum_{i=k-n}^{m-n-1} |U_n x_{i+1} - x_{i+1} - U_n x_i + x_i|$$

The last equality comes from the fact that

$$Ux_i - x_i = Ux_k - x_k \quad \text{for} \quad i = k, \ldots, k+n-1$$

and zero elsewhere in the interval $[k-n, \ldots, m-n-1]$, and then the variation of this block pulse is simply twice the height.

(ii) Suppose $m - n - 1 \geq w$. Then

$$\sum_{i=k}^{m-n-1} |\bigvee^{n} x_{i+1} - \bigvee^{n} x_i| = x_w - x_{k+n} + x_w - x_{m-n}$$

or

$$\sum_{i=k}^{m-n-1} |\bigvee^{n} x_{i+1} - \bigvee^{n} x_i| = x_w - x_k + x_w - x_{m-n} + x_k - x_{k+n}.$$

This again yields

$$\mu_j = \tau_j - \frac{2}{n} \sum_{i=k-n}^{m-n-1} |U_n x_i - x_i| = \tau_j - 2(U_n x_k - x_k).$$

In both cases therefore

$$\begin{aligned} T(\bigvee^n x) = \sum_{j=-\infty}^{\infty} \mu_j &= \sum_{j=-\infty}^{\infty} \tau_j - \sum_{j=-\infty}^{\infty} \frac{2}{n} \sum_{i=t_j-n}^{t_{j+1}-n-1} |U_n x_i - x_i| \\ &= T(x) - \sum_{j=-\infty}^{\infty} \sum_{i=t_j-n}^{t_{j+1}-n-1} |(U_n x - x)_{i+1} - (U_n x - x)_i| \\ &= T(x) - T(U_n x - x) \end{aligned}$$

or $T(x) = T(\bigvee^n x) + T(x - U_n x)$. By a similar argument, or by the usual duality argument, $T(x) = T(\bigwedge^n x) + T(x - L_n x)$, completing the proof. □

Theorem 4.4 *For $x \in M_{n-1}$, $n \geq 1$;*

$$T(x) = T(U_n x) + T(x - U_n x)$$

and

$$T(x) = T(L_n x) + T(x - L_n x).$$

Proof. The proof is simple since $\bigwedge^n$ is variation diminishing so that $T(x) \geq T(\bigwedge^n \bigvee^n x) + T(x - U_n x)$ and the equality follows from the subadditivity of T, since $U_n = \bigwedge^n \bigvee^n$. A similar proof, or the usual duality argument yields the other equality. □

Theorem 4.5 *For $x \in M_{n-1}$, $n \geq 1$;*

$$T(x) = T(L_n U_n x) + T(x - L_n U_n x)$$

and

$$T(x) = T(U_n L_n x) + T(x - L_n U_n x).$$

Proof. $T(x) \leq T(L_n U_n x) + T(x - L_n U_n x)$, by the usual subadditivity of T. But by the first equality of the previous theorem $T(x) = T(U_n x) + T(x - U_n x)$, and by the second $T(x) = T(L_n U_n x) + T(U_n x - L_n U_n x) + T(x - U_n x)$.

By the usual subadditivity the last two terms are not smaller than $T(U_n x - L_n U_n x + x - U_n x)$ and therefore $T(x) \geq T(L_n U_n x) + T(x - L_n U_n x)$. This argument can be repeated, or to illustrate the usual duality argument;

$$\begin{aligned} T(x) &= T(-x) = T(L_n U_n(-x) + T(-x - L_n U_n(-x)) \\ &= T(-U_n L_n(x)) + T(-x + U_n L_n x) \\ &= T(U_n L_n x) + T(x - U_n L_n x) \end{aligned}$$

The important point is however that the total variation, and the local variation, can allocate proportional weight to constituent parts of the decomposition. A fractional part of the measurement vector length can be allocated to each resolution level. This is useful. □

5 Comparison of Haar- and $LULU$-decompositions.

In the previous comparisons the complications at endpoints of a finite sequence are avoided for the sake of simplicity. These problems can be treated satisfactorily in a variety of different ways, usually by letting both beginning and end of a finite sequence move to zero in a variety of different reasonable ways. Furthermore, there is an arguably unfair advantage that is easily overlooked if only the first decompositions are considered.

If the sequences are finite, then a Haar-decomposition projects the sequence x onto a nested set of subspaces, each being of half the dimension of the previous until a constant is reached. The $LULU$-operators map, in a way that is projection-like, (roughly speaking, as near to a projection as can be expected with a non-linear operator) onto nested subsets $M_1 \supset M_2 \supset M_3 \supset \cdots$ etc. Since these are not subspaces, there is no question of dimension, but to be fair it should be realized that there is not a decomposition into octaves, but rather a linear decrease in "flexibility" (for lack of a better word). A fairer comparison would be if the decomposition were from M_0 to M_1 to M_3 to M_5 etc., since, if the range of the projections P_j of the Haar-decomposition are R_j we have that $R_j \subset M_{2^j-1}$. The proof of this is obvious if we observe that a sequence in R_j is a sampled B-spline of order 1, which is $(2^j - 1)$-monotone, since it is a sampling of a piecewise constant function. The sequence is therefore made up of successive sections of 2^j equal values, and clearly every $2^j + 1$ successive values are monotone.

At this stage there has been no attempt to suggest comparison in a significant advantage of the wavelet-decomposition, namely the economising in representation and computation by the fact that the smoother component and the wavelet component can be economically stored by the respective spline and wavelet basis. This yields an effective representation in no more numbers than the original number of element of the sequence. These coefficients can furthermore be quantized without major distortion in the reconstructed sequence.

In the $LULU$-decompositon there is no basis that is generally useful for economization. There are several possibilities for savings by coding the noise components, which can be progressively more sparse. Similarly the smoother part could have progressively larger constant sections, permitting some economising. This whole issue is not addressed here, and it is generally complicated, like in the case of the two-dimensional image processing, with both wavelet decompositions and median decompositions. The primary comparisons to be made here are more fundamental in nature. The $LULU$-decompositions are an alternative to the prevalent median transform of the same type. These seem generally to be good in the two-dimensional case of image processing (1). A disadvantage listed is the computational complexity. A more serious disadvantage seems to be the lack of theory. Wavelet theory has a comprehensive and beautiful theory for analysis. For linear operators, Fourier transforms and their inverses, provide a framework for analysis. For a large class of operators, mathematical morphology provides a framework of analysis, and is well developed, especially in the two-dimensional case. Starck,

Murtagh and Bijaoni state that median transforms are considered better than morphological filters. This may be when considering only the simpler compositions like U_n and L_n, in the one-dimensional case, when $L_n \leq M_n \leq U_n$ is true, but the interval $[L_n, U_n]$ is too large and L_n and U_n are only comparable to M_n in approximating properties when the significant features in the noise are one-sided. The inequality $U_nL_n \leq M_n \leq L_nU_n$, which is not general in morphological filters, is much sharper and the operators U_nL_n and L_nU_n are good approximations to the median. This is visible in the foregoing examples where the first decompositions with L_nU_n and U_nL_n are compared. Since this difference can be argued to be within the fundamental uncertainty interval of the concept "impulse" and an associated concept of "resolution". They must both bc expected to be equivalent to the median for the purpose at hand.

The advantage argued here is in the application to analysis of measurements. Significant features of measurements that are fundamental are "edges", "local trend", and "pulses". The words "edges", "trend", "impulse", "pulse" and "resolution" are all widely used in science and technology and are very often not clearly defined in the context. For a multiresolution decomposition to be called such, it may be prudent to be more precise.

At this stage it is good to argue clearly a fundamental ambiguity in the pulse-decomposition with $LULU$-operators. (It must be noted that the Haar-decomposition also gives ambiguous decompositions at each level, depending on the choice of the knots.) Two sufficiently separated single pulses would both be removed. If they are next to each other they will not be. If separated by one index an ambiguity results (5). After the next two decomposition no pulse of lower resolution will be present. The only activity is in the first three layers indicating the fundamental resolution interval $[1,3]$. Two Haar decompositions of the same pulse smear activity into all levels, making a stricter use of the term "resolution" difficult.

It is not the purpose here to expand on all the various advantages, and disadvantages, over wavelet transforms, but to indicate the type of use that has been, and can be made in the detection of detail at various resolution levels. The method of analysis is relatively unknown, and based on the strong mathematical structure on the basic operators. Having defined a pure n-impulse p_n it may be useful to define a pulse p of resolution between m and n if $p_m \leq p \leq p_n$. If superimposed noise is comparatively small, it is clear that, by the syntoneness of the operators involved in the $LULU$-decomposition, the pulse will appear in the m-th level and disappear in the n-th level. Thus it may be a more precise way of defining what is meant by a pulse at a resolution level (or interval). If two such pulses are separated sufficiently they can be resolved. When they are superimposed, or close enough, the fundamental uncertainty at that resolution level can be recognised by the difference between the decomposition with L_nU_n and U_nL_n. Clearly other interpretations are also possible, but the ambiguity is at least apparent from the different sequences $x - L_1U_1x$ and $x - U_1L_1x$.

At each level n the operators L_n, U_n, U_nL_n and L_nU_n are all idempotent and co-idempotent and all compositions are one of these four. This near-ring of operators has a complete order given by $L_n \le U_nL_n \le L_nU_n \le U_n$. In the decomposition procedure several layers of such operators are used, and since they are all syntone operators, relative orders are inherited and remain useful for analysis. The well-established median operators, popularly used for smoothing out impulsive noise and accepted to be very good for preservation of edges and other significant features are contained between such $LULU$-outputs. When decompositions with L_nU_n and U_nL_n are compared, the difference indicates an "amount of ambiquity, which turns out to be very useful. Using the "commutators" like $L_nU_n - U_nL_n$ resulted in the design of recording instruments for times of arrival of shock pulses for location purposes. In the first application attempted, accuracy of such time of arrival estimates of a shock wave have exceeded the best previous designs of an international company by a large margin.

6 Computational considerations and characterisation of the resolution levels

From the definition of $LULU$-operators they appear computationally expensive, but for the purpose of successive decompositions by L_nU_n (or U_nL_n) there are considerable computational (and conceptual-) simplifications. The example with L_nU_n can be chosen to illustrate. Given a sequence $x = \{x_{ij}\ i = 1, 2, \dots, N\}$ in $\mathcal{M}_0$ the first decomposition is with U_1, followed by L_1 to yield a sequence L_1U_1x in $\mathcal{M}_1$ and a residual $R_1x = x - L_1U_1x$. At each stage, the sequence that is decomposed by L_nU_n is in $\mathcal{M}_{n-1}$, which permits a simplified calculation. Consider $(U_nx)_i = \min\{\max\{x_{i-n}, \dots, x_i\}, \dots, \max\{x_i, \dots, x_{i+n}\}$. Near endpoints, where left (or right-) neighbours are not defined, it is simple to simply omit them in the maximum (or minimum) calculators. This is clearly equivalent to appending sufficient values equal to x_1 to the left (and equal to x_N at the right). Given that $n \ge 1$ and x is $(n-1)$-monotone the first (and the last) values of U_nx can be copied from x. This is since the following arguments hold.

$\{x_k, x_{k+1} \dots, x_{k+n}\}$ is monotone for each k.

(i) Assume $x_1 \le x_2 \le \dots \le x_{n+1}$ then for $j = 1, \dots, n$

$$\begin{aligned} U_nx_j &= \min\{\max\{x_1, \dots, x_j\}, \max\{x_2, \dots, x_j, x_{j+1}\}, \dots \\ &\quad \dots, \max\{x_j, \dots, x_{j+n}\} \\ &= x_j, \end{aligned}$$

since $x_j = \max\{x_1, \dots, x_j\}$ and x_j is not larger than all the others since they are upper bounds.

(ii) Assume $x_1 \ge x_2 \ge \dots \ge x_{n+1}$.
Noting that $x_n \ge x_{n+1}$ implies that $x_n \ge x_{n+1} \ge \dots \ge x_{n+n}$, it is clear that

$\max\{x_j, x_{j+1}, \dots, x_{j+n}\} = x_j$, so that

$$\begin{aligned} U_n x_j &= \min\{\max\{1, \dots, x_j\}, \dots, \max\{x_j, x_{j+1}, \dots, x_{j+n}\} \\ &= x_j, \end{aligned}$$

since the last maximum is x_j and the previous cannot be less.
Thus $U_n x_j = x_j$ for $j \in \{1, \dots, n\}$.

A similar argument also holds for $j \in \{N-n+1, \dots, N\}$ so that $2n$ values of $U_n x_j$ are already calculated. The following sequences are calculated successively.
$e_i = \max\{x_i, x_{i+1}, \dots, x_{i+n}\} = \max\{x_i, x_{i+n}\}$, $i = 2$ to $N - n - 1$, the simplification being due to the fact that $x_i, x_{i+1}, \dots, x_{i+n}$ are monotone since $x \in M_{n-1}$. The maximum operator does not preserve the monotonicity, so that a similar simplification fails in; $t_i = \min\{e_{i-n}, \dots, e_i\}$, for $i = n+1$ to $N - n - 1$

This is followed by a similar process for $L_n(U_n x)$, where $U_n x$ is again in M_{n-1}.

It is therefore clear that if $p \in P_n$ and $p = x - U_n x$, where $P_n = M_{n-1} \setminus M_n$, that

$$p_i = 0 \quad \text{for} \quad 1 \le i \le n \quad \text{and} \quad N - n + 1 \le i \le N.$$

Therefore if $p \neq 0$, then the first index j where $p_j \neq 0$ is the first value of a constant region such that $p_j = p_{j+1} = \dots p_{j+n-1}$ followed by a point $p_{j+n} = 0$. This is a block pulse of negative amplitude $\alpha_{n_j} = p_j$. Noting that $p \in M_{n-1}$ it is clear that $p_{j+n-1} \le p_{j+n}$. It follows that $p_i \ge 0$ for $j + n \le i \le j + 2n - 1$, and since $p = x - U_n x$ is a non-positive sequence $p_i = 0$ for $j+n \le i \le j+2n-1$, the first index following $j + n - 1$ where p can differ from zero is thus $j + 2n$.

There are thus no more than $\dfrac{N-2n}{2n}$ such pulses in p, and each has some negative amplitude. Applying L_n to $U_n x \in M_{n-1}$ similarly yields no more than $\dfrac{N-2n}{2n}$ pulses of positive amplitude in $U_n x - L_n U_n x$. Noting that L_n preserves the pulses in $U_n x$, since those sections are n-monotone, the positive and negative pulses cannot overlap. In total therefore $x - L_n U_n x = x - U_n x + U_n x L_n U_n x$ cannot have more than $\dfrac{N-2n}{n}$ pulses and can be fully specified by a sequence of no more than $\dfrac{N-2n}{n}$ amplitudes α_{j_n}.

To compute the residual $x - L_n U_n x$ at each level, only values different from 0 need a subtraction. (The minority of points, definitely less than $\frac{N}{n}$, but generally much less.) There is no point in letting n be more than $N - 1$, since then $x - L_n U_n x = 0$ and $L_n U_n x$ a constant. There are indications that further saving can be achieved, as the above estimates are not sharp. Certainly it is clear that a great amount of parallelisation is possible, depending upon the equipment available. Some simple examples demonstrate very economic coding if some structures are present. Such economising is likely to occur often when quantising (and/or thresholding) is introduced for the amplitude, as is the case with measurements in fixed bit format It should be noted that in such a case none but the finite number of possible numbers available are sufficient for all decompositions, since the only operations performed are selections on those values present. This is an important economisation, compared to other values that are introduced by wavelet decompositions.

Example 6.1 *Let* $N = 2k$ *and* $x = \{x_i = |i - k|\}$

At level 1. $(x - L_1U_1x)_i = \delta_{ik}$ *(The Kronecker delta)*

At level 3. $(x - L_3U_3)L_2U_2L_1U_1x_i = \delta_{ik-1}$

At level 5 $\qquad = \delta_{kk-2}$

$\vdots$

At level $2m + 1 \qquad = \delta_{ik-m}.$

Thus there are precisely N pulses. Each is of width $2m - 1$ *and constant amplitude, costing requires N starting indexes and N amplitudes (in this case* 1*). (Here symmetry economises.)*

Example 6.2 *When random noise from a cubic-B-spline distribution was added to three uniform level pulses of duration* 1, 2 *and* 3 *the visually significant counts were* 23, 6 *and* 5 *out of a sequence length of* 50. *Quantizing and thresholding appropriately, and reconstituting from only* 3 *significant pulses yielded an image that substantially contained the significant features from* 3 *indexes and amplitudes. The thresholds can be chosen from estimates corresponding to amplitudes and distribution of the random noise. Impulsive noise of a prescribed width can be removed by omitting higher levels altogether.*

From the above examples, and others, it seems as if a sequence can always be decomposed into a total of not more than N pulses. This looks provable by a very simple argument, but is being thoroughly investigated, together with other computational simplifications, and remarkable shape preserving properties of the $LULU$-approximation.

7 Conclusion

The theory of $LULU$-operators, and the more general theory of a mathematical morphology, provide the possibility of a specific stricter concept of "resolution" coupled to the concepts of "pulse" and "local trend". This permits systematic decomposition of a sequence into components at resolution levels 1 to any n in more than one way. Choices can be made that provide for a multiresolution analysis that is effective, efficient, consistent and stable and permits economical coding for storage and transmission of information contained in the sequence. The total variation of the sequence is preserved at each separation stage, providing a quantitative measure of relative contribution of each resolution level for automatic decisions on thresholding, quantising, truncation and economisation in storage and transmission. Coding of the information at each resolution level is simple. Reconstruction, fully or partially, is straightforward, and all computations are simple and extensively parallisable. The theory that supports the procedure is furthermore indirectly applicable to explain and predict some properties of median-decompositions, which are generally regarded to be very good for image processing, as these components are related to those of two equivalent $LULU$-operators used in the decompositions developed here.

The idea that morphological filters are not as good as the medians for decomposition can be shown to be a result of a suboptimal choice amongst these. There are indications that the ideas based on the properties of $LULU$-operators can be generalised extensively to other morphological filters, and promising generalisations have already been discovered.

There are further remarkable shape preservation properties of the $LULU$-separators discovered recently, which the author intends publishing soon.

References

[1] A. Bijaoni, F. Murtagh, J.L. Starck, *Image Processing and Data Analysis – The Multiscale Approach*, Cambridge University Press, 1998.

[2] C.K. Chui, *Wavelets: A Mathematical Tool for Signal Analysis,* SIAM, Philadelphia, 1997.

[3] C.L. Mallows, *Some Theory of Nonlinear Smoothers,* (Ann. Statist. 8, no. 4 (1980), 695–715).

[4] P. Maragos, R.W. Schafer, *Morphological Filters – Part II: Their relations to Median, Order-Statistic, and Stack Filters,* IEEE Transactions on Acoustics, Speech and Signal Processing, Vol. ASSP-35 (1987), 1170–1184.

[5] C.H. Rohwer, *Idempotent One-Sided Approximation of Median Smoothers,* Journal of Approximation Theory. Vol 58, No. 2 (1989).

[6] C.H. Rohwer, and L.M. Toerien, *Locally Monotone Robust Approximation of Sequences*, Journal of Computational and Applied Mathematics **36** (1991), 399–408.

[7] C.H. Rohwer, *Projections and Separators,* Quaestiones Mathematicae **22** (1999), 219–230.

[8] C.H. Rohwer, *Fast Approximation with Locally Monotone Sequences.* Proceedings 4th FAAT Conference, Maratea, (to appear).

[9] H.L. Royden, *Real Analysis,* Macmillan, N.J., 1969.

[10] J. Serra, *Image Analysis and Mathematical Morphology,* Acad. Press, London 1982.

[11] P.F. Velleman, ***Robust nonlinear data smoothers: Definitions and recommendations***, Proc. Natl. Acad. Sci. USA. **74**, No. 2 (1977), 434–436.

[12] Zhou Xing-Wei, Yang De-Yun, Ding Run-Tao, ***Infinite length roots of median filters,*** Science in China, Ser. A, Vol. 35 (1992), 1496–1508.

Department of Mathematics
University of Stellenbosch
Private Bag X1
Matieland, 7602
South Africa
Email address: chr@maties.sun.ac.za

Advanced Problems in Constructive Approximation
(Eds.) M.D. Buhmann and D.H. Mache
International Series of Numerical Mathematics Vol. 142, 187–201

H-splines and Quasi-interpolants on a Three Directional Mesh

Paul Sablonnière

Abstract

Let τ (resp. τ^*) be the uniform three-directional mesh of the plane generated by the vectors $e_1 = (1,0), e_2 = (0,1), e_3 = (-1,-1)$ (resp. $e_1^* = (1,0), e_2^* = (-\frac{1}{2}, \frac{\sqrt{3}}{2}), e_3^* = (-\frac{1}{2}, -\frac{\sqrt{3}}{2})$). Let $P_n^s(\tau)$ and $P_n^s(\tau^*)$ be the spaces of piecewise polynomial functions of degree n and smoothness s on these meshes. There exist two interesting families of B-splines, respectively in the spaces $P_{3r+1}^{2r}(\tau), r \geq 0$ and $P_{3r}^{2r-1}(\tau), r \geq 1$. In the first space, B-splines with minimal support are simultaneously box-splines and H_{r+1}-splines, i.e., their support is the hexagon H_{r+1}, centered at the origin, whose sides are composed of $r+1$ edges of triangles of the mesh. In the second space, there exist three types of box-splines whose supports are non regular hexagons. Generalizing examples given in [18] and [19], we construct H_{r+1}-splines in the space $P_{3r}^{2r-1}(\tau)$ as linear combinations of translates of three box-splines. Then we construct various differential and discrete quasi-interpolants (QI) which have the best possible approximation order, for degrees (resp. smoothness orders) ranging from 3 to 10 (resp. from 1 to 6). Their computation is made easier thanks to the symmetry properties of H-splines. Finally, we give some examples of QI with nearly minimal infinite norms, which we call near-best quasi-interpolants.

1 Introduction and notations

Let τ (resp. τ^*) be the uniform three-directional mesh of the plane generated by the vectors $e_1 = (1,0), e_2 = (0,1), e_3 = (-1,-1)$ (resp. $e_1^* = (1,0), e_2^* = (-\frac{1}{2}, \frac{\sqrt{3}}{2}), e_3^* = (-\frac{1}{2}, -\frac{\sqrt{3}}{2})$). Let $P_n^s(\tau)$ and $P_n^s(\tau^*)$ be the spaces of piecewise polynomial functions (abbr. ppf) of degree n and smoothness s respectively defined on these meshes. It is well known (see e.g. [1], [2], [3], [6]) that there exist interesting families of B-splines, respectively in the two spaces $P^{2r}_{3r+1}(\tau), r \geq 0$ and $P^{2r-1}_{3r}(\tau), r \geq 1$. In the first space, B-splines M_{r+1} with minimal support (abbr. ms-splines) are simultaneously box-splines and H_{r+1}-splines, i.e., their support is the hexagon H_{r+1} (centered at the origin) whose sides are composed of $r+1$ edges of the mesh. They already appeared in the seventies (e.g. in [11],[12],[14],[15]) and later in Bézier form ([17], [18], [19], [20]). In the second space, the situation is more complex: there exist two types of ms-splines (whose support is a non regular hexagon) and three types of box-splines with a larger support (which is another kind of non regular hexagon), see e.g. [2]. Generalizing examples given in [18] and [19], we construct H_{r+1}-splines Λ_{r+1} as linear combinations of translates of the three types of box-splines. Similarly, there exist H^*_{r+1}-splines M^*_{r+1} (resp. Λ^*_{r+1}) in the spaces $P^{2r}_{3r+1}(\tau^*)$ (resp. $P^{2r-1}_{3r}(\tau^*)$). Then, using the inverses of Fourier transforms of H-splines and the results of [9], we construct *differential quasi-interpolants* (abbr. DQI), associated with the H- or H^*-splines previously defined, having the best possible approximation order. Their computation is made easier thanks to the symmetry properties of hexagonal supports. Moreover, the examples studied for the first degrees show that they seem to be particularly suited to the approximation of harmonic functions and polynomials. Next we recall results of [21] and [22] in order to give examples of *discrete quasi-interpolants* (abbr. dQI). The degrees (resp. smoothness orders) of our examples range from 3 to 10 (resp. 1 to 6). Finally, we construct examples of dQI with *quasi-minimal infinite norms*, which we call *near-best* QI. We thus provide a partial answer to the problem raised in Chapter III of [3] about the determination of discrete quasi-interpolants with minimal infinite norm (see Section 5.2 for details). For additional informations on QI, see for example [1], Chapter 11, [3], Chapter 3, [6], Chapter 8 and also the papers [4],[10]. Most of the results presented below are only sketched and more detailed proofs will be given elsewhere in a more complete paper. New results are essentially those of Sections 2.2 and 3.2 on H_{r+1}-splines Λ_{r+1} and Λ^*_{r+1}, and of Sections 5 and 6 on dQI, specifically those concerning near-best dQI.

For a box-spline $B(\cdot|X)$ associated with a set of directions $X \subset \mathbf{R}^2$, we use the following notations (see e.g.[1], Chapter 11):

$$\mathcal{Y} = \mathcal{Y}(X) = \{Y \subset X \,|\, \langle X \backslash Y \rangle \neq \mathbf{R}^2\}$$

$$\mathcal{D} = \mathcal{D}(X) = \{f : D_Y f = 0 \ for\ all\ Y \in \mathcal{Y}\} \ where\ D_Y f = (\Pi_{y \in Y} D_y) f$$

$$d = d(X) = \min\{|Y| : Y \in \mathcal{Y}\} - 1$$

$$\mathbb{B}(X) = \{V \subset X : |V| = \dim\langle V \rangle = 2\}, \ \dim \mathcal{D}(X) = |\mathbb{B}(X)|$$

For a box-spline $\phi = B(\cdot|X)$ or more generally for a B-spline ϕ, we denote by

$$\mathcal{S}(X) = \mathcal{S}(\phi) = \langle \{\phi(\cdot - \alpha), \alpha \in \mathbf{Z}^2\} \rangle$$

the *spline space* generated by integer translates of ϕ.

We also recall some important properties of box-splines: denoting by Π_d the space of bivariate polynomials of total degree at most d, we have $\Pi_d \subseteq \mathcal{D}(X)$, but Π_{d+1} is not a subset of $\mathcal{D}(X)$, therefore the approximation order of smooth functions in $\mathcal{S}(X)$ is $O(h^{d+1})$ for a triangulation with meshsize h.

Defining $sc(t) = \frac{\sin(t/2)}{(t/2)}$ and $\hat{f}(y) = \int_{\mathbf{R}^2} \exp(-ix^T y) f(x) dx$, then the Fourier transform of the box-spline $B(\cdot|X)$ can be written

$$\hat{B}(y|X) = \Pi_{x \in X} sc(x^T y)$$

Finally, for any invertible matrix A and $X^* = AX$, one has (see e.g. [10]):

$$B(x|X^*) = B(x|AX) = \frac{1}{\det(A)} B(A^{-1}x|X), \;\; \hat{B}(y|X^*) = \hat{B}(A^T y|X)$$

Setting $V = \mathbf{Z}^2$ and $V^* = AV$, we easily get

$$\sum_{\beta \in V^*} B^*(x - \beta) = \frac{1}{\det(A)}$$

So, if $\det(A) \neq 1$, box-splines on the triangulation τ^* with vertices V^* have to be normalized in order to get a partition of unity.

2 H-splines on the triangulation τ

2.1 H-splines in $P_{3r+1}^{2r}(\tau)$

In the space $P_{3r+1}^{2r}(\tau), r \geq 0$, there exists a unique box-spline M_{r+1} with hexagonal support H_{r+1}. It is defined as the convolution product

$$M_{r+1} = M_1 * M_1 * \ldots * M_1 \; (r+1 \text{ times})$$

of the piecewise affine pyramid M_1 with support the unit hexagonal cell H_1 centered at the origin. The set of directions defining this box-spline is

$$X_{r+1} = \{e_1(r+1), e_2(r+1), e_3(r+1)\}$$

where the parentheses mean that each direction is taken with multiplicity $r+1$. In this case, one can check that $d(X_{r+1}) = d_{r+1} = 2r+1$, the minimal cardinal of a subset $Y \in \mathcal{Y}$. Since $\mathcal{D}_{r+1} = \mathcal{D}(X_{r+1})$ contains Π_{2r+1}, but not Π_{2r+2}, therefore, the approximation order of smooth functions in this space is $O(h^{2r+2})$ for a triangulation of meshsize h. Moreover dim $\mathcal{D}_{r+1} = 3(r+1)^2$ =number of triangles in H_{r+1} having the same orientation. The Fourier transform of M_{r+1} is $\hat{M}_{r+1}(y) = [sc(y_1)sc(y_2)sc(y_3)]^{r+1}$, where $y_3 = -(y_1 + y_2)$.

2.2 H-splines in $P_{3r}^{2r-1}(\tau)$

For $s = 1, 2, 3$, there exist three box-splines $B_{r+1}^{(s)} = B(\cdot|X_{r+1}^{(s)})$ generated by the subsets of vectors defined respectively by $X_{r+1}^{(s)} = X_{r+1} \backslash \{e_s\}$. Assuming that their supports are centered at the origin, we define the H_{r+1}-spline $\Lambda_{r+1} \in P_{3r}^{2r-1}(\tau)$ by

$$\Lambda_{r+1}(x) = \frac{1}{3}\{B_{r+1}^{(1)}(x + \frac{1}{2}e_1) + B_{r+1}^{(2)}(x + \frac{1}{2}e_2) + B_{r+1}^{(3)}(x + \frac{1}{2}e_3)\}$$

Theorem 2.1 (i) *The Fourier transform of* Λ_{r+1} *is equal to*

$$\hat{\Lambda}_{r+1}(y) = \hat{\Lambda}_1(y)[sc(y_1)sc(y_2)sc(y_3)]^r$$

where Λ_1 *is the characteristic function of the hexagon* H_1 *and*

$$\hat{\Lambda}_1(y) = -\frac{2}{3}\{y_1cos(y_1) + y_2cos(y_2) + y_3cos(y_3)\}/y_1y_2y_3$$

(ii) *The spline space* $\mathcal{S}(\Lambda_{r+1})$ *contains* Π_{2r}*, therefore the approximation order of smooth functions in this space is* $O(h^{2r+1})$ *for a triangulation of meshsize* h*.*

Proof. since $\mathcal{D}(X_{r+1}^{(s)})$ contains Π_{2r} for $s = 1, 2, 3$, it can be verified, by using Strang-Fix conditions, that Π_{2r} is also included in the spline space $\mathcal{S}(\phi)$. □

Remark 2.2 *The Bernstein (abbr. B)-coefficients of* $M_2 \in P_4^2$*,* $M_3 \in P_7^4$*,* $M_4 \in P_{10}^6$ *can be found in [19], which is available from the author. A general algorithm for the computation of B-coefficients of box-splines is given in [7]. The B-coefficients of* $\Lambda_2 \in P_3^1$*,* $\Lambda_3 \in P_6^3$*, and* $\Lambda_4 \in P_9^5$ *are also in [19]. In general, they can be computed by using [7] for box-splines in the above definition of* Λ_{r+1}*.*

3 H-splines on the triangulation τ^*

In this section, $X_1^* = \{e_1^*, e_2^*, e_3^*\} = AX_1$, whith $X_1 = \{e_1, e_2, e_3\}$ and

$$A = \begin{pmatrix} 1 & -\frac{1}{2} \\ 0 & \frac{\sqrt{3}}{2} \end{pmatrix}, \quad A^{-1} = \begin{pmatrix} 1 & -\frac{1}{\sqrt{3}} \\ 0 & \frac{2}{\sqrt{3}} \end{pmatrix}, \quad A^T = \begin{pmatrix} 1 & 0 \\ -\frac{1}{2} & \frac{\sqrt{3}}{2} \end{pmatrix}$$

In particular, $\det(A) = \frac{\sqrt{3}}{2}$.

3.1 H-splines in $P_{3r+1}^{2r}(\tau^*)$

As the set of directions is now

$$X_{r+1}^* = AX_{r+1} = \{\, e_1^*(r+1), e_2^*(r+1), e_3^*(r+1)\,\}$$

the box-spline in this space is

$$M_{r+1}^*(x) = B(x|X_{r+1}^*) = \frac{2}{\sqrt{3}}M_{r+1}(A^{-1}x)$$

and its Fourier transform is

$$\hat{M}^*_{r+1}(y) = \hat{M}_{r+1}(A^T y) = [sc(y_1)sc(-\frac{1}{2}y_1 + \frac{\sqrt{3}}{2}y_2)sc(\frac{1}{2}y_1 + \frac{\sqrt{3}}{2}y_2)]^{r+1}$$

There are two possible normalizations: M^*_{r+1} or $M^{\#}_{r+1} = \frac{2}{\sqrt{3}}M^*_{r+1}$ satisfying respectively

$$\int_{\mathbf{R}^2} M^*_{r+1}(x)dx = \hat{M}^*_{r+1}(0) = 1, \quad \sum_{\beta\in V^*} M^{\#}_{r+1}(x-\beta) = 1$$

3.2 H-splines in $P^{2r-1}_{3r}(\tau^*)$

Similarly, the H^*_{r+1}-spline is defined by

$$\Lambda^*_{r+1}(x) = \frac{2}{\sqrt{3}}\Lambda_{r+1}(A^{-1}x)$$

or through its Fourier transform

$$\hat{\Lambda}^*_{r+1}(y) = \hat{\Lambda}_{r+1}(A^T y) = \hat{\Lambda}_1(A^T y)[sc(y_1)sc(-\frac{1}{2}y_1 + \frac{\sqrt{3}}{2}y_2)sc(\frac{1}{2}y_1 + \frac{\sqrt{3}}{2}y_2)]^r$$

As in the preceding case, we can choose two normalizations: Λ^*_{r+1} or $\Lambda^{\#}_{r+1} = \frac{2}{\sqrt{3}}\Lambda^*_{r+1}$ satisfying respectively

$$\int_{\mathbf{R}^2} \Lambda^*_{r+1}(x)dx = \hat{\Lambda}^*_{r+1}(0) = 1, \quad \sum_{\beta\in V^*} \Lambda^{\#}_{r+1}(x-\beta) = 1$$

4 Differential quasi-interpolants (DQI)

4.1 Differential quasi-interpolants on τ

We use results of [9]: given some H-spline ϕ, take the inverse of its Fourier transform

$$\frac{1}{\hat{\phi}(y)} = \sum_{\alpha\in V} a_\alpha y^\alpha$$

Assuming that Π_d is the maximal space of polynomials of type Π_n included in the spline space generated by ϕ, define the differential operator

$$\mathbb{D}_\phi = \sum_{|\alpha|\le d} (-i)^{|\alpha|} a_\alpha D^\alpha$$

Then the *differential quasi-interpolant* (abbr. DQI) Q_ϕ associated with ϕ is defined, for a given smooth function f, by

$$Q_\phi f(x) = \sum_{\alpha\in V} \mathbb{D}_\phi f(\alpha)\phi(x-\alpha)$$

The operator Q_ϕ is *exact* on Π_d, i.e., $Q_\phi p = p$ for all $p \in \Pi_d$, therefore the approximation order of a smooth function f by the associated DQI on the triangulation with meshsize h is maximal, i.e., $O(h^{d+1})$.

In order to illustrate this method, we give the DQIs associated with the H_r-splines M_r and Λ_r for $r = 2, 3, 4$. For sake of simplicity, we set

$$\mathbb{D}_r = \mathbb{D}_{M_r}\ ,\ \tilde{\mathbb{D}}_r = \mathbb{D}_{\Lambda_r}\ ,\ Q_r = Q_{M_r}\ ,\ \tilde{Q}_r = Q_{\Lambda_r}$$

Introducing the auxiliary differential operator ($D_1 = \frac{\partial}{\partial x}$ and $D_2 = \frac{\partial}{\partial y}$)

$$\mathbf{D} = D_1^2 + D_1 D_2 + D_2^2$$

then, with the help of a computer algebra system, one obtains respectively the coefficients of the differential quasi-interpolants Q_r, for $r = 2, 3, 4$:

$$\mathbb{D}_2 = I - \frac{1}{6}\mathbf{D},\ \ \mathbb{D}_3 = I - \frac{1}{4}\mathbf{D} + \frac{1}{30}\mathbf{D}^2,$$

$$\mathbb{D}_4 = I - \frac{1}{3}\mathbf{D} + \frac{7}{120}\mathbf{D}^2 - \frac{1}{140}\mathbf{D}^3 - \frac{1}{15120}\mathsf{D}_6$$

where D_6 is the differential operator of order 6 defined by

$$\mathsf{D}_6 = (D_1 D_2 D_3)^2 = D_1^2 D_2^2 (D_1 + D_2)^2$$

For the coefficients of the differential quasi-interpolants $\tilde{Q}_r$, $r = 2, 3, 4$, we obtain respectively

$$\tilde{\mathbb{D}}_2 = I - \frac{2}{9}\mathbf{D},\ \ \tilde{\mathbb{D}}_3 = I - \frac{11}{36}\mathbf{D} + \frac{83}{1620}\mathbf{D}^2$$

$$\tilde{\mathbb{D}}_4 = I - \frac{7}{18}\mathbf{D} + \frac{131}{1620}\mathbf{D}^2 - \frac{2453}{204120}\mathbf{D}^3 - \frac{1}{10080}\mathsf{D}_6$$

4.2 Differential quasi-interpolants on τ^*

For box-splines $B(\cdot|X^*)$, let us define

$$\mathcal{Y}^* = \mathcal{Y}(X^*) = \{Y^* \subset X^* \mid \langle X^* \backslash Y^* \rangle \neq \mathbf{R}^2\}$$

$$\mathcal{D}^* = \mathcal{D}(X^*) = \{g : D^*_Y g = 0 \ for\ all\ Y^* \in \mathcal{Y}^*\}\ where\ D^*_Y g = (\Pi_{y^* \in Y^*} D_y) g$$

With any given polynomial $p \in \mathcal{D}(X)$, one can associate the polynomial $q(x) = p(A^{-1}x)$. Reciprocally, for any polynomial $q \in \mathcal{D}(X^*)$, there exists a unique polynomial $p \in \mathcal{D}(X)$ defined by $p(x) = q(Ax)$. Using the notation $D = (D_1, D_2)^T$, this can be verified by observing that for any $y^* = Ay$, there holds

$$\begin{aligned} D_{y^*} q(x) &= (y^*)^T D q(x) = y^T A^T D q(x) = y^T A^T D p(A^{-1}x) \\ &= (y^T A^T) A^{-T} D p(x) = y^T D p(x) = D_y p(x) \end{aligned}$$

For $p \in \mathcal{D}(X)$, $D_Y p = 0$ for all $Y \in \mathcal{Y}(X)$, therefore the above equations imply $D_{Y^*} q = 0$ for all $Y^* \in \mathcal{Y}(X^*)$, and $q \in \mathcal{D}(X^*)$. The converse is also true: the fact that $q \in \mathcal{D}(X^*)$ implies immediately that $p \in \mathcal{D}(X)$. So, to each box-spline ϕ on τ corresponds a unique box-spline ϕ^* on τ^* whose associated DQI is defined, for any smooth function g, by

$$Q_{\phi^*} g(x) = \sum_{\beta \in V^*} \mathbb{D}_{\phi^*} g(\beta) \phi^*(x - \beta)$$

The differential operator $\mathbb{D}_{\phi^*}$ and its coefficients are respectively given by

$$\mathbb{D}_{\phi^*} = \sum_{\beta \in V^*} (-i)^{|\beta|} b_\beta (D^*)^\beta, \ \textit{with} \ \frac{1}{\hat{\phi}^*(y)} = \sum_{\beta \in V^*} b_\beta y^\beta$$

$$D^* = A^T D = (D_1^*, D_2^*) = (D_1, -\frac{1}{2} D_1 + \frac{\sqrt{3}}{2} D_2)$$

As in Section 4.1, Q_{ϕ^*} is exact on Π_d and the approximation order of a smooth function g by the associated DQI on the triangulation with meshsize h is maximal, i.e., $O(h^{d+1})$.

In order to illustrate this method, let us give the DQIs associated with the H_r^*-splines M_r^* and Λ_r^* for $r = 2, 3, 4$. For sake of simplicity, we set

$$\mathbb{D}_r^* = \mathbb{D}_{M_r}^* \ , \ \tilde{\mathbb{D}}_r^* = \mathbb{D}_{\Lambda_r}^* \ , \ Q_r^* = Q_{M_r}^* \ , \ \tilde{Q}_r^* = Q_{\Lambda_r}^*$$

Using the Laplace operator $\boldsymbol{\Delta} = D_1^2 + D_2^2$ then one obtains respectively the coefficients of the differential quasi-interpolants Q_r^*, for $r = 2, 3, 4$:

$$\mathbb{D}_2^* = I - \frac{1}{8} \boldsymbol{\Delta}, \ \mathbb{D}_3^* = I - \frac{3}{16} \boldsymbol{\Delta} + \frac{3}{160} \boldsymbol{\Delta}^2$$

$$\mathbb{D}_4^* = I - \frac{1}{4} \boldsymbol{\Delta} + \frac{21}{640} \boldsymbol{\Delta}^2 - \frac{1459}{483840} \boldsymbol{\Delta}^3 + \frac{1}{483840} \mathsf{D}_6^*$$

where D_6^* is the 6th-order differential operator defined by

$$\mathsf{D}_6^* = D_1^6 - 15 D_1^4 D_2^2 + 15 D_1^2 D_2^4 - D_2^6$$

For the coefficients of the differential quasi-interpolants $\tilde{Q}_r^*$, $r = 2, 3, 4$, we obtain respectively

$$\mathbb{D}_2^* = I - \frac{1}{6} \boldsymbol{\Delta}, \ \tilde{\mathbb{D}}_3^* = I - \frac{11}{48} \boldsymbol{\Delta} + \frac{83}{2880} \boldsymbol{\Delta}^2$$

$$\tilde{\mathbb{D}}_4^* = I - \frac{7}{24} \boldsymbol{\Delta} + \frac{131}{2880} \boldsymbol{\Delta}^2 - \frac{4909}{967680} \boldsymbol{\Delta}^3 + \frac{1}{322560} \mathsf{D}_6^*$$

Remark 4.1 *Approximation of harmonic functions or polynomials.*

Interesting simplifications occur when we assume that the function f to be approximated is harmonic, i.e., satisfies $\Delta f = 0$. From the preceding calculations, we deduce that the operators Q_r^* and $\tilde{Q}_r^*$, for $r = 2, 3$ coincide with the simple Schoenberg operators:

$$S_r^* f(x) = \sum_{\beta \in V^*} f(\beta) M_r^\#(x - \beta) \text{ and } \tilde{S}_r^* f(x) = \sum_{\beta \in V^*} f(\beta) \Lambda_r^\#(x - \beta)$$

They are exact on harmonic polynomials in the spaces Π_r, $r = 2, 3, 4, 5$. For higher degrees, there appear in the expression of $\mathbb{D}_{\phi^*}$ some sparse differential operators of high orders with very small rational coefficients. For box-splines $M_4^\#$ of degree 10 and class C^6, we obtain for example

$$Q_4^* f(x) = \sum_{\beta \in V^*} (f + \frac{1}{483840} \mathsf{D}_6^* f)(\beta) M_4^\#(x - \beta)$$

This QI is exact on the subspace of harmonic polynomials in Π_7. Similarly, for H_4^*-splines $\Lambda_4^\#$ of degree 9 and class C^5, we obtain

$$\tilde{Q}_4^* f(x) = \sum_{\beta \in V^*} (f + \frac{1}{322560} \mathsf{D}_6^* f)(\beta) \Lambda_4^\#(x - \beta)$$

This QI is exact on the subspace of harmonic polynomials in Π_6. This kind of situation also holds for H^*-splines and DQI of higher degrees.

Now, let us consider a polynomial $f(z) = p(x, y) + iq(x, y)$ of one complex variable $z = x + iy$. Thanks to Cauchy-Riemann conditions, it is well known that the real and imaginary parts $p(x, y)$ and $q(x, y)$ of f are harmonic functions. Therefore defining, for any H^*-spline ϕ^*

$$Q_{\phi^*} f(z) = Q_{\phi^*} p(x, y) + i Q_{\phi^*} q(x, y)$$

we see that the operators Q_r^* and $\tilde{Q}_r^*$, for $r = 2, 3$ coincide with the Schoenberg operators, and that, for higher degrees, the expression of $Q_{\phi^*} f(z)$ is greatly simplified. This property is very interesting and we plan to develop this point with some applications in a further paper.

Remark 4.2 *Two families of simple and efficient DQIs.*

For $r \geq 2$, the first two terms of differential operators $\mathbb{D}_\phi$ or $\mathbb{D}_{\phi^*}$ are the following

$$\mathbb{D}_r = I - \frac{r}{12}\mathbf{D}, \ \tilde{\mathbb{D}}_r = I - \frac{3r+2}{36}\mathbf{D}, \ \mathbb{D}_r^* = I - \frac{r}{16}\mathbf{\Delta}, \ \tilde{\mathbb{D}}_r^* = I - \frac{3r+2}{48}\mathbf{\Delta}$$

The corresponding DQI are exact on Π_3 (except $\tilde{Q}_2$ which is exact on Π_2), therefore, their approximation order is $O(h^4)$ on a triangulation of meshsize h. As a *by-product, all monomials $x^\alpha = x_1^{\alpha_1} x_2^{\alpha_2}, |\alpha| \leq 3$ have simple expansions in series of translates of B-splines.*

Similarly, for $r \geq 3$, the first three terms of differential operators $\mathbb{D}_\phi$ or $\mathbb{D}_{\phi^*}$ are the following

$$\mathbb{D}_r = I - \frac{r}{12}\mathbf{D} + \frac{r(5r+1)}{1440}\mathbf{D}^2, \quad \tilde{\mathbb{D}}_r = I - \frac{3r+2}{36}\mathbf{D} + \frac{45r^2+69r+52}{12960}\mathbf{D}^2$$

$$\mathbb{D}_r^* = I - \frac{r}{16}\mathbf{\Delta} + \frac{r(5r+1)}{2560}\mathbf{\Delta}^2, \quad \tilde{\mathbb{D}}_r^* = I - \frac{3r+2}{48}\mathbf{\Delta} + \frac{45r^2+69r+52}{23040}\mathbf{\Delta}^2$$

The corresponding DQIs are exact on Π_5 (except $\tilde{Q}_3$ which is exact on Π_4), therefore, their approximation order is $O(h^6)$ on a triangulation of meshsize h and all monomials $\{x^\alpha = x_1^{\alpha_1} x_2^{\alpha_2}, |\alpha| \leq 5\}$ have simple expansions in series of translates of B-splines.

Remark 4.3 *In [5], the Taylor expansions of functions $(sc(y))^{-r}$ are given in terms of central factorial numbers. They can be used for explicit formal expressions of coefficients of DQI.*

5 Discrete quasi-interpolants (dQI)

In this section, we use notations and results of [21] and [22]. As it is easier to work with the triangulation τ^*, we only study dQIs on this mesh. They are easily extended to the triangulation τ with minor modifications. Let $\mathbf{H}_\mathbf{r}^*$ denote the space of *hexagonal sequences* with support $W_r^* = H_r^* \cap V^*$, namely the set of vertices of the hexagonal mesh V^* which lie inside and on the boundary of the hexagon H_r^*. By definition, these hexagonal sequences are *invariant with respect to the group of symmetries of the hexagon.* Such a sequence, for example $\mathbf{a} \in \mathbf{H}_\mathbf{4}^*$, can be represented as follows

$$\mathbf{a} = [a_0|a_1|a_2, a_3|a_4, a_5|a_6, a_7, a_8]$$

since $\dim \mathbf{H}_4^* = 9$ (see Theorem 2.2 of [20]). With this sequence is associated a unique *central difference operator* $\delta_\mathbf{a}$ defined by

$$\delta_\mathbf{a}(f) = a_0 f(0) + a_1 \sum_{s=1}^{3} [f(e_s^*) + f(-e_s^*)] + a_3 \sum_{s=1}^{3} [f(2e_s^*) + f(-2e_s^*)$$

$$+ a_2 \sum_{s \neq t} f(e_s^* - e_t^*) + a_4 \sum_{s \neq t} [f(e_s^* - 2e_t^*) + f(2e_s^* - e_t^*)] + a_5 \sum_{s=1}^{3} [f(3e_s^*) + f(-3e_s^*)$$

$$+ a_6 \sum_{s \neq t} f(2e_s^* - 2e_t^*) + a_7 \sum_{s \neq t} [f(e_s^* - 3e_t^*) + f(3e_s^* - e_t^*)] + a_8 \sum_{s=1}^{3} [f(4e_s^*) + f(-4e_s^*)].$$

The set $\mathbb{H}^* = \cup_{r\geq 0}\mathbf{H}_r^*$ is a convolution algebra for the usual convolution of sequences ($\mathbf{H}_0^* = \{\delta_0\}$ is reduced to the unit sequence $\delta_0 = [1]$ associated with the Dirac linear functional $\delta_0(f) = f(0)$). This algebra is isomorphic to the composition algebra $\mathbb{L}^*$ of all central difference operators (abbr. CdOs). Here $\mathbb{L}^*$ stands for Laplace because CdOs associated with basic sequences are discrete approximations of the Laplace operators $\mathbf{\Delta}$ or $\mathbf{\Delta}^2$. It is the case for the following basic hexagonal sequences whose associated CdOs are denoted $\delta_1, \delta_2, \delta_3, \delta_4$ and are used later in Section 5.2

$$\mathbf{a_1} = [-6|1], \quad \mathbf{a_2} = [-6|0|0,1], \quad \mathbf{a_3} = [-48|9|-1,0], \quad \mathbf{a_4} = [12|-3|1,0]$$

According to [8], p.544-546, δ_1, δ_2, δ_3 and δ_4 are respectively approximations of $\frac{3}{2}\mathbf{\Delta}$, $6\mathbf{\Delta}$, $9\mathbf{\Delta}$ and $\frac{9}{16}\mathbf{\Delta}^2$.

For any H^*-spline ϕ^* generating a spline space $\mathcal{S}(\phi^*)$ containing Π_d, the Schoenberg operator S_{ϕ^*} coincide on Π_d with the CdO $\delta_{\mathbf{a}}$ associated with the hexagonal sequence $\mathbf{a} = \phi^*|V^*$ of values of ϕ^* at the meshpoints lying inside supp(ϕ^*) (see Theorem of [20]). For example, if supp$(\phi^*) = H_4^*$, then we have respectively

$$a_0 = \phi^*(0),\ a_1 = \phi^*(e_1^*),\ a_2 = \phi^*(e_1^* - e_3^*),\ a_3 = \phi^*(2e_1^*),\ a_4 = \phi^*(2e_1^* - e_3^*)$$

$$a_5 = \phi^*(3e_1^*),\ a_6 = \phi^*(2e_1^* - 2e_3^*),\ a_7 = \phi(3e_1^* - e_3^*),\ a_8 = \phi^*(4e_1^*)$$

Both operators S_{ϕ^*} and $\delta_{\mathbf{a}}$ are isomorphisms of Π_d and the inverse of $\delta_{\mathbf{a}}$ in $\mathbb{L}^*$ can be formally expressed as $\delta_{\mathbf{b}} = \delta_{\mathbf{a}}^{-1}$ where $\mathbf{b} \in \mathbb{H}^*$ is an *infinite hexagonal sequence.* In order to get a *finite* hexagonal sequence, we associate with the difference operator $\bar{\delta}_{\mathbf{b}} = \delta_{\mathbf{b}}|\Pi_d$ the dQI defined by

$$\mathrm{Q}_{\phi^*} f(x) = \sum_{\beta\in V^*} \bar{\delta}_{\mathbf{b}}(f(\cdot + \beta))\phi^*(x - \beta)$$

We have proved in ([20], Theorem 3.1) that this dQI is exact on Π_d.

5.1 Examples

Example 5.1 *The hexagonal sequence* $\mathbf{a}$ *and the corresponding CdO* $\delta_{\mathbf{a}}$ *associated with the* C^2 *quartic box-spline* $M_2^\#$ *are respectively* $\mathbf{a} = [\frac{1}{2}|\frac{1}{12}]$ *and* $\delta_{\mathbf{a}} = \delta_0 + \frac{1}{12}\delta_1$. *Therefore* $\bar{\delta}_{\mathbf{b}} = \delta_0 - \frac{1}{12}\delta_1$ *with* $\mathbf{b} = [\frac{3}{2}|\frac{1}{12}]$ *and the associated dQI defined by*

$$\mathrm{Q}_2^* f(x) = \sum_{\beta\in V^*} [f - \frac{1}{12}\delta_1 f](\beta) M_2^\#(x - \beta)$$

is exact on Π_3. *The approximation order for a smooth function* f *is* $O(h^4)$ *on a triangulation with meshsize* h. *Moreover, the uniform norm of* Q_2^* *is obviously* *bounded by the sum of absolute values of coefficients of* **b** *(see Section* 6 *below). Since we have*

$$[f - \frac{1}{12}\delta_1 f](\beta) = \frac{3}{2}f(\beta) - \frac{1}{12}\sum_{s=1}^{3}[f(\beta + e_s) + f(\beta - e_s)]$$

we deduce that $\|\mathrm{Q}_2^*\|_\infty \leq 2$. *The exact value of this norm is equal to the Chebyshev norm of the standard Lebesgue function associated with the operator. Of course, it is smaller than* 2.

Example 5.2 *The hexagonal sequence* **a** *and the CdO* $\delta_{\mathbf{a}}$ *associated with the* C^1 *cubic* H_2*-spline* $\Lambda_2^\#$ *are respectively* $\mathbf{a} = [\frac{1}{3}|\frac{1}{9}]$ *and* $\delta_{\mathbf{a}} = \delta_0 + \frac{1}{9}\delta_1$. *Therefore* $\bar{\delta}_{\mathbf{b}} = \delta_0 - \frac{1}{9}\delta_1$ *with* $\mathbf{b} = [\frac{5}{3}|\frac{1}{9}]$ *and the associated dQI defined by*

$$\tilde{\mathrm{Q}}_2^* f(x) = \sum_{\beta \in V^*} [f - \frac{1}{9}\delta_1 f](\beta)\Lambda_2^\#(x - \beta)$$

is exact on Π_2. *The approximation order for a smooth function* f *is* $O(h^3)$ *on a triangulation with meshsize* h. *As in the previous case, the uniform norm of* $\tilde{\mathrm{Q}}_2^*$ *is bounded by the sum of absolute values of coefficients of* **b**. *Since we have*

$$[f - \frac{1}{9}\delta_1 f](\beta) = \frac{5}{3}f(\beta) - \frac{1}{9}\sum_{s=1}^{3}[f(\beta + e_s) + f(\beta - e_s)]$$

we deduce that $\|\tilde{\mathrm{Q}}_2^*\|_\infty \leq \frac{7}{3} \approx 2.33$ *For the computation of dQIs associated with* H^**-splines of higher degrees, we need to define bases of hexagonal sequences (or of corresponding CdOs).*

5.2 Some bases in $\mathbb{H}^*$ and $\mathbb{L}^*$

Let us consider the three pairs (δ_1, δ_2), (δ_1, δ_3), and (δ_1, δ_4) of CdOs defined at the beginning of this section. Consider the three families $\mathcal{B}_s$ of CdOs consisting of all powers $\delta_1^p \delta_s^q$, for positive integers p and q and for $s \in \{2, 3, 4\}$ fixed (here, the product means, of course, the composition of CdOs). We have proved in [20] and [21] that these families are bases of $\mathbb{L}^*$, i.e., that each CdO can be expressed as a linear combination of elements of any of the three bases. Therefore, we obtain three expansions for each CdO.

Example 5.3 *Box-spline of degree* 7 *and class* C^4

The hexagonal sequence **a** *and the three CdOs* ${\delta_{\mathbf{a}}}^{(s)}, s = 2, 3, 4$ *associated with the* C^4 *box-spline* $M_3^\#$ *of degree* 7 *with support* H_3^*, *are respectively (see e.g.* [19], [21]*).*

$$\mathbf{a} = \frac{1}{840}[288|86|5, 1], \quad \delta_{\mathbf{a}}^{(2)} = \delta_0 + \frac{37}{280}\delta_1 + \frac{5}{1680}\delta_1^2 - \frac{1}{560}\delta_2$$

$$\delta_{\mathbf{a}}^{(3)} = \delta_0 + \frac{41}{280}\delta_1 + \frac{1}{840}\delta_1^2 - \frac{1}{280}\delta_3, \quad \delta_{\mathbf{a}}^{(4)} = \delta_0 + \frac{33}{280}\delta_1 + \frac{1}{840}\delta_1^2 + \frac{1}{280}\delta_4$$

w.r.t. the three bases $\mathcal{B}_2$, $\mathcal{B}_3$ and $\mathcal{B}_4$. The corresponding truncated formal inverses on Π_5 and the associated hexagonal sequences are the following

$$\bar{\delta}_{\mathbf{b}}^{(2)} = \delta_0 - \frac{37}{280}\delta_1 + \frac{3407}{235200}\delta_1^2 - \frac{1}{560}\delta_2 - \frac{37}{78400}\delta_1\delta_2 + \frac{1}{313600}\delta_2^2$$

$$\mathbf{b}^{(2)} = [\frac{372193}{156800}| - \frac{12919}{47040}|\frac{103}{3675}, \frac{8971}{470400}| - \frac{37}{78400}, -\frac{37}{78400}|\frac{1}{156800}, 0, \frac{1}{313600}]$$

$$\bar{\delta}_{\mathbf{b}}^{(3)} = \delta_0 - \frac{41}{280}\delta_1 + \frac{4763}{235200}\delta_1^2 + \frac{1}{280}\delta_3 - \frac{41}{39200}\delta_1\delta_3 + \frac{1}{78400}\delta_3^2$$

$$\mathbf{b}^{(3)} = [\frac{87637}{39200}| - \frac{3473}{14700}|\frac{1781}{117600}, \frac{397}{29400}|\frac{1}{1225}, \frac{1}{39200}|\frac{1}{78400}, 0, 0]$$

$$\bar{\delta}_{\mathbf{b}}^{(4)} = \delta_0 - \frac{33}{280}\delta_1 + \frac{2987}{235200}\delta_1^2 - \frac{1}{280}\delta_4$$

$$\mathbf{b}^{(4)} = [\frac{12307}{5600}| - \frac{5507}{23520}|\frac{2567}{117600}, \frac{2987}{235200}]$$

For $s = 2, 3, 4$, the associated dQI are

$$\mathsf{Q}_3^{*(s)} f(x) = \sum_{\beta \in V^*} \bar{\delta}_{\mathbf{b}}^{(s)} f(\cdot + \beta) M_3^{\#}(x - \beta)$$

and we obtain the following bounds for their uniform norms

$$\|\mathsf{Q}_3^{*(2)}\|_\infty \le \frac{5283}{1225} \approx 4.31, \quad \|\mathsf{Q}_3^{*(3)}\|_\infty \le \frac{4698}{1225} \approx 3.83, \quad \|\mathsf{Q}_3^{*(4)}\|_\infty \le \frac{7467}{1960} \approx 3.81$$

Other examples can be found in [21], *in particular on the 4-directional mesh.*

Remark 5.4 *In* [22], *we proposed a first method of solving the problem of dQI with minimal infinite norm. It consisted in trying to determine good bases of the algebras of hexagonal sequences or of associated CdOs in order to get small rational coefficients in formal inverses and consequently small norms for the corresponding dQI. In the next section, we propose a second method which seems more effective.*

6 Near-best discrete quasi-interpolants

Given a dQI

$$\mathsf{Q}_{\phi^*} f(x) = \sum_{\alpha \in V^*} \{ \sum_{\beta \in W^*} b_\beta f(\alpha + \beta) \} \phi^*(x - \beta)$$

where $\phi^* = M_r^{\#}$ or $\Lambda_r^{\#}$ and $W^* = \mathrm{supp}(\phi^*) \cap V^*$, we define respectively the fundamental function and the Lebesgue function

$$L^*(x) = \sum_{\beta \in W^*} b_\beta \phi^*(x - \beta), \quad \lambda^*(x) = \sum_{\alpha \in V^*} |\phi^*(x - \alpha)|$$

Since the dQI can be written in the form

$$\sum_{\alpha \in V^*} f(\alpha) L^*(x-\alpha)$$

its infinite (or uniform) norm is equal to the Chebyshev norm of λ^*, hence it is bounded by the l^1-norm of the hexagonal sequence $\mathbf{b}$

$$\|Q_{\phi^*}\|_\infty = |\lambda^*|_\infty \leq \nu_1(\mathbf{b}) = \sum_{\beta \in W^*} |b_\beta|$$

In all examples studied previously, the sequence $\mathbf{b}$ was defined as a truncated formal inverse of $\mathbf{a}$ (resp. $\delta_{\mathbf{a}}$) for the convolution product (resp. the composition product). We had almost no flexibility in the choice of its coefficients, except the possibility of using different bases of $\mathbb{L}^*$, therefore the values of $\nu_1(\mathbf{b})$ were not in general minimal.

Here we change our strategy: *we choose a priori a sequence* $\mathbf{b}$, *with a larger hexagonal support*, and *we minimize* $\nu_1(\mathbf{b})$ *under the linear constraints consisting of reproducing all monomials in* Π_d (in fact, we have also the possibility of reproducing monomials of lower degrees). This problem is well known to be equivalent to a linear programming one (see e.g. [13], Section 4.2.3, l^1 problem).

Example 6.1 C^1 *cubic* H_2*-splines*

In Section 5.1 *(Example 2), we got the sequence* $\mathbf{b} = [\frac{5}{3}|\frac{1}{9}]$ *with support* H_1 *satisfying* $\nu_1(\mathbf{b}) = \frac{7}{3} \approx 2.33$. *Choosing an hexagonal sequence* $\mathbf{b}' \in \mathbf{H}_2^*$, *i.e., of type* $[b_0|b_1|b_2, b_3]$, *we obtain an optimal solution* $\mathbf{b}' = [\frac{7}{6}|0|0, -\frac{1}{36}]$ *with a much smaller norm* $\nu_1(\mathbf{b}') = \frac{4}{3} \approx 1.33$. *Of course, the associated dQI is also exact on* Π_2.

Example 6.2 C^2 H_2*-splines=quartic box-splines*

In Section 5.1 *(example 1), we got the sequence* $\mathbf{b} = [\frac{3}{2}|\frac{1}{12}]$ *with support* H_1 *satisfying* $\nu_1(\mathbf{b}) = 2$. *Choosing an hexagonal sequence* $\mathbf{b}' \in \mathbf{H}_2^*$, *we obtain an optimal solution* $\mathbf{b}' = [\frac{9}{8}|0|0, -\frac{1}{48}]$ *with a much smaller norm* $\nu_1(\mathbf{b}') = \frac{5}{4} = 1.25$. *Of course, the associated dQI is also exact on* Π_3.

It is interesting to observe that the nonzero coefficients of the two new sequences $\mathbf{b}'$ are at the center and at the vertices of the hexagon H_2. This situation also occurs for other types of H-splines with higher degrees and smoothness orders.

Example 6.3 C^4 H_3*-splines=box-splines of degree* 7

The sequence $\mathbf{b}' = [\frac{123}{80}|0| - \frac{7}{30}, -\frac{23}{160}]$ *is one of the optimal solutions in* $\mathbf{H}_2^*$ *(there is an infinite set of solutions in that case) and* $\nu_1(\mathbf{b}') = \frac{19}{5} = 3.8$. *It is simpler and slightly better than the solution* $\mathbf{b}^{(4)}$ *given in the example of Section* 5.2.

A systematic study of these problems is still under investigation. Some partial results will be given at the congress *Curves and Surfaces* in Saint-Malo (June 26-July 4, 2002).

References

[1] B.D. Bojanov, H.A. Hakopian, A.A. Sahakian, *Spline functions and multivariate interpolation*, Kluwer, Dordrecht, 1993.

[2] C. de Boor, K. Höllig, *Bivariate box-splines and smooth pp functions on a three direction mesh*, J. Comput. Appl. Math. **9** (1983), 13–28.

[3] C. de Boor, K. Höllig, S. Riemenschneider, *Box-splines*, Springer, New-York, 1993.

[4] C. de Boor, *Quasiinterpolants and approximation power of multivariate splines* in *Computation of curves and surfaces*, W. Dahmen, M. Gasca, C.A. Micchelli (eds), Kluwer, Dordrecht (1990), 313–345.

[5] P.L. Butzer, M. Schmidt, E.L. Stark, L. Voigt, *Central factorial numbers; their main properties and some applications*, Numer. Funct.Anal. and Optimiz. **10**, 5&6 (1989), 419–488.

[6] C.K. Chui, *Multivariate splines*, CBMS-NSF Regional Conference Series in Appl. Math., Vol. 54, SIAM, Philadelphia, 1988.

[7] C.K. Chui, M.J. Lai, *Algorithms for generating B-nets and graphically displaying spline surfaces on three and four directional meshes*, Comp. Aided Geom. Design **8** (1991), 479–493.

[8] L. Collatz, *The numerical treatment of differential equations*, Springer, Berlin, 1966.

[9] W. Dahmen, C.K. Micchelli, *Translates of multivariate splines*, Linear Algebra Appl. **52/53** (1983), 217–234.

[10] W. Dahmen, C.K. Micchelli, *Recent progress in multivariate splines*, in *Approximation Theory IV*, C.K. Chui, L.L. Schumaker, J.D. Ward (eds), Academic Press (1983), 27–121.

[11] P.O. Frederickson, *Triangular spline interpolation*, Report 6-70, Lakehead University, 1970.

[12] P.O. Frederickson, *Generalized triangular splines* , Report 7-71, Lakehead University, 1971.

[13] P.E. Gill, W. Murray, M.H. Wright, *Practical optimization*, Academic Press, London, 1997.

[14] F. di Guglielmo, *Construction d'approximations des espaces de Sobolev sur des réseaux en simplexes*, Calcolo **6** (1969), 279–331.

[15] F. di Guglielmo, *Méthode des éléments finis: une famille d'approximations des espaces de Sobolev par les translatées de p fonctions*, Calcolo **7** (1970), 185–233.

[16] M.J. Lai, *Fortran subroutines for B-nets of box-splines on three- and four-directional meshes*, Numer. Algorithms **2** (1992), 33–38.

[17] M.A. Sabin, *The use of piecewise forms for the numerical representation of shape*, Thesis, Budapest, 1977.

[18] P. Sablonnière, *Bases de Bernstein et approximants splines*, Thèse de Doctorat, Université de Lille, 1982.

[19] P. Sablonnière, *A catalog of B-splines of degree* ≤ 10 *on a three direction mesh*, Publication ANO 132, Université de Lille, 1984.

[20] P. Sablonnière, *Bernstein-Bézier methods for the construction of bivariate spline approximants*, Comput. Aided Geom. Design **2** (1985), 29–36.

[21] P. Sablonnière, *Quasi-interpolants associated with H-splines on a three direction mesh* , J. Comput. Appl. Math. **66** (1996), 433–442.

[22] P. Sablonnière, *On some families of B-splines on the uniform four-directional mesh of the plane*, Conference on multivariate approximation and interpolation with applications in CAGD, signal and image processing. Eilat, Israel, September 7-11, 1998 (unpublished).

[23] P. Sablonnière, *Quasi-interpolantes splines sobre particiones uniformes*, Meeting on Approximation Theory, Ubeda, Spain, July 2000. Prépublication 00-38, Institut de Recherche Mathématique de Rennes.

Centre de Mathématiques
INSA de Rennes
20, Avenue des Buttes de Coësmes
35043-Rennes cedex, France
Email address: Paul.Sablonniere@insa-rennes.fr

Advanced Problems in Constructive Approximation
(Eds.) M.D. Buhmann and D.H. Mache
International Series of Numerical Mathematics Vol. 142, 203–221

Approximation by Positive Definite Kernels

Robert Schaback and Holger Wendland

Abstract

This contribution extends earlier work [16] on interpolation/approximation by positive definite basis functions in several aspects. First, it works out the relations between various types of kernels in more detail and more generality. Second, it uses the new generality to exhibit the first example of a discontinuous positive definite function. Third, it establishes the first link from (radial) basis function theory to n-widths, and finally it uses this link to prove quasi-optimality results for approximation rates of interpolation processes and decay rates for eigenvalues of integral operators having smooth kernels.

1 Kernel functions

Let Ω be a domain in $\mathbb{R}^d$. We want to work with large-dimensional data-dependent spaces of functions on Ω. A simple way to do this is to consider functions of the form

$$s_{\alpha,X} := \sum_{j=1}^{M} \alpha_j \Phi(x_j, \cdot) \tag{1}$$

for "data sets" $X = \{x_1, \ldots, x_M\} \subseteq \Omega \subseteq \mathbb{R}^d$, coefficients $\alpha \in \mathbb{R}^M$ and a "kernel" function

$$\Phi : \Omega \times \Omega \to \mathbb{R}, \quad \Omega \subseteq \mathbb{R}^d. \tag{2}$$

We start with a short review of the basic features of kernels, but we do not follow the standard path. The functions (1) form a finite-dimensional space

$$S_{X,\Phi} := \text{ span } \{\Phi(x, \cdot) \ : \ x \in X\} \tag{3}$$

of dimension at most M. These spaces are the ones we want to work with. The union of these spaces is

$$\mathcal{S}_\Phi := \text{span}\ \{\Phi(x,\cdot)\ :\ x \in \Omega\}. \tag{4}$$

Of course, everything is useless if $\Phi = 0$, and at least one would like to have linear independence of functions $\Phi(x,\cdot)$ for $x \in \Omega$.

Definition 1.1 *We call a kernel* (2) nondegenerate *on* Ω, *if for all finite data sets* $X = \{x_1, \ldots, x_M\}$ *the functions* $\Phi(x_j,\cdot)$, $x_j \in X = \{x_1, \ldots, x_M\}$ *are linearly independent over* Ω.

There are plenty of such kernels, e.g. $\Phi(x,y) := \exp(x^T y)$, $x, y \in \mathbb{R}^d$ is nondegenerate on every subset of $\mathbb{R}^d$ that contains at least an interior point.

But we also want to have a norm structure on the space (4). The simplest axiomatic way to do this is to use the kernel itself:

Definition 1.2 *A function* (2) *on* $\Omega \subseteq \mathbb{R}^d$ *that generates an inner product of the form*

$$(\Phi(x,\cdot), \Phi(y,\cdot))_\Phi = \Phi(x,y) \text{ for all } x, y \in \Omega \tag{5}$$

on the space $\mathcal{S}_\Phi$ *will be called a* reproducing kernel *on* Ω.

Clearly, a reproducing kernel has simple properties like

$$\begin{array}{rcll} \Phi(x,x) & \geq & 0 & \text{for all } x \in \Omega, \\ \Phi(x,y) & = & \Phi(y,x) & \text{for all } x, y \in \Omega, \\ \Phi(x,y)^2 & \leq & \Phi(x,x)\Phi(y,y) & \text{for all } x, y \in \Omega, \end{array}$$

but we just note them in passing. Equation (5) turns $\mathcal{S}_\Phi$ into a pre-Hilbert space, and it allows to write

$$(f(\cdot), \Phi(y,\cdot))_\Phi := f(y) \text{ for all } y \in \Omega, f \in \mathcal{S}_\Phi,$$

because the equation holds for all functions $f_x(y) := \Phi(x,y)$ and thus on the whole space $\mathcal{S}_\Phi$.

The formal closure $\mathcal{N}_\Phi$ of $\mathcal{S}_\Phi$ under the inner product $(.,.)_\Phi$ will be a Hilbert space, and an abstract element f of $\mathcal{N}_\Phi$ can be interpreted as a function on Ω by

$$(f, \Phi(y,\cdot))_\Phi =: f(y) \text{ for all } y \in \Omega, f \in \mathcal{N}_\Phi, \tag{6}$$

because the left-hand side makes sense on the closure. Equation (6) is the reason why a kernel Φ is usually called reproducing with respect to a specific Hilbert space of functions: it allows to recover the function values of an element f of the Hilbert space by (6). Standard sources for results on reproducing kernel Hilbert spaces are [1, 8], while results on native spaces are compiled in [7, 15].

Definition 1.3 *If Φ is a reproducing kernel on $\Omega \subseteq \mathbb{R}^d$, we call the space*

$$\mathcal{N}_\Phi := \text{clos}_{(.,.)_\Phi} \mathcal{S}_\Phi := \text{clos}_{(.,.)_\Phi} \text{ span } \{\Phi(x,\cdot) \, : \, x \in \Omega\}$$

the native space *for Φ.*

If the kernel is just reproducing and possibly degenerate or even zero, we cannot get a rich native Hilbert space. But in many situations we have both properties, and then we get another useful notion:

Theorem 1.4 *If a kernel* (2) *is reproducing and nondegenerate on $\Omega \subseteq \mathbb{R}^d$, it is* (strictly) positive definite *there. This means that for all finite data sets $X = \{x_1, \dots, x_M\}$ the matrices*

$$A_{X,\Phi} := (\Phi(x_j, x_k))_{1 \le j,k \le M}$$

are symmetric and positive definite. The converse is also true: a positive definite kernel is nondegenerate and reproducing.

Proof. If we have a reproducing kernel Φ, the matrices $A_{X,\Phi}$ are Gramians and thus positive semidefinite. If the kernel is nondegenerate, the matrices must be positive definite, because Gramians of linearly independent functions are positive definite.

For the converse, we start with a positive definite kernel Φ and consider functions $s_{\alpha,X}$ of the form (1). We have

$$s_{\alpha,X}(x_k) = \alpha^T A_{X,\Phi} e_k, \; 1 \le k \le M$$

using the k-th unit vector $e_k \in \mathbb{R}^M$. By positive definiteness we can conclude that such a function can only vanish on X if the coefficients are zero. This proves that the kernel is nondegenerate, and it implies that finitely generated functions $s_{\alpha,X}$ from (4) are uniquely determined by α and X. Thus we can define a bilinear form by (5) on the functions $\Phi(x,\cdot)$ that generate $\mathcal{S}_\Phi$ and use (1) again to write

$$\begin{aligned}\alpha^T A_{X,\Phi} \alpha = \sum_{j,k=1}^M \alpha_j \alpha_k \Phi(x_j, x_k) &= \left(\sum_{j=1}^M \alpha_j \Phi(x_j,\cdot), \sum_{k=1}^M \alpha_k \Phi(x_k,\cdot) \right)_\Phi \\ &= \|s_{\alpha,X}\|_\Phi^2 \ge 0\end{aligned}$$

for all $\alpha \in \mathbb{R}^M, X = \{x_1, \dots, x_M\} \subseteq \Omega \subseteq \mathbb{R}^d$ to conclude the definiteness of the bilinear form. □

There are other equivalent formulations for positive definiteness of a kernel Φ on $\Omega \subseteq \mathbb{R}^d$:

Theorem 1.5 *If a kernel Φ is reproducing on $\Omega \subseteq \mathbb{R}^d$, then the following properties are equivalent:*

1. *The functions $\Phi(x_j, \cdot)$ are linearly independent on Ω for all finite data sets $X = \{x_1, \dots, x_M\} \subseteq \Omega$.*
2. *For all finite data sets $X = \{x_1, \dots, x_M\}$ the matrices $A_{X,\Phi}$ are positive definite.*
3. *All point evaluation functionals for distinct points in Ω are linearly independent in the dual of $\mathcal{N}_\Phi$.*
4. *The native space $\mathcal{N}_\Phi$ separates points of Ω, i.e., for all finite data sets $X = \{x_1, \dots, x_M\} \subseteq \Omega$ and all points $x_j \in X$ there is a function $f_j \in \mathcal{N}_\Phi$ such that $f_j(x_k) = \delta_{jk}$, $1 \le j, k \le M$.*

Proof. We already know the equivalence of properties 1 and 2. In the dual of the native space, we can use (6) to see that the Riesz representer of the point evaluation functional $\delta_x : f \mapsto f(x)$ is the function $\Phi(x, \cdot)$, and thus

$$(\delta_x, \delta_y)_{\mathcal{N}_\Phi^*} = \Phi(x, y) \text{ for all } x, y \in \Omega$$

holds in the dual of the native space. Thus the matrices $A_{X,\Phi}$ are Gramians of the point evaluation functionals δ_{x_j}, $x_j \in X$ in the dual of the native space, and linear independence of the functionals is equivalent to positive definiteness of the matrix.

If we have property 4, we can easily see that the point evaluation functionals for any finite point set are linearly independent, since for a vanishing linear combination we get

$$0 = \left(\sum_{k=1}^{M} \alpha_k \delta_{x_k} \right) f_j = \sum_{k=1}^{M} \alpha_k f_j(x_k) = \alpha_j, \ 1 \le j \le M.$$

Conversely, property 4 follows from property 2 by interpolation. We define vectors and functions

$$a_j := A_{X,\Phi}^{-1} e_j, \ f_j := s_{a_j,X}, \ 1 \le j \le M$$

via (1). This gives

$$f_j(x_k) = s_{a_j,X}(x_k) = a_j^T A_{X,\Phi} e_k = e_j^T A_{X,\Phi}^{-1} A_{X,\Phi} e_k = e_j^T e_k = \delta_{jk}, \ 1 \le j, k \le M.$$

□

For completeness, we add a standard observation that goes the other way round:

Theorem 1.6 *If $\mathcal{H}$ is a Hilbert space of functions on Ω such that all point evaluation functionals for distinct points in Ω are linearly independent in the dual of $\mathcal{H}$, then $\mathcal{H}$ is the native space of a nondegenerate reproducing kernel.*

Proof. We define Φ as the Riesz representer for the point evaluation functionals, i.e., by (6) for all $f \in \mathcal{H}$. Then we get (5) by putting $f_x(\cdot) := \Phi(x, \cdot)$ into (6), and the previous theorems yield that Φ is a positive definite reproducing kernel on Ω

with its native space $\mathcal{N}_\Phi$ being necessarily a closed subspace of $\mathcal{H}$. But we can use (6) to show that an element f of $\mathcal{H}$ which is orthogonal to all $\Phi(y, \cdot)$ must vanish on Ω, and thus the spaces $\mathcal{H}$ and $\mathcal{N}_\Phi$ coincide. □

This result shows that reproducing positive definite kernels are not exotic. They automatically arise for any Hilbert space of functions where point evaluation is a continuous and nondegenerate operation.

We list a series of important special forms of kernels: $\Phi(x, y) =$

Radial Basis Functions	$\phi(\|x-y\|_2)$	$\forall x, y \in \mathbb{R}^d$
Translation-invariant Kernels on $\mathbb{R}^d$	$\Psi(x-y)$	$\forall x, y \in \mathbb{R}^d$
Zonal Kernels on Spheres	$\phi(x^T y)$	$\forall x, y \in S^{d-1}$
Periodic Kernels on Tori	$\Psi(x-y)$	$\forall x, y \in [0, 2\pi]^d$
Convolution Kernels	$\int_\Sigma \Psi(x, s)\Psi(y, s)d\mu(s)$	$\forall x, y \in \Omega$
	$\Psi : \Omega \times \Sigma \to \mathbb{R}$	
Hilbert-Schmidt Kernels	$\sum_{i \in I} \lambda_i \varphi_i(x) \varphi_i(y)$	$\forall x, y \in \Omega$
	$\varphi_i : \Omega \to \mathbb{R},\ \lambda_i > 0$	$\forall i \in I$

This paper focuses on Hilbert-Schmidt kernels, because it turns out that they are quite general, though they look rather special. This will be topic of the next section. But we should add some remarks on the other cases. Translation-invariant kernels occur as reproducing kernels of translation-invariant Hilbert spaces of functions on $\mathbb{R}^d$. They allow Fourier transform methods and are positive definite in $\mathbb{R}^d$, if their Fourier transform exists and is positive almost everywhere. Radial basis functions additionally have rotational symmetry. By replacing Fourier transforms by other transforms, one can deal with the other cases. Zonal kernels $\phi(x^T y)$ are positive definite, if their symmetrized spherical transform, i.e., their expansion into Legendre polynomials as functions of the cosine of the angle θ between x and y has positive coefficients. For periodic kernels on tori, one simply uses positivity of the coefficients of the Fourier series representation.

These observations immediately show that many kernels have series expansions with positive coefficients, and thus they come close to the Hilbert-Schmidt kernel form that we want to study in the next section.

2 Hilbert-Schmidt kernels

Before we delve into the standard way of looking at those kernels, i.e., by introducing an integral operator in $L_2(\Omega)$, we want to focus on a somewhat more abstract view that does not require a link to embeddings into L_2 spaces.

Definition 2.1 *For each index i from a countable index set I let there be a positive weight λ_i and a function $\varphi_i : \Omega \to \mathbb{R}$ such that for all $x \in \Omega$ the condition*

$$\sum_{i \in I} \lambda_i \varphi_i^2(x) < \infty \tag{7}$$

is satisfied and such that any finite subset of the φ_i is linearly independent over Ω. Then the function

$$\Phi(x,y) = \sum_{i \in I} \lambda_i \varphi_i(x) \varphi_i(y) \ : \ \Omega \times \Omega \to \mathbb{R} \tag{8}$$

is called a Hilbert-Schmidt kernel.

Theorem 2.2 *Any Hilbert-Schmidt kernel Φ is a reproducing kernel on the native space*

$$\mathcal{N}_\Phi := \left\{ \sum_{i \in I} c_i \varphi_i \ : \ c_i \in \mathbb{R}, \ \sum_{i \in I} \frac{c_i^2}{\lambda_i} < \infty \right\}. \tag{9}$$

Proof. Note first that our summability Condition (7) implies that the kernel series is summable. Furthermore, the functions in $\mathcal{N}_\Phi$ are well defined because of

$$\sum_{i \in I} |c_i \varphi_i(x)| = \sum_{i \in I} \frac{|c_i|}{\sqrt{\lambda_i}} \sqrt{\lambda_i} |\varphi_i(x)|| \le \sqrt{\sum_{i \in I} \frac{c_i^2}{\lambda_i}} \sqrt{\sum_{i \in I} \lambda_i \varphi_i^2(x)}.$$

By our assumption on linear independence, all finite linear combinations of the φ_i have unique coefficients, and we can define the inner product

$$\begin{aligned} (\varphi_i, \varphi_j)_\Phi &:= \frac{\delta_{ij}}{\lambda_i} \\ \left(\sum_{i \in I} c_i \varphi_i, \sum_{j \in I} d_j \varphi_j \right)_\Phi &:= \sum_{i \in I} \frac{c_i d_i}{\lambda_i} \end{aligned}$$

on these functions. We get a pre-Hilbert space whose closure is $\mathcal{N}_\Phi$. By easy calculations, all $\Phi(x, \cdot)$ are in $\mathcal{N}_\Phi$ and both (6) and (5) hold. □

Unfortunately, the linear independence assumptions of Definitions 1.1 and 2.1 differ, and we cannot conclude that a Hilbert-Schmidt kernel is nondegenerate in general. For example, if all φ_i have a common zero, the nondegeneracy fails.

Theorem 2.3 *If the space of all finite linear combinations of the generating functions φ_i of a Hilbert-Schmidt kernel Φ of the form (8) separates points of Ω in the sense of assertion 4 of Theorem 1.5, the kernel is nondegenerate.*

Proof. Assume there is a vanishing linear combination $s_{\alpha,X}$ for some

$$X = \{x_1, \ldots, x_M\} \subseteq \Omega.$$

Then

$$0 = \|s_{\alpha,X}\|_\Phi^2 = \sum_{j,k=1}^M \alpha_j \alpha_k \Phi(x_j, x_k) = \sum_{i \in I} \lambda_i \left(\sum_{j=1}^M \alpha_j \varphi_i(x_j) \right)^2$$

implies that all sums $\sum_{j=1}^M \alpha_j \varphi_i(x_j)$ are zero. Taking linear combinations with the coefficients of point-separating functions, we can conclude that α vanishes. □

We now know that under mild assumptions all Hilbert-Schmidt kernels are positive definite reproducing kernels of some Hilbert space. We now assert the converse, but we need some tool to proceed from a fairly general kernel Φ, e.g. a radial basis function on $\mathbb{R}^d$, to certain functions φ_i and positive weights λ_i that allow to rewrite Φ in the form (8). This will be done by going back to the origin of Hilbert-Schmidt theory, i.e., eigenfunction expansions of kernels of compact integral operators.

Definition 2.4 *Let* $\Phi : \Omega \times \Omega \to \mathbb{R}$ *be a kernel. If the integral operator*

$$\mathcal{I}_\Phi(f) := \int_\Omega f(t)\Phi(t, \cdot)dt \tag{10}$$

maps $L_2(\Omega)$ *into itself and is* **c***ompact,* **i***njective,* **p***ositive, and* **s***elfadjoint, we say that* Φ *is a* CIPS *kernel on* $L_2(\Omega)$.

Theorem 2.5 *Any* **CIPS** *kernel on* $L_2(\Omega)$ *has an absolutely and uniformly convergent representation* (8) *with* $I := \mathbb{N}$*and*

$$\lambda_1 \geq \lambda_2 \geq \cdots > 0 \text{ and } \lambda_i \to 0 \text{ for } i \to \infty$$

and a complete orthonormal system $\{\varphi_i\}_{i \in \mathbb{N}}$ *in* $L_2(\Omega)$ *of eigenfunctions, i.e.,*

$$\mathcal{I}_\Phi(\varphi_i) = \lambda_i \varphi_i \text{ for all } i \in \mathbb{N}.$$

Proof. The existence of the eigenfunctions and the series representation is a consequence of standard ([12]) spectral theory of selfadjoint compact operators on $L_2(\Omega)$. Uniform convergence of the series follows from the theorem of Mercer, and we get (7). □

Definition 2.6 *A Hilbert-Schmidt kernel on* Ω *that has the properties asserted in Theorem* 2.5 *will be called a* positive *Hilbert-Schmidt kernel (***PHS***) on* $L_2(\Omega)$.

Note that positivity and injectivity of the integral operator means that

$$(f, g)_{\mathcal{I}_\Phi} := (\mathcal{I}_\Phi(f), g)_2 = (f, \mathcal{I}_\Phi(g))_2 \text{ for all } f, g \in L_2(\Omega)$$

is an inner product on $L_2(\Omega)$. The notion of positive definiteness of a kernel is different, and it does not seem easy to connect these properties. We further note that for PHS kernels we also have

$$(f, \mathcal{I}_\Phi(g))_\Phi = (f, g)_2 \text{ for all } f \in \mathcal{N}_\Phi,\ g \in L_2(\Omega), \tag{11}$$

and the native space (9) is embedded into $L_2(\Omega)$ as

$$\mathcal{N}_\Phi = \left\{ f \in L_2(\Omega) : \sum_{i \in \mathbb{N}} \frac{(f, \varphi_i)_2^2}{\lambda_i} < \infty \right\} \tag{12}$$

with the inner product taking the form

$$(f, g)_\Phi = \sum_{i \in \mathbb{N}} \frac{(f, \varphi_i)_2 (g, \varphi_i)_2}{\lambda_i} \text{ for all } f, g \in \mathcal{N}_\Phi. \tag{13}$$

Theorem 2.7 *The following are equivalent:*

1. *The kernel Φ is* **PHS** *in $L_2(\Omega)$.*
2. *The kernel Φ is reproducing on Ω with the above native space $\mathcal{N}_\Phi \subseteq L_2(\Omega)$ and a complete L_2-orthonormal system of functions φ_i such that (7) holds.*
3. *The kernel Φ is a* **CIPS** *kernel on $L_2(\Omega)$.*

Proof sketch. The implication $3 \Rightarrow 1$ is Theorem 2.5, while the implication $1 \Rightarrow 2$ follows from Theorem 2.2. If 2 holds, the integral operator is the limit of integral operators whose kernels are the finite partial sums of Φ, and thus is compact. Injectivity and positivity follow easily, because all λ_i are positive. □

Theorem 2.8 *If Φ is a reproducing kernel on Ω such that*

$$\begin{aligned} \int_\Omega \Phi(y, y) dy &< \infty \\ \int_\Omega \int_\Omega \Phi(x, y)^2 dx dy &< \infty \\ \int_\Omega \Phi(x, y) f(y) dy &= 0 \text{ for all } x \in \Omega \text{ implies } f = 0 \text{ in } L_2(\Omega) \end{aligned}$$

then Φ is a **CIPS** *kernel on $L_2(\Omega)$.*

Proof sketch. The first additional hypothesis guarantees that the native space of Φ can be embedded into $L_2(\Omega)$. The second ensures compactness of the integral operator in $L_2(\Omega)$. Then spectral theory [12] allows to conclude the existence of an expansion (8) with L_2-orthogonal φ_i and rather general weights, but the reproduction property implies that the weights are nonnegative. The third additional hypothesis guarantees injectivity of the integral operator, positivity of all weights, and completeness of the system of orthogonal eigenfunctions. Details are in [16]. □

Note that injectivity of $\mathcal{I}_\Phi$ is essential here, but the nondegeneracy of the kernel and the separation property are not mentioned at all. Theorem 2.8 shows that very many kernels have a positive Hilbert-Schmidt form, and this motivates our concentration on those kernels in the remaining sections. We close this section by noting that we are still lacking useful conditions that allow to relate properties of Φ like positive definiteness or nondegeneracy to properties of $\mathcal{I}_\Phi$ like positivity or injectivity.

3 A discontinuous example

The techniques of the previous section allow to construct new kernels from expansions. These expansions may be based on a complete set of L_2-orthonormal functions, but they can also be quite general as in Definition 2.1 and Theorem 2.2. So far, all known kernels are at least continuous, but we can use the new technique to present a discontinuous case as an example. We modify an approach due to Fabien Hinault (private communication, 2000).

Let us mimic part of a Haar basis on $\mathbb{R}$ by taking scaled and shifted characteristic functions

$$H_k^j(x) := \chi_{[0,1)}\left(2^j x - k\right) = \chi_{[k2^{-j},(k+1)2^{-j})}(x) \text{ for all } k \in \mathbb{Z}, j \geq 0,\ x \in \mathbb{R}.$$

They have the properties

$$\begin{array}{rcll} H_k^j(x) & = & 1 \text{ iff } k = \lfloor 2^j x \rfloor & \text{else } = 0, \\ H_k^j(x) H_k^j(y) & = & 1 \text{ iff } k = \lfloor 2^j x \rfloor = \lfloor 2^j y \rfloor & \text{else } = 0. \end{array}$$

With a summable sequence of positive weights ρ_j, $j \geq 0$ we define

$$\begin{array}{rcl} \Phi(x,y) & := & \displaystyle\sum_{j=0}^{\infty} \rho_j \sum_{k=-\infty}^{\infty} H_k^j(x) H_k^j(y) \\ & = & \displaystyle\sum_{\substack{j=0 \\ \lfloor 2^j x \rfloor = \lfloor 2^j y \rfloor}}^{\infty} \rho_j \end{array}$$

for all $x, y \in \mathbb{R}$. Note now that $\lfloor 2^j x \rfloor = \lfloor 2^j y \rfloor$ for some $j \geq 0$ can hold only if x and y are of the same sign and do not differ by 1 or more. Moreover, the identity $\lfloor 2^j x \rfloor = \lfloor 2^j y \rfloor$ means that x and y coincide in their binary expansions in all of the pre-period digits and in the first j post-period digits. This means

$$\Phi(x,y) = \begin{cases} \displaystyle\sum_{j=0}^{m} \rho_j & x, y \text{ coincide in sign and all leading binary digits} \\ & \text{up to the } m\text{-th after the period} \\ 0 & \text{else} \end{cases}$$

and in particular

$$\Phi(x,x) = \sum_{j=0}^{\infty} \rho_j.$$

Thus the kernel is piecewise constant and has a finite evaluation scheme, if the sum over the ρ_j has a known value.

Theorem 3.1 *The kernel* Φ *is positive definite.*

Proof. In view of Theorems 2.2 and 2.3 we only have to show that the functions H_k^j separate points. Take a set $X = \{x_1, \ldots, x_M\} \subseteq \mathbb{R}$, pick an arbitrary index $s \in \{1, \ldots, M\}$ and a $j > 0$ such that

$$|x_r - x_s| > 2^{-j} \text{ for all } r \neq s,\ 1 \leq r \leq M.$$

This implies $\lfloor 2^j x_s \rfloor \neq \lfloor 2^j x_r \rfloor$ for all $r \neq s$. Then we pick $k = \lfloor 2^j x_s \rfloor$ and find that $H_k^j(x_s) = 1$ while $H_k^j(x_r) = 0$ for all $r \neq s$, and we get the separation. □

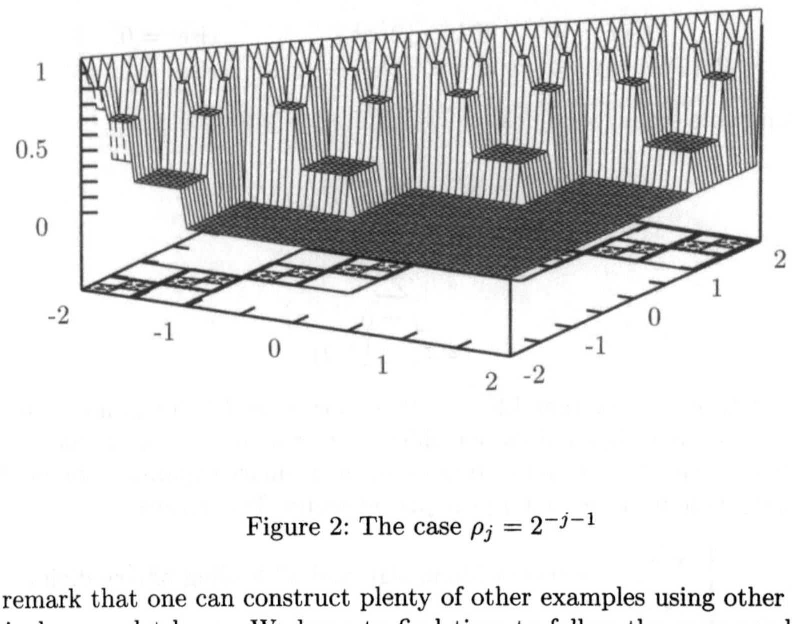

Figure 2: The case $\rho_j = 2^{-j-1}$

We remark that one can construct plenty of other examples using other bases, in particular wavelet bases. We hope to find time to follow the open road towards "refinable kernels" elsewhere.

4 Native space and range

From here on we always assume a positive definite kernel Φ that is a positive Hilbert-Schmidt kernel on $L_2(\Omega)$, and in particular we consider the native space (12) and the inner product (13) there. Note that the action of the integral operator $\mathcal{I}_\Phi$ of (10) on a function f with expansion coefficients $(f,\varphi_i)_2$ just consists of a multiplication of the coefficients by λ_i.

The range of the integral operator $\mathcal{I}_\Phi$ of (10) then is

$$\mathcal{R}_\Phi := \left\{ f \in L_2(\Omega) \ : \ \sum_{i\in\mathbb{N}} \frac{(f,\varphi_i)_2^2}{\lambda_i^2} < \infty \right\},$$

and it is the native space of the convolution kernel

$$\begin{aligned}(\Phi * \Phi)(x,y) \quad &:= \int_\Omega \Phi(x,t)\Phi(y,t)dt \\ &= \sum_{i=1}^{\infty} \lambda_i^2 \varphi_i(x)\varphi_i(y)\end{aligned}$$

Consequently we have the inclusions

$$\mathcal{R}_\Phi \quad = \quad \mathcal{N}_{\Phi*\Phi} \quad \subseteq \quad \mathcal{N}_\Phi \quad \subseteq \quad L_2(\Omega).$$

The subspace $\mathcal{R}_\Phi$ of the native space $\mathcal{N}_\Phi$ is of quite some importance. For completeness, we add a result from [16] that generalizes [14]:

Theorem 4.1 *The convergence order of interpolants to functions from $\mathcal{R}_\Phi$ is twice the convergence order of functions from the native space $\mathcal{N}_\Phi$.*

Proof. The interpolant $s_{f,X,\Phi} \in S_{X,\Phi}$ of (3) to a function f from $\mathcal{N}_\Phi$ in data locations $X = \{x_1, \ldots, x_M\}$ with *fill distance*

$$h_X := \sup_{x\in\Omega} \min_{1\le j\le M} \|x - x_j\|_2$$

has a standard [13] error bound

$$\|f - s_{f,X,\Phi}\|_2^2 \le F_\Phi(h_X)\|f - s_{f,X,\Phi}\|_{\mathcal{N}_\Phi}^2 \le F_\Phi(h_X)\|f\|_{\mathcal{N}_\Phi}^2 \tag{14}$$

for all $x \in \Omega$ with a certain function F_Φ that depends on the smoothness of Φ. We assert that for $f = \mathcal{I}_\Phi(g) \in \mathcal{R}_\Phi$ there is an improved bound

$$\|f - s_{f,X,\Phi}\|_2^2 \le F_\Phi(h_X)^2\|f\|_{\Phi*\Phi}^2 = F_\Phi(h_X)^2\|g\|_2^2.$$

To this end, we use the standard [15] orthogonality relation

$$(f - s_{f,X,\Phi}, s_{f,X,\Phi})_{\mathcal{N}_\Phi} = 0$$

and the property (11) of $\mathcal{I}_\Phi$ for $f = \mathcal{I}_\Phi(g) \in \mathcal{R}_\Phi = \mathcal{N}_{\Phi*\Phi}$ to find

$$\begin{array}{rcl} \|f - s_{f,X,\Phi}\|^2_{\mathcal{N}_\Phi} & = & (f - s_{f,X,\Phi}, f)_{\mathcal{N}_\Phi} \\ & = & (f - s_{f,X,\Phi}, \mathcal{I}_\Phi(g))_{\mathcal{N}_\Phi} \\ & = & (f - s_{f,X,\Phi}, g)_2 \\ & \le & \|f - s_{f,X,\Phi}\|_2 \|g\|_2 \\ & \le & \sqrt{F_\Phi(h_X)} \|f - s_{f,X,\Phi}\|_{\mathcal{N}_\Phi} \|g\|_2 \\ \|f - s_{f,X,\Phi}\|_{\mathcal{N}_\Phi} & = & \sqrt{F_\Phi(h_X)} \|g\|_2 \end{array}$$

and we can plug this into the standard error bound (14) to arrive at

$$\begin{array}{rcl} \|f - s_{f,X,\Phi}\|^2_2 & \le & F_\Phi(h_X) \|f - s_{f,X,\Phi}\|^2_{\mathcal{N}_\Phi} \\ & \le & F_\Phi(h_X)^2 \|g\|^2_2 \end{array}$$

with

$$\|g\|_2^2 = (g, g)_2 = (\mathcal{I}_\Phi(g), g)_{\mathcal{N}_\Phi} = (\mathcal{I}_\Phi(g), \mathcal{I}_\Phi(g))_{\Phi*\Phi} = \|f\|^2_{\Phi*\Phi}. \qquad \square$$

If we ask somewhat more than (7), i.e.,

$$\sum_{i \in I} \sqrt{\lambda_i} \varphi_i^2(x) < \infty \tag{15}$$

we can define the *convolution square-root* of Φ by the kernel

$$\sqrt{\Phi}(x, y) := \sum_{i=1}^{\infty} \sqrt{\lambda_i} \varphi_i(x) \varphi_i(y)$$

and get

$$\mathcal{R}_\Phi \;=\; \mathcal{N}_{\Phi*\Phi} \;\subseteq\; \mathcal{R}_{\sqrt{\Phi}} \;=\; \mathcal{N}_\Phi \;\subseteq\; \mathcal{N}_{\sqrt{\Phi}} \;\subseteq\; L_2(\Omega).$$

5 n-widths

From now on we let Φ be a positive Hilbert-Schmidt kernel on $L_2(\Omega)$ and assume (15) to play safe. We make use of the fact that we have integral operators related to $\sqrt{\Phi}(x, y)$ or $\Phi(x, y)$ that map $L_2(\Omega)$ into $\mathcal{N}_\Phi$ or $\mathcal{R}_\Phi$. This opens the road for applications of the theory of n-widths [11]. For the convenience of the reader, we will review that part that is of interest for us. For a subset A of a Hilbert space H, the Kolmogorov n-width is defined by

$$d_n(A; H) := \inf_{V_n} \sup_{f \in H} \inf_{s \in V_n} \|f - s\|_H.$$

Here, the outer infimum is taken over all n-dimensional subspaces V_n of H. An n-dimensional space V_n^* is called optimal if

$$E(A; V_n^*) := \sup_{f \in A} \inf_{s \in V_n^*} \|f - s\|_H = d_n(A; H).$$

In our case, the Hilbert space H will always be $H = L_2(\Omega)$ and the set A will essentially be either $\mathcal{N}_\Phi$ or $\mathcal{R}_\Phi$. Actually, to avoid problems with scaling we will take A rather to be the unit ball in that space, i.e., $A = S(\mathcal{N}_\Phi)$ or $A = S(\mathcal{R}_\Phi)$, where we used the general notation $S(H) = \{h \in H : \|h\|_2 \leq 1\}$. This perfectly fits into the theory of n-width of compact operators, where A is the image of the unit ball of the linear space H under a continuous mapping T. In our case, the mapping is given by $\mathcal{I}_{\sqrt{\Phi}}$ and $\mathcal{I}_\Phi$, respectively.

Lemma 5.1 *The unit ball of the native space $\mathcal{N}_\Phi$ is the image of the unit ball of $L_2(\Omega)$ under the operator $\mathcal{I}_{\sqrt{\Phi}}$, i.e., $S(\mathcal{N}_\Phi) = \mathcal{I}_{\sqrt{\Phi}}(S(L_2(\Omega)))$. Similarly, we have for $\mathcal{R}_\Phi$ that $S(\mathcal{R}_\Phi) = \mathcal{I}_\Phi(S(L_2(\Omega)))$.*

Proof. If $f = \mathcal{I}_{\sqrt{\Phi}}v$ with $v \in S(L_2(\Omega))$, then, by definition of the native space norm, $\|f\|_\Phi = \|v\|_2$. The same holds in the second case. □

The results of Pinkus' book [11], in particular, Corollary 2.6 of Chapter IV yield:

Theorem 5.2 *Let Φ be a positive Hilbert-Schmidt kernel on $L_2(\Omega)$ with* (15). *Then, the n-widths for the unit ball in $\mathcal{N}_\Phi$ and $\mathcal{R}_\Phi$ are given by*

$$\begin{aligned} d_n(S(\mathcal{N}_\Phi); L_2(\Omega)) &= \sqrt{\lambda_{n+1}}, \\ d_n(S(\mathcal{R}_\Phi); L_2(\Omega)) &= \lambda_{n+1}, \end{aligned}$$

respectively. In both cases, the subspace

$$V_n^* := \ span\ \{\varphi_1, \ldots, \varphi_n\}$$

is optimal. The associated optimal data functionals have the form

$$\rho_k(f) := (f, \varphi_k)_2 \text{ for all } f \in L_2(\Omega).$$

As said before, the proof can be found in Pinkus' book, but it is also not too difficult. For example, to see that V_n^* is optimal for $S(\mathcal{N}_\Phi)$ we simply use $f_n = \sum_{j=1}^n (f, \varphi_j)_2 \varphi_j \in V_n^*$ as the approximant to $f \in S(\mathcal{N}_\Phi)$ to get

$$\|f - f_n\|_2^2 = \sum_{j=n+1}^\infty (f, \varphi_j)_2^2 = \sum_{j=n+1}^\infty \lambda_j \frac{(f, \varphi_j)_2^2}{\lambda_j} \leq \lambda_{n+1} \sum_{j=n+1}^\infty \frac{(f, \varphi_j)_2^2}{\lambda_j} \leq \sqrt{\lambda_{n+1}},$$

since $\|f\|_\Phi = \sum \frac{(f, \varphi_j)_2^2}{\lambda_j} \leq 1$. □

The good news here is that we have found best rates for n-term approximation. The bad news is that for standard radial cases neither the φ_i nor the λ_i are known. Furthermore, the optimal functionals are not easily accessible numerically. Thus the next section tries to compare the optimal n-width errors with the behaviour of standard interpolation in n data locations or with simple approximation schemes.

6 Quasi-optimal processes

Here, we shall look at approximation or interpolation schemes to see whether they realize the optimal behaviour outlined in Theorem 5.2 or not. Since the eigenfunctions are not accessible in many cases, and since the inner products with eigenfunctions are not practically relevant as data functionals, we have to be satisfied with quasi-optimal subspaces instead of optimal subspaces.

Definition 6.1 *An n-dimensional subspace $V_n \subseteq H$ is called quasi-optimal for $A \subseteq H$ if there exists a constant $C > 0$, independent of n, such that*

$$E(A; V_n) \leq C d_n(A; H).$$

Since $E(A; V_n) \geq d_n(A; H)$ is always satisfied, both quantities are equivalent, which we will also denote by $E(A; V_n) \sim d_n(A; H)$.

We now look at some special cases from the literature, and we start with approximation on the sphere $S^{d-1} = \{x \in \mathbb{R}^d : \|x\|_2 = 1\}$. Here, things are generally presented upside down, i.e., one starts with a family of orthonormal functions, namely spherical harmonics and defines the kernel Φ by its expanding series so that the eigenvalues of the corresponding integral operator are the Fourier coefficients of the kernel. To be more precise, let $\{Y_{\ell,k} : 1 \leq k \leq N(d, \ell)\}$ denote the usual orthonormal basis for the space of spherical harmonics of degree ℓ (cf. [10]), where

$$N(d, 0) = 1, \quad \text{and} \quad N(d, \ell) = \frac{2\ell + d - 2}{\ell} \binom{\ell + d - 3}{\ell - 1}, \; \ell > 0.$$

Then the kernel has an expansion of the form

$$\Phi(p, q) = \sum_{\ell=0}^{\infty} \sum_{k=1}^{N(d,\ell)} a_{\ell,k} Y_{\ell,k}(p) Y_{\ell,k}(q). \tag{16}$$

For simplicity, we will assume that the kernel is radial or zonal, which is equivalent to the fact that for a fixed ℓ all coefficients $a_{\ell,k}$, $1 \leq k \leq N(d, \ell)$, are the same, i.e., $a_\ell := a_{\ell,k}$, $1 \leq k \leq N(d, \ell)$.

Under this assumption, it is actually more natural to look at the space of spherical harmonics up to order ℓ,

$$V_\ell := \text{span}\{Y_{\lambda,k} : 0 \leq \lambda \leq \ell, 1 \leq k \leq N(d, \lambda)\},$$

which is the restriction of the space of d-variate polynomials of degree at most ℓ to the sphere and has dimension $\dim V_\ell = N(d + 1, \ell)$. The n-width theory gives

Corollary 6.2 *If the coefficients $\alpha_\ell = \alpha_{\ell,k}$, $1 \le k \le N(d,\ell)$, of the kernel* (16) *form a sufficiently fast decaying, nonincreasing, and positive sequence, then*

$$d_n(S(\mathcal{N}_\Phi), L_2(S^{d-1})) = \sqrt{a_\ell},$$

for $N(d+1,\ell) \le n < N(d+1,\ell+1)$.

This is the result to which we have to compare the known estimates for interpolation by positive definite kernels. In the latter context it is usual to assume that

$$N(d,\ell)a_\ell \le C(1+\ell)^{-\alpha}$$

which is, since $N(d,\ell)$ grows like $\mathcal{O}(\ell^{d-2})$, equivalent to $a_\ell = \mathcal{O}(\ell^{-\alpha-d+2})$. The reason for looking at $N(d,\ell)a_\ell$ rather than a_ℓ is that this number appears naturally for "radial" kernels, since the addition theorem (cf. [10]) yields

$$\Phi(p,q) = \sum_{\ell=0}^{\infty} \frac{N(d,\ell)a_\ell}{\omega_{d-1}} P_\ell(p \cdot q),$$

where ω_{d-1} denotes the surface area of S^{d-1} and P_ℓ is the Legendre polynomial of degree ℓ in d dimensions, normalized by $P_\ell(1) = 1$.

In case of interpolation by positive definite kernels it is usual to measure the approximation orders in terms of the so-called fill distance, which is in this context $h_X := \sup_{x \in S^{d-1}} \min_{x_j \in X} \operatorname{dist}(x, x_j)$. Here, dist is the usual spherical distance.

The following result comes from Dyn/Narcowich/Ward [4], Jetter/Stöckler/Ward [6], and Morton/Neamtu [9].

Theorem 6.3 *Suppose Φ is a radial positive definite kernel on the sphere with $a_\ell = \mathcal{O}(\ell^{-\alpha})$, $\ell \to \infty$, with $\alpha > d$. Then, the interpolation error can be bounded by*

$$\|f - s_{f,X}\|_\infty \le C h_X^{\frac{\alpha-1}{2}} \|f\|_\Phi.$$

The L_∞-error bound leads immediately to an L_2-error bound, which we now want to compare with the results from n-width theory. To achieve this, we have to relate h_X to ℓ, since by Corollary 6.2 the n-width is rather related to ℓ than to n in this situation,

$$d_n(S(\mathcal{N}_\Phi); L_2(S^{d-1})) = \mathcal{O}(\ell^{-\frac{\alpha-d-2}{2}}).$$

This is hopeless in the general case, but the situation changes in case of quasi-uniform data sets. A set $X \subseteq S^{d-1}$ of n points is said to be quasi-uniform if $h_X^{d-1} \sim 1/n$. Since we also know that $n \sim N(d+1,\ell) \sim \ell^{d-1}$ we can conclude

$$\|f - s_{f,X}\|_2 = \mathcal{O}(\ell^{-\frac{\alpha-1}{2}}).$$

Corollary 6.4 *Interpolation of function values in quasi-uniform data locations by positive definite "radial" kernels on the sphere may fail to be quasi-optimal by order at most $\frac{d-1}{2}$ if the kernel has eigenvalues with algebraic decay.*

Our formulation of the corollary just poses an upper bound on the deviation from quasi-optimality, but we think that we actually have a quasi-optimal approximation scheme. The reason for our optimistic point of view is the following. We gained the L_2 approximation error simply by integrating the L_∞-error. In the light of the $\mathbb{R}^d$ theory, this seems to be too naive. In the $\mathbb{R}^d$ case it is, in a similar situation, possible to gain an additional $d/2$ in the order by using a localization trick, which dates back to Duchon's initial work on thin-plate splines (cf. [2, 3]). This trick should also work in the sphere setting, but so far nobody has ever tried it.

Note that in the just described situation the native space is actually the Sobolev space $H^s(S^{d-1})$ with $s = \frac{\alpha+d}{2} - 1$.

For Euclidean space $\mathbb{R}^d$ and bounded domains Ω therein, we usually do not know the orthogonal Hilbert-Schmidt expansions in $L_2(\Omega)$. Thus we cannot assess the optimality of the known error bounds. The state-of-the-art in results on optimality of rates of approximation provided by interpolation is in [17, 20]. Instead of optimality results for approximations, we here get upper bounds on the decay of the unknown eigenvalues. Curiously enough, this means that approximation theory provides results on the spectrum of integral operators.

On $\mathbb{R}^d$ we make the following assumptions:

- the kernel $\Phi(x,y) = \phi(x-y)$ is symmetric and Fourier-transformable,
- we consider interpolation by translates of ϕ on n asymptotically quasi-uniform data locations in a bounded domain $\Omega \subseteq \mathbb{R}^d$, which has a sufficiently smooth boundary.

Let us look at the case of limited smoothness (e.g. [13]) first. For

$$\hat{\phi}(\omega) \sim (1 + \|\omega\|_2)^{-d-\beta}, \ \|\omega\|_2 \to \infty, \tag{17}$$

there is an error bound

$$\|f - s_{f,X}\|_\infty \le Ch^{\beta/2}\|f\|_\Phi$$

This error bound can be improved by Duchon's localization trick as mentionened earlier (see for example [19]) to

$$\|f - s_{f,X}\|_2 \le Ch^{(\beta+d)/2}\|f\|_\Phi,$$

provided that the boundary of Ω is sufficiently smooth.

In case of quasi-uniform data, which now becomes $h^d_{X,\Omega} \sim 1/n$, the latter means in terms of n,

$$\|f - s_{f,X}\|_2 \leq Cn^{-(\beta+d)/2d}\|f\|_\Phi.$$

The error of the optimal process must be asymptotically smaller, and this implies

Theorem 6.5 *The eigenvalues of the Hilbert-Schmidt operator $\mathcal{I}_\Phi$ with kernel Φ on $L_2(\Omega)$ and Fourier transform satisfying (17) for a bounded domain $\Omega \subseteq \mathbb{R}^d$ satisfy*

$$\lambda_{n+1} \leq Cn^{-(\beta+d)/d}$$

for $n \to \infty$. □

Again, as in the case of the sphere, the native space is a Sobolev space $H^s(\Omega)$, $s = (\beta + d)/2$. For Sobolev spaces, the optimal n-widths are known (Jerome 1970 [5]):

$$d_n(S(H^s(\Omega); L_2(\Omega)) = \sqrt{\lambda_{n+1}} = \mathcal{O}(n^{-s/d}) \text{ for } n \to \infty$$

and we can compare with the interpolation error bounds for $H^s(\Omega)$ with $\Omega \subseteq \mathbb{R}^d$. They have the form (14) with $s = (\beta + d)/2 > 0$, and we get

Theorem 6.6 *Interpolation in quasi-uniform locations by translates of reproducing kernels that generate Sobolev spaces is quasi-optimal.*

Since Sobolev kernels and Wendland functions [18, 19] reproduce spaces that are norm-equivalent to Sobolev spaces, we have

Corollary 6.7 *Interpolation in asymptotically regular data locations by translates of Sobolev kernels or Wendland functions is quasi-optimal.* □

Generalizations to other radial basis functions are not known, but would be welcome.

The case of unlimited smoothness occurs for inverse multiquadrics and Gaussians, and it leads to Fourier transforms with a decay like

$$\hat{\phi}(\omega) \leq C\exp(-c\|\omega\|_2), \ \|\omega\|_2 \to \infty. \tag{18}$$

Then there is an error bound [13]

$$\|f - s_{f,n}\|_\infty \leq C\exp(-c/h)\|f\|_\Phi \leq C\exp(-cn^{1/d})\|f\|_\Phi.$$

Theorem 6.8 *For a kernel Φ with exponential decay (18) of its Fourier transform, the eigenvalues of the integral operator $\mathcal{I}_\Phi$ in $L_2(\Omega)$ for a bounded domain $\Omega \subseteq \mathbb{R}^d$ satisfy*

$$\lambda_{n+1} \leq C\exp(-cn^{1/d})$$

for $n \to \infty$. □

References

[1] M. Atteia. *Hilbertian kernels and spline functions.* North-Holland, Amsterdam, 1992. Studies in Computational Mathematics (4).

[2] J. Duchon. Interpolation des fonctions de deux variables suivant le principe de la flexion des plaques minces. *Rev. Française Automat. Informat. Rech. Opér. Anal. Numer.*, 10:5–12, 1976.

[3] J. Duchon. Sur l'erreur d'interpolation des fonctions de plusieurs variables pas les D^m-splines. *Rev. Française Automat. Informat. Rech. Opér. Anal. Numer.*, 12(4):325–334, 1978.

[4] N. Dyn, F.J. Narcowich, and J.D. Ward. Variational principles and Sobolev-type estimates for generalized interpolation on a riemannian manifold. *Constructive Approximation*, 15(2):174–208, 1999.

[5] J.W. Jerome. On n-widths in Sobolev spaces and applications to elliptic boundary value problems. *Journal of Mathematical Analysis and Applications*, 29:201–215, 1970.

[6] K. Jetter, J. Stöckler, and J.D. Ward. Error estimates for scattered data interpolation on spheres. *Mathematics of Computation*, 68:733–747, 1999.

[7] L.-T. Luh. *Characterizations of native spaces.* Dissertation, Universität Göttingen, 1998.

[8] H. Meschkowski. *Hilbertsche Räume mit Kernfunktion.* Springer, Berlin, 1962.

[9] Tanya M. Morton and M. Neamtu. Error bounds for solving pseudodifferential equations on spheres by collocation with zonal kernels. Preprint Dept. of Math. Vanderbilt University, Nashville TN 37240, 2000.

[10] C. Müller. *Spherical Harmonics.* Springer, Berlin, 1966.

[11] A. Pinkus. *n-widths in Approximation Theory.* Springer, 1985.

[12] D. Porter and D.G. Stirling. *Integral equations: a practical treatment from spectral theory to applications.* Cambridge Texts in Applied Mathematics. Cambridge University Press, 1990.

[13] R. Schaback. Error estimates and condition numbers for radial basis function interpolation. *Advances in Computational Mathematics*, 3:251–264, 1995.

[14] R. Schaback. Improved error bounds for scattered data interpolation by radial basis functions. *Mathematics of Computation*, 68:201–216, 1999.

[15] R. Schaback. Native Hilbert spaces for radial basis functions i. In M.D. Buhmann, D.H. Mache, M. Felten, and M.W. Müller, editors, *New Developments in Approximation Theory*, number 132 in International Series of Numerical Mathematics, pages 255–282. Birkhäuser Verlag, 1999.

[16] R. Schaback. A unified theory of radial basis functions. *J. Comp. Appl. Math.*, pages 165–177, 2000.

[17] R. Schaback and H. Wendland. Inverse and saturation theorems for radial basis function interpolation. to appear in Math. Comp.

[18] H. Wendland. Piecewise polynomial, positive definite and compactly supported radial functions of minimal degree. *Advances in Computational Mathematics*, 4:389–396, 1995.

[19] H. Wendland. Sobolev-type error estimates for interpolation by radial basis functions. In A. LeMéhauté, C. Rabut, and L.L. Schumaker, editors, *Surface Fitting and Multiresolution Methods*, pages 337–344. Vanderbilt University Press, Nashville, TN, 1997.

[20] H. Wendland. Optimal approximation orders in L_p for radial basis functions. *East J. Approx.*, 6:87–102, 2000.

Universität Göttingen
Institut für Numerische und Angewandte Mathematik
Lotzestr. 16–18
D-37083 Göttingen (Germany)

Advanced Problems in Constructive Approximation
(Eds.) M.D. Buhmann and D.H. Mache
International Series of Numerical Mathematics Vol. 142, 223–236

Inequalities for Polynomials With Weights Having Infinitely many Zeros on the Real Line

József Szabados

Abstract

We prove infinite-finite range, as well as Bernstein–Markov type inequalities for generalized algebraic polynomials on the real line when the weight is the product of a Freud-type weight and of another function which has infinitely many roots on the real line. This kind of investigation is an analogue of the so-called genralized Jacobi weights on finite intervals.

1 Introduction

Let $u(x) = e^{-Q(x)}$ be a Freud-type weight function on $\mathbf{R} := (-\infty, \infty)$, i.e., let $Q : \mathbf{R} \to \mathbf{R}$ be even, continuous, Q'' continuous, $Q' > 0$ in $(0, \infty)$ and

$$0 < \alpha := \inf_{x \in \mathbf{R}^+} \frac{(xQ'(x))'}{Q'(x)} \le \sup_{x \in \mathbf{R}^+} \frac{(xQ'(x))'}{Q'(x)} < \infty. \tag{1}$$

(A typical example is $Q(x) = |x|^\alpha$, $\alpha > 0$.) Further let $\mathcal{P}_\nu$ be the set of generalized algebraic polynomials

$$P(x) = \omega \prod_{j=1}^{s} |x - z_j|^{\nu_j}$$

of degree $\nu = \sum_{j=1}^{s} \nu_j$, where $\omega > 0$, z_j, $j = 1, \ldots, s$ are pairwise different complex numbers, and the real numbers $\nu_j \ge 1$, $j = 1, \ldots, s$ are the corresponding multiplicities. This is an obvious generalization of the absolute value of ordinary polynomials (where all the ν_j's are positive integers).

For such weights and generalized polynomials the infinite-finite range inequality

$$||Pu||_{L_p(\mathbf{R})} \le c||Pu||_{L_p(|x| \le a_\nu(1 - K\nu^{-2/3}))}, \qquad P \in \mathcal{P}_\nu \tag{2}$$

as well as the Markov–Bernstein inequality

$$||P'u||_{L_p(\mathbf{R})} \le cM_\nu ||Pu||_{L_p(\mathbf{R})}, \qquad M_\nu = \begin{cases} 1 & 0<\alpha<1, \\ \log \nu & \alpha = 1, \\ \frac{\nu}{a_\nu} & \alpha > 1 \end{cases}, \qquad P \in \mathcal{P}_\nu \quad (3)$$

hold for any generalized polynomial $P \in \mathcal{P}_\nu$ and $0 < p \le \infty$, where $c > 0$ depends only on the weight and on $K > 0$,[3] and a_ν is the so-called Mhaskar–Rahmanov–Saff number satisfying

$$||Pu||_{L_\infty(\mathbf{R})} = ||Pu||_{L_\infty(|x|\le a_\nu)}, \qquad \text{for all} \quad P \in \mathcal{P}_\nu \quad (4)$$

(cf. Levin–Lubinsky [3], Theorem 1.8). Inequalities (2)–(3), in a much more general form, can be found in [3], Theorem 1.9(a) and Corollary 1.16.

Our purpose here is to generalize inequalities (2)–(3) for weights which have infinitely many zeros on the real line. The motivation for such investigations is the following. On a finite interval, in case of Jacobi weights, polynomial inequalities, properties of orthogonal polynomials, estimates for the Christoffel functions, etc., were generalized for weights having *finitely many* algebraic type roots inside the interval considered (see works of Badkov, Mastroianni, Vértesi and others). A natural analogue of these investigations is when the weights on $\mathbf{R}$ have *infinitely many* zeros. A first (and most important) step in this subject can be the establishing of (2)–(3) for such weights.

So let

$$0 < t_1 < t_2 < \cdots, \quad \lim -k \to \infty t_k = \infty \quad (5)$$

and

$$m_k > 0, \ k = 1, \dots, \quad \liminf_{k\to\infty} m_k > 0 \quad (6)$$

be two sequences of real numbers. Under some growth condition on the sequences (5)–(6), we will construct a weight function on $\mathbf{R}$ whose *real* roots are exactly the numbers $\pm t_k$ with corresponding multiplicities m_k, $k = 1, 2, \dots$. In other words, we consider the case when the roots are symmetric with respect to zero. It would be easy to handle the general case, but since the formulas are too involved, we restrict ourselves to this situation. Besides, we will associate Freud-type weights to this weight which is, by definition, symmetric to zero. Our construction will be such that, besides the prescribed roots $\pm t_k$, $k = 1, 2, \dots$, the weight function will contain complex roots as well. We could have avoided this by making a more delicate construction, but the ideas we use can be seen much better if we do not care about non-real roots.

Our assumption on the sequences (5)–(6) is the following: there exists a $\varrho \ge 0$ such that

$$\sum_{k=1}^{\infty} \frac{m_k}{t_k^{\varrho+\varepsilon}} < \infty, \quad \text{but} \quad \sum_{k=1}^{\infty} \frac{m_k}{t_k^{\varrho-\varepsilon}} = \infty \qquad \text{for all} \quad \varepsilon > 0. \quad (7)$$

[3]In what follows, $c > 0$ will denote constants independent of ν and p, not necessarily the same at each occurrence.

Typical examples for such sequences are $t_k = k^\beta$, $\beta > 0$ (when $\varrho = 1/\beta$), or $t_k = 2^k$ (when $\varrho = 0$). (Here we took $m_k = 1$, $k = 1, 2, \ldots$.)

Now consider the function

$$v(x) = \prod_{k=1}^{\infty} \left| 1 - \left(\frac{x}{t_k} \right)^{2q} \right|^{m_k}, \qquad q = [\varrho/2] + 1. \tag{8}$$

Here the infinite product converges uniformly in each compact subset of **R**. This will be seen from the following

Lemma 1.1 *Suppose the conditions* (7) *hold with some* $\varrho \geq 0$ *and for all* $\varepsilon > 0$. *Then*

(a) $v(x) \leq e^{c|x|^{\varrho+\varepsilon}}$ *for all* $x \in \mathbf{R}$, $\varepsilon > 0$, *and* (b) *with* $t_0 = 0$,

(b) $v(x) \geq e^{-c|x|^{\varrho+\varepsilon}}$ *for* $|x| \in \cup_{j=1}^{\infty} I_j := \cup_{j=1}^{\infty} \left[t_{j-1} + \frac{m_{j-1}}{t_{j-1}^{\varrho+\varepsilon}}, t_j - \frac{m_j}{t_j^{\varrho+\varepsilon}} \right]$,

where in both cases $c > 0$ *depends on* v *and* ε.[4]

Proof. (a) By symmetry, we may assume that $x \geq 0$. Let

$$N(x) = \sum_{t_k \leq x} m_k. \tag{9}$$

By the first relation in (7), evidently

$$N(x) \leq c(\varepsilon) x^{\varrho+\varepsilon} \qquad \text{for all} \quad \varepsilon > 0. \tag{10}$$

Using the geometric-arithmetic means inequality as well as the first condition in (7) we obtain

$$\begin{aligned} v(x) &\leq \prod_{t_k \leq x} \left| 1 - \frac{x^{2q}}{t_k^{2q}} \right|^{m_k} \leq \left\{ \prod_{t_k \leq x} \left(\frac{2x}{t_k} \right)^{m_k} \right\}^{2q} \leq \left\{ (2x)^{N(x)} \prod_{t_k \leq x} \frac{1}{t_k^{m_k}} \right\}^{2q} \\ &\leq (2x)^{2qN(x)} \left\{ \prod_{t_k \leq x} \left(\frac{1}{t_k^{\varrho+\varepsilon}} \right)^{m_k} \right\}^{2q} \varrho + \varepsilon \\ &\leq (2x)^{2qN(x)} \left(\frac{\sum_{t_k \leq x} \frac{m_k}{t_k^{\varrho+\varepsilon}}}{N(x)} \right)^{\frac{2q}{\varrho+\varepsilon} N(x)} \leq \left(\frac{c x^{\varrho+\varepsilon}}{N(x)} \right)^{\frac{2q}{\varrho+\varepsilon} N(x)}. \end{aligned}$$

Hence, using that the function $(a/u)^u$ attains its maximum in the interval $[1, \infty)$ at $u = a/e$ we obtain

$$v(x) \leq (ce)^{\frac{2qc}{e(\varrho+\varepsilon)} x^{\varrho+\varepsilon}},$$

which proves the first statement of the lemma.

---

[4]If $a > b$, then let $[a, b] = \emptyset$

(b) First we show that $I_j \neq \emptyset$ for infinitely many j's, thus the statement is not about the empty set. Namely, if $I_j = \emptyset$, then

$$\frac{m_{j-1}}{t_{j-1}^{\varrho+\varepsilon}} + \frac{m_j}{t_j^{\varrho+\varepsilon}} > t_j - t_{j-1}.$$

If this were true for sufficiently large j's then adding these inequalities we would arrive at a contradiction with the first relation in (7).

Now we may assume that $x \geq 1$, since for $0 \leq x \leq 1$ the statement is trivial. We have

$$v(x) \geq \prod_{t_k \leq x} \left|1 - \frac{x}{t_k}\right|^{m_k} \prod_{x < t_k \leq 2x} \left|1 - \frac{x}{t_k}\right|^{m_k} \prod_{t_k > 2x} \left|1 - \frac{x^{2q}}{t_k^{2q}}\right|^{m_k} = P_1 P_2 P_3.$$

Here by (6) and (9)–(10)

$$P_1 \geq \left(\frac{t_{j-1} + \frac{m_{j-1}}{t_{j-1}^{\varrho+\varepsilon}}}{t_{j-1}} - 1 \right)^{N(x)} \geq t_{j-1}^{-(1+\varrho+\varepsilon)N(x)} \geq x^{-(1+\varrho+\varepsilon)c(\varepsilon/2)x^{\varrho+\varepsilon/2}}$$

$$\geq e^{-cx^{\varrho+\varepsilon}}, \qquad x \in I_j.$$

Next,

$$P_2 \geq \left(1 - \frac{x}{t_j}\right)^{N(2x)}, \qquad x \in I_j.$$

Here we distinguish two cases. If $t_j \geq 2x$, then

$$P_2 \geq 2^{-N(x)} \geq e^{-cx^{\varrho+\varepsilon}}, \qquad x \in I_j;$$

while if $t_j \leq 2x$ then

$$P_2 \geq \left(1 - \frac{t_j - \frac{m_j}{t_j^{\varrho+\varepsilon}}}{t_j}\right)^{N(x)} \geq t_j^{-(1+\varrho+\varepsilon)N(x)} \geq (2x)^{-(1+\varrho+\varepsilon)N(x)}, \qquad x \in I_j,$$

and here the estimate can be continued as for P_1.

Finally, using the inequality $1 - u \geq e^{-2u}$, $0 \leq u \leq 1/2$ we obtain

$$P_3 \geq e^{-2x^{2q} \sum_{t_k > 2x} \frac{m_k}{t_k^{2q}}} \geq e^{-2^{\varrho+\varepsilon+1-2q} x^{\varrho+\varepsilon} \sum_{k=1}^{\infty} \frac{m_k}{t_k^{\varrho+\varepsilon}}} \geq e^{-cx^{\varrho+\varepsilon}}, \qquad x \in I_j,$$

provided $\varepsilon \leq 2q - \varrho$. (Evidently, it suffices to prove the lower estimate for sufficiently small ε's. □

Now we return to the definition of our weight function. Lemma 1.1 tells us that in order to have a weight decaying exponentially at infinity, we have to multiply v by a proper Freud-type weight. Our weight with the infinitely many zeros (5) will be

$$w(x) = u(x)|v(x)|. \tag{11}$$

Under some condition on α, this new weight function will behave at infinity similarly to $u(x)$.

2 The infinite-finite range inequality

Theorem 2.1 *With the previous notations, if*

$$0 \leq \varrho < \alpha, \tag{12}$$

then we have

$$||P_\nu w||_{L_p(\mathbf{R})} \leq c||P_\nu w||_{L_p(|x|\leq \mu_\nu a_\nu)} \qquad \text{for all} \quad P_\nu \in \mathcal{P}_\nu, \tag{13}$$

where a_ν is the Mhaskar–Rahmanov–Saff number associated with u (see (4)), $0 < p \leq \infty$, and

$$\mu_\nu = 1 + c_1\nu^{\varrho/\alpha-1+\varepsilon} - c_2\nu^{-2/3}, \qquad \varepsilon > 0 \quad \text{arbitrary}, \tag{14}$$

with $c, c_1, c_2 > 0$ absolute constants.

Remark. (14) shows that by (12), we always have $\lim_{\nu\to\infty}\mu_\nu = 1$. In fact, a sharp inequality similar to (2) can be obtained if $\alpha > 3\varrho$. (13) shows that the presence of the weight v causes no significant change in the situation: the finite interval in (13) is more or less the same as for the Freud weight u.

Proof. We present the proof for the case $0 < p < \infty$; the case $p = \infty$ runs along parallel lines, but everything is much simpler. We have to estimate the quantity

$$\int_{|x|\geq \mu_\nu a_\nu} (P_\nu w)^p = \int_{\mu_\nu a_\nu \leq |x| \leq \lambda a_\nu} (P_\nu w)^p + \int_{|x|\geq \lambda a_\nu} (P_\nu w)^p := \mathcal{I}_1 + \mathcal{I}_2,$$

where the constant $\lambda > 1$ will be chosen later.

Estimate of $\mathcal{I}_1$. Let

$$R_\nu(x) = \prod_{t_k \leq \lambda a_\nu} \left|1 - \left(\frac{x}{t_k}\right)^{2q}\right|^{m_k}, \quad S_\nu(x) = \prod_{t_k > \lambda a_\nu} \left|1 - \left(\frac{x}{t_k}\right)^{2q}\right|^{m_k}. \tag{15}$$

Then by (10) $R_\nu \in \mathcal{P}_\tau$, where

$$\tau = 2qN(\lambda a_\nu) \leq c\nu^{\varrho/\alpha+\varepsilon}, \qquad \varepsilon > 0,$$

since

$$a_\nu \leq c\nu^{1/\alpha} \tag{16}$$

(cf. [5], (5.8)). On the other hand, evidently S_ν is monotone decreasing in the interval $[0, \lambda a_\nu]$. Thus applying (2) with $P_\nu R_\nu \in \mathcal{P}_{\nu+\nu^{\varrho/\alpha+\varepsilon}}$ instead of P_ν, and using the above mentioned monotonicity twice we obtain

$$
\begin{aligned}
\mathcal{I}_1 &= \int_{\mu_\nu a_\nu \le |x| \le \lambda a_\nu} (P_\nu R_\nu S_\nu u)^p \le S_\nu(\mu_\nu a_\nu)^p \int_{|x|\ge \mu_\nu a_\nu} (P_\nu R_\nu u)^p \\
&\le S_\nu(\mu_\nu a_\nu)^p \int_{|x|\ge a_{\nu+\nu^{\varrho/\alpha+\varepsilon}}(1-K(\nu+\nu^{\varrho/\alpha+\varepsilon})^{-2/3})} (P_\nu R_\nu u)^p \\
&\le cS_\nu(\mu_\nu a_\nu)^p \int_{|x|\le a_{\nu+\nu^{\varrho/\alpha+\varepsilon}}(1-K(\nu+\nu^{\varrho/\alpha+\varepsilon})^{-2/3})} (P_\nu R_\nu u)^p \\
&\le cS_\nu(\mu_\nu a_\nu)^p \int_{|x|\le \mu_\nu a_\nu} (P_\nu R_\nu u)^p \le c \int_{|x|\le \mu_\nu a_\nu} (P_\nu w)^p,
\end{aligned}
$$

since

$$
\begin{aligned}
a_{\nu+\nu^{\varepsilon/\alpha+\varepsilon}} &\{1 - K(\nu + \nu^{\varrho/\alpha+\varepsilon})^{-2/3}\} \\
&\le a_\nu \left(1 + c_1 \left|\frac{\nu + \nu^{\varrho/\alpha+\varepsilon}}{\nu} - 1\right|\right)(1 - K\nu^{-2/3}) \\
&\le a_\nu \left(1 + c_1 \nu^{\varrho/\alpha - 1 + \varepsilon} - c_2 \nu^{-2/3}\right) = \mu_\nu a_\nu
\end{aligned}
$$

(cf. [5], (5.9)).

Estimate of $\mathcal{I}_2$. This quantity will turn out to be exponentially smaller than $\mathcal{I}_1$, but the proof is more involved. Let

$$
\tilde{I}_k = \begin{cases} I_k, & |I_k| \ge 1/\nu, \\ \emptyset & \textit{otherwise}, \end{cases} \qquad k = 1, 2, \ldots .
$$

Let k_ν be defined by

$$
\tilde{I}_\nu = \begin{cases} t_{k_\nu} + \frac{m_{k_\nu}}{t_{k_\nu}^{\varrho+\varepsilon}}, & \mu_\nu a_\nu - t_{k_\nu} - \frac{m_{k_\nu}}{t_{k_\nu}^{\varrho+\varepsilon}} > 1/\nu, \\ \emptyset & \textit{otherwise}. \end{cases}
$$

Define $J_\nu = \cup_{I_k \subset [-\mu_\nu a_\nu, \mu_\nu a_\nu]} I_k \cup I_\nu$. Then, by Lemma, Part (b),

$$
|v(x)| \ge e^{-c|x|^{\varrho+\varepsilon}} \ge e^{-c|x|^\alpha} \ge e^{-cQ(x)}, \qquad |x| \in J_\nu,\ 0 < \varepsilon \le \alpha - \varrho, \tag{17}
$$

where in the last step we used the inequality

$$
Q(x) \ge c|x|^\alpha, \qquad x \in \mathbf{R}, \tag{18}
$$

which is easily obtained from (1) by integration. Hence by the monotonicity of $Q(x)$ in $\mathbf{R}^+$ we get

$$
A_\nu := ||P_\nu w||_{L^p(|x|\le \mu_\nu a_\nu)} \ge e^{-cQ(a_\nu)} ||P_\nu||_{L^p(J_\nu)},
$$

i.e., by

$$
Q(a_\nu) \sim \nu \tag{19}
$$

(cf. [5], relation (5.5)) we get $||P_\nu||_{L^p(J_\nu)} \leq e^{c\nu} A_\nu$. Thus using the Nikolski-type inequality

$$||P_\nu||_{L_\infty(I_k)} \leq c\left(\frac{\nu^2}{|I_k|}\right)^{1/p} ||P_\nu||_{L_p(I_k)} \leq c\nu^{3/p}||P_\nu||_{L_p(I_k)}, \qquad I_k \neq \emptyset$$

(cf. [1], Theorem A.4.4) we get

$$||P_\nu||_{L^\infty(J_\nu)} \leq \nu^{3/p} e^{c\nu} A_\nu \leq e^{c\nu} A_\nu.$$

Since by construction, for $K_\nu := [-a_\nu, a_\nu] \setminus J_\nu$ we have (see (15) and (16))

$$|K_\nu| \leq 2\sum_{k=1}^{k_\nu}\left(\frac{m_{k-1}}{t_{k-1}^{\varrho+\varepsilon}} + \frac{m_k}{t_k^{\varrho+\varepsilon}}\right) + \frac{2N(a_\nu)+2}{\nu} + \frac{2Ka_\nu}{\nu^{2/3}} \leq c\left(1+\frac{a_\nu}{\nu^{2/3}}\right),$$

the Remez inequality (cf. [1], Theorem A.4.1) applied to the interval $[-a_\nu, a_\nu]$ yields

$$||P_\nu||_{L^\infty[-a_\nu,a_\nu]} \leq e^{\nu\sqrt{\frac{|K_\nu|}{2a_\nu}}} A_\nu \leq e^{c\nu/\sqrt{a_\nu}+c\nu^{1/3}} A_\nu \leq e^{c\nu} A_\nu.$$

Thus by Chebyshev's inequality

$$P_\nu(x) \leq \left(\frac{2|x|}{a_\nu}\right)^\nu e^{c\nu} A_\nu, \qquad |x| \geq a_\nu.$$

(This is a weaker version of the Chebyshev inequality, which can be proved for generalized polynomials first with rational ν_j's, and then with a limiting process for arbitrary ν_j's.) Hence

$$P_\nu(x)u(x) \leq e^{c\nu - \frac{1}{2}Q(x) - \varphi_\nu(x)} A_\nu,$$

where

$$\varphi_\nu(x) = \frac{1}{2}Q(x) - \nu \log\frac{2|x|}{a_\nu}.$$

Here by

$$Q(\lambda x) \geq c\lambda^\alpha Q(x) \tag{20}$$

(cf. (5.3) in [5]), by (19) for sufficiently large λ

$$\varphi_\nu(\lambda a_\nu) = \frac{1}{2}Q(\lambda a_\nu) - \nu\log(2\lambda) \geq c\lambda^\alpha Q(a_\nu) - \nu\log(2\lambda) \geq c\nu(\lambda^\alpha - \log(2\lambda)) > 0, \tag{21}$$

Moreover,

$$\begin{aligned}\varphi_\nu'(x) = \frac{1}{2}Q'(x) - \frac{\nu}{|x|} &\geq \frac{\alpha Q(x) - \nu}{|x|} \geq \frac{\alpha\lambda^\alpha Q(x/\lambda) - \nu}{|x|} \\ &\geq \frac{\alpha\lambda^\alpha Q(a_\nu) - 2\nu}{2|x|} \geq \frac{c\nu(\lambda^\alpha - 1)}{2|x|} > 0, \qquad |x| \geq \lambda a_\nu,\end{aligned}$$

provided λ is large enough. This together with (21) implies $\varphi_\nu(x) > 0$ if $|x| \geq \lambda a_\nu$. Hence by Lemma (a) and (18)

$$v(x) \leq e^{c|x|^{\varrho+\varepsilon}} \leq e^{c\mu|x|^\alpha} \leq e^{c\mu Q(x)}, \qquad |x| \geq \lambda a_\nu$$

holds with any small enough $\mu > 0$ for sufficiently large ν's, and thus we obtain

$$P_\nu(x)w(x) \leq e^{c\nu - cQ(x)} A_\nu \qquad (|x| \geq \lambda a_\nu).$$

Integrating the pth power of both sides, using (20) we get

$$\begin{aligned}\mathcal{I}_2 &\leq e^{cp\nu} A_\nu^p \int_{|x|\geq\lambda a_\nu} e^{-\frac{p}{2}Q(x)}\,dx \leq e^{cp\nu} A_\nu^p \int_{|x|\geq\lambda a_\nu} |x|Q'(x)e^{-\frac{p}{2}Q(x)}\,dx \\ &\leq e^{cp\nu} A_\nu^p \int_{|x|\geq\lambda a_\nu} Q'(x)e^{-\frac{p}{3}Q(x)}\,dx \leq e^{cp\nu} A_\nu^p 6cpe^{-p3Q(\lambda a_\nu)} \\ &\leq e^{cp\nu(1-\lambda^\alpha)} A_\nu^p \leq e^{-cp\nu} A_\nu^p\end{aligned}$$

for sufficiently large λ, which together with the estimate for $\mathcal{I}_1$ proves the theorem. □

Remark. Instead of (7), we could have used the function

$$v(x) = \prod_{k=1}^{\infty} \left\{ \left|1 - \left(\frac{x}{t_k}\right)^2\right|^{m_k} \exp\left(m_k \sum_{j=1}^{q-1} \frac{1}{j}\left(\frac{x}{t_k}\right)^{2j}\right)\right\}$$

(for $q = 1$ the sum is counted as zero). The advantage of this function is that its roots are *exactly* the numbers $\pm t_k$ with multiplicities m_k, $k = 1, 2, \ldots$. The disadvantage is the presence of the convergence factors which causes technical difficulties.

3 Markov–Bernstein inequalities

We now wish to generalize (3) for the weights (11). Unfortunately, in the generality of (7) we cannot do this for the sequences (5)–(6). What we need here is a slightly stronger condition than (7): from now on we will assume that, with the notation $\Delta t_k := t_k - t_{k-1}$, $k = 1, 2, \ldots$, we have

$$\begin{cases} \gamma := \lim_{k\to\infty} \frac{\log \Delta t_k}{\log t_k} = \lim_{k\to\infty} \frac{\log \Delta t_k}{\log t_{k-1}} > -\infty, \\ m_k \leq M, \ k = 1, 2, \ldots \end{cases} \tag{22}$$

(of course, $\gamma \leq 1$). The examples $t_k = k^\beta$, $\beta > 0$ and $t_k = 2^k$ mentioned after (7) still satisfy this (with $\gamma = 1 - 1/\beta$ and $\gamma = 1$, respectively). The novelty in condition (23) is that it restricts the distance between adjacent t_k's (from below).

That (23) implies (7) with $\varrho = 1 - \gamma$ is easy to show. Namely, if $\varepsilon > 0$ is arbitrary, then (23) yields that

$$c \max(t_{k-1}^{\gamma-\varepsilon}, t_k^{\gamma-\varepsilon}) \le \Delta t_k \le c \min(t_{k-1}^{\gamma+\varepsilon}, t_k^{\varrho+\varepsilon}), \qquad k = 1, 2, \dots, \tag{23}$$

whence

$$\sum_{k=1}^{\infty} \frac{m_k}{t_k^{\varrho+2\varepsilon}} \le c \sum_{k=1}^{\infty} \frac{\Delta t_k}{t_k^{\gamma+\varrho+\varepsilon}} = c \sum_{k=1}^{\infty} \frac{\Delta t_k}{t_k^{1+\varepsilon}} \le c \int_{t_1}^{\infty} \frac{dt}{t^{1+\varepsilon}} < \infty,$$

while

$$\sum_{k=1}^{\infty} m_k \frac{t_k^{\varrho-2\varepsilon}}{\geq} c \sum_{k=1}^{\infty} \frac{\Delta t_{k+1}}{t_k^{\gamma+\varrho-\varepsilon}} = c \sum_{k=1}^{\infty} \frac{\Delta t_{k+1}}{t_k^{1-\varepsilon}} \ge c \int_{t_1}^{\infty} \frac{dt}{t^{1-\varepsilon}} = \infty.$$

Obviously, (23) is stronger than (7): for the sequence

$$t_{2k} = 2k, \quad t_{2k+1} = 2k + 2^{-k}, \quad m_k = 1, \qquad k = 0, 1, \dots,$$

(7) holds with $\varrho = 1$, while (23) fails to hold (the limits do not exist).

Thus, for any sequences (5)–(6) with the property (23) we form the function (8) with

$$0 \le \varrho = 1 - \gamma. \tag{24}$$

Then, with the Freud-type weight $u(x) = e^{-Q(x)}$ we consider our weight w defined in (11) and state

Theorem 3.1 *With the previous notations, if the characteristic α of the Freud-type weight u defined in* (1) *satisfies*

$$\alpha > \frac{(3-\gamma)(1-\gamma)}{2\left[\frac{1-\gamma}{2}\right] + 1 + \gamma}, \tag{25}$$

then we have

$$||P_\nu' w||_{L_p(\mathbf{R})} \le e^{cM} M_\nu ||P_\nu w||_{L_p(\mathbf{R})} \qquad \textit{for all} \quad P_\nu \in \mathcal{P}_\nu \quad \text{and} \quad 0 < p \le \infty. \tag{26}$$

Note that condition (26) is very delicate: if γ is approaching an odd negative integer from above then the right-hand side of (26) tends to infinity.

Proof. First we prove the theorem for $p = \infty$. (The condition (12), i.e.,

$$\alpha > 1 - \gamma \tag{27}$$

evidently holds, cf. (25) and (26).)

For convenience of notation, let $m = \nu^1 1 - \gamma + \varepsilon$ with a small $\varepsilon > 0$ to be determined later. We show that

$$0 < e^{-cM} \le \prod_{t_k > m} \left| 1 - \left(\frac{x}{t_k}\right)^{2q} \right|^{m_k} \le 1, \qquad |x| \le \lambda a_\nu,\ q = \left[\frac{1-\gamma}{2}\right] + 1, \tag{28}$$

where the constant $\lambda > 2$ will be determined later. By (3.6), we have for

$$|x| \le \lambda a_\nu \le c\nu^{1/\alpha} \le 12\nu^{\frac{1}{1-\gamma+\varepsilon}} = \frac{m}{2}, \qquad 0 < \varepsilon < \alpha - 1 + \gamma$$

(see (16)) the inequality

$$\prod_{t_k > m} \left| 1 - \left(\frac{x}{t_k} \right)^{2q} \right|^{m_k} \le 1, \quad |x| \le \lambda a_\nu.$$

On the other hand, by $1 - t \ge e^{-2t}$ $(0 \le t \le 1/2)$, as well as (16), (23) and (24) we obtain

$$\prod_{t_k > m} \left| 1 - \left(\frac{x}{t_k} \right)^{2q} \right|^{m_k} \ge \exp\left(-2x^{2q} \sum_{t_k \ge m} \frac{m_k}{t_k^{2q}} \right)$$

$$\ge \exp\left(-cMa_\nu^{2q} \sum_{t_k \ge m} \frac{\Delta t_k}{t_k^{2q+\gamma-\varepsilon}} \right) \ge \exp\left(-cM\nu^{2q/\alpha} \int_m^\infty \frac{dt}{\nu^{2q+\gamma-\varepsilon}} \right)$$

$$\ge \exp\left(-cM\nu^{2q/\alpha} m^{1-2q-\gamma+\varepsilon} \right) \ge \exp\left(-cM\nu^{\frac{2q}{\alpha} - \frac{2q+\gamma-1-\varepsilon}{1-\gamma+\varepsilon}} \right) \ge e^{-cM}, \quad |x| \le \lambda a_\nu$$

provided

$$0 < \varepsilon \le \frac{\alpha(2q+\gamma-1) - 2q(1-\gamma)}{2q+\alpha}.$$

The latter inequality can be satisfied because of (26). Hence (3.7) is proved.

Besides (3.7), we will need the existence of polynomials $Q_\nu(x) \in \mathcal{P}_{c\nu}$ such that

$$u(x) \sim Q_\nu(x), \qquad |x| \le \lambda a_\nu \tag{29}$$

(cf. [4], Theorem 1.3).

Now let

$$T_\nu(x) = \prod_{t_k \le m} \left| 1 - \left(\frac{x}{t_k} \right)^{2q} \right|^{m_k}. \tag{30}$$

Then by (9)–(10) (using the latter with $\varepsilon/2$ instead of ε), $T_\nu \in \mathcal{P}_\nu$. Thus Theorem 1 (applied in a weaker form) yields with $P_\nu T_\nu \in \mathcal{P}_{2\nu}$ (by utilizing (3.7) and (3.8) as well)

$$\begin{aligned} ||P_\nu' w||_{L_\infty(\mathbf{R})} &\le c||P_\nu' w||_{L_\infty(|x|\le 2a_\nu)} \le c||P_\nu' T_\nu u||_{L_\infty(|x|\le 2a_\nu)} \\ &\le c||(P_\nu T_\nu)' u||_{L_\infty(|x|\le 2a_\nu)} + c||P_\nu T_\nu' u||_{L_\infty(|x|\le 2a_\nu)} \\ &\le cM_{2\nu}||P_\nu T_\nu u||_{L_\infty(|x|\le a_{2n})} + 2q\left\| P_\nu T_\nu Q_\nu \sum_{t_k \le m} \frac{m_k x^{2q-1}}{x^{2q} - t_k^{2q}} \right\|_{L_\infty(|x|\le 2a_\nu)} \\ &\le e^{cM} M_\nu ||P_\nu w||_{L_\infty(\mathbf{R})} + 2q\left\| P_\nu T_\nu Q_\nu \sum_{t_k \le m} \frac{m_k x^{2q-1}}{x^{2q} - t_k^{2q}} \right\|_{L_\infty(|x|\le 2a_\nu)}, \end{aligned} \tag{31}$$

provided $\lambda > 2$ is chosen such that $a_{2\nu} \le \lambda a_\nu$ so that (3.7) is applicable.

Here we have to estimate the second term on the right-hand side. By symmetry, we may assume $0 \le x \le 2a_\nu$. First,

$$\begin{aligned}
&\left\| P_\nu T_\nu Q_\nu \sum_{t_k \le m} \frac{m_k x^{2q-1}}{x^{2q} - t_k^{2q}} \right\|_{L_\infty(0 \le x \le t_1/2)} \\
&\le cM \|P_\nu T_\nu Q_\nu\|_{L_\infty(0 \le x \le t_1/2)} \sum_{k=1}^{\infty} \frac{1}{t_k^{2q}} \qquad (32)\\
&\le cM \|P_\nu w\|_{L_\infty(\mathbf{R})} \sum_{k=1}^{\infty} \frac{1}{t_k^{1-\gamma+\varepsilon}} \le cM \|P_\nu w\|_{L_\infty(\mathbf{R})},
\end{aligned}$$

where $\varepsilon = 2q - 1 + \gamma > 0$.

Now if $t_k \ge 2a_\nu + 1M_\nu$, then

$$\left\| \frac{P_\nu T_\nu Q_\nu m_k}{x - t_k} \right\|_{L_\infty(|x| \le 2a_\nu)} \le cMM_\nu \|P_\nu w\|_{L_\infty(|x| \le 2a_\nu)} \le cMM_\nu \|P_\nu w\|_{L_\infty(\mathbf{R})}, \tag{33}$$

while for $t_k \le 2a_\nu + 1M_\nu$ we have

$$\left\| \frac{P_\nu T_\nu Q_\nu m_k}{x - t_k} \right\|_{L_\infty(|x| \le \lambda a_\nu,\ |x - t_k| \ge 1/M_\nu)} \le MM_\nu \|P_\nu w\|_{L_\infty(\mathbf{R})}.$$

Hence using Remez inequality (cf. [1], Theorem A.4.1) on the interval $|x| \le \lambda a_\nu$ we get

$$\begin{aligned}
\left\| \frac{P_\nu T_\nu Q_\nu m_k}{x - t_k} \right\|_{L_\infty(|x| \le \lambda a_\nu)} &\le \exp\left(\frac{c\nu}{a_\nu M_\nu}\right) M_\nu \|P_\nu w\|_{L_\infty(\mathbf{R})} \qquad (34)\\
&\le cMM_\nu \|P_\nu w\|_{L_\infty(\mathbf{R})}.
\end{aligned}$$

This shows that (3.12) holds for all t_k's.

In order to estimate the norm on the right-hand side of (3.10) in the interval

$$u_{j-1} := \frac{t_{j-1} + t_j}{2} \le x \le \frac{t_j + t_{j+1}}{2} =: u_j, \qquad j \ge 1 \text{ fixed}, \tag{35}$$

we use (3.12) to estimate the term $k = j$ on the right-hand side of (3.10):

$$\begin{aligned}
\left\| \frac{P_\nu T_\nu Q_\nu m_j x^{2q-1}}{x^{2q} - t_j^{2q}} \right\|_{L_\infty(u_{j-1}, u_j)} &\le M \left\| \frac{P_\nu T_\nu Q_\nu}{x - t_j} \right\|_{L_\infty(|x| \le 2a_\nu)} \\
&\le cMM_\nu \|P_\nu w\|_{L_\infty(\mathbf{R})}.
\end{aligned}$$

Next, we estimate the sum for $k \ne j$ on the right-hand side of (3.10). Using (24) we obtain for the x's specified in (3.14) (some of the sums below may be empty)

$$
\begin{aligned}
&\sum_{k\neq j}\frac{m_k x^{2q-1}}{|x^{2q}-t_k^{2q}|}\\
&\le 2M\sum_{t_k\le x/2}\frac{1}{x}+M\sum_{\substack{x/2\le t_k\le t_{j-2}\\ \text{or } t_{j+2}\le t_k\le 2x}}\frac{1}{|x-t_k|}\\
&\quad+\frac{2M}{\Delta t_j}+\frac{2M}{\Delta t_{j+1}}++cMx^{2q-1}\sum_{t_k>2x}\frac{1}{t_k^{2q}}\\
&\le cM\frac{N(x/2)}{x}+cMx^{-\gamma+\varepsilon}\left(\sum_{x/2\le t_k\le t_{j-2}}\frac{\Delta t_{k+1}}{x-t_k}++\sum_{t_{j+2}\le t_k\le 2x}\frac{\Delta t_k}{t_k-x}\right)\\
&\quad+\frac{cM}{t_{j-1}^{\gamma-\varepsilon}}+\frac{cM}{t_j^{\gamma-\varepsilon}}+cMx^{2q-1}\sum_{t_k>2x}\frac{\Delta t_k}{t_k^{2q+\gamma-\varepsilon}}\\
&\le cMx^{-\gamma+\varepsilon}\left(1+\int_{t_1}^{t_{j-1}}\frac{dt}{x-t}+\int_{t_{j+1}}^{2x}\frac{dt}{t-x}\right)+cM\max(1,x^{-\gamma+\varepsilon})\\
&\quad+cMx^{2q-1}\int_{2x}^{\infty}\frac{dt}{\nu^{2q+\gamma-\varepsilon}}\\
&\le cMx^{-\gamma+\varepsilon}\left(1+\log\Delta t_j+\log\frac{2x}{\Delta t_{j+1}}\right)+cM\max(1,x^{-\gamma+\varepsilon})+cMx^{-\gamma+\varepsilon}\\
&\le cM\max(1,x^{-\gamma+\varepsilon}\log x)\\
&\le cM\max(1,x^{-\gamma+2\varepsilon})
\end{aligned}
$$

provided $0<\varepsilon<2q+\gamma-1$. If $0\le\gamma\le 1$, then this is bounded. Otherwise we obtain

$$
\begin{aligned}
\sum_{k\neq j}\frac{m_k x^{2q-1}}{|x^{2q}-t_k^{2q}|}&\le cMx^{-\gamma+2\varepsilon}\le cMa_\nu^{-\gamma+2\varepsilon}\\
&\le cMna_\nu\le cMM_\nu,\qquad u_{j-1}\le x\le u_j,
\end{aligned}
\tag{36}
$$

provided $a_\nu^{1-\gamma+2\varepsilon}\le c\nu$. With respect to (16), the latter inequality is satisfied if $\alpha\ge 1-\gamma+2\varepsilon$. But this is true because of (3.6), provided $\varepsilon>0$ is small enough.

Now (3.15) yields

$$
\begin{aligned}
\left\|P_\nu T_\nu Q_\nu\sum_{t_k\le m}\frac{m_k x^{2q-1}}{x^{2q}-t_k^{2q}}\right\|_{L_\infty(u_{j-1}\le x\le u_j)}&\le cMM_\nu\|P_\nu T_\nu Q_\nu\|_{L_\infty(u_{j-1}\le x\le u_j)}\\
&\le cMM_\nu\|P_\nu w\|_{L_\infty(u_{j-1}\le x\le u_j)}.
\end{aligned}
$$

Summing up the pth power of these inequalities for j's and taking into account

(3.11) we obtain

$$\left\| P_\nu T_\nu Q_\nu \sum_{t_k \le m} \frac{m_k x^{2q-1}}{x^{2q} - t_k^{2q}} \right\|_{L_\infty(|x|\le 2a_\nu)} \le cMM_\nu ||P_\nu w||_{L_\infty(\mathbf{R})},$$

which with respect to (3.10) proves Theorem 2 for $p = \infty$.

Now let $0 < p < \infty$. According to the theory of Christoffel functions applied for the Legendre polynomials (see e.g. [2], Lemma 3.1 on p. 100), there exist (ordinary) polynomials $K_\nu(x) \in \mathcal{P}_\nu$ such that

$$\int_{-2}^{2} K_\nu(x)\,dx \le \frac{c}{\nu} \quad \text{and} \quad K_\nu(0) = 1. \tag{37}$$

The other tool we are going to use is

$$|v(x)| \sim T_\nu(x) \in \mathcal{P}_\nu, \qquad |x| \le \lambda a_\nu \tag{38}$$

(cf. (3.7) and (3.9)).

Let t be a fixed parameter, and apply the just proved part of the theorem to

$$P_\nu(x) K_\nu\left(\frac{x-t}{da_\nu}\right)$$

instead of $P_\nu(x)$ (where $d > 1$ is a constant to be specified later, as well as Theorem 1, (3.17) and Nikolski's inequality between L_∞ and L_1 spaces

$$||P_\nu u||_{L_\infty(\mathbf{R})} \le \frac{c\nu}{a_\nu} ||P_\nu u||_{L_1(\mathbf{R})}$$

(cf. [1], Theorem 6.2.10) to get

$$\begin{aligned}
\left|\left[P_\nu(x)K_\nu\left(\frac{x-t}{da_\nu}\right)\right]'\right| w(x) &\le e^{cM} M_\nu \left\| P_\nu(\xi) K_\nu\left(\frac{\xi-t}{da_\nu}\right) w(\xi)\right\|_{L_\infty(\mathbf{R})} \\
&\le e^{cM} M_\nu \left\| P_\nu(\xi) K_\nu\left(\frac{\xi-t}{da_\nu}\right) w(\xi)\right\|_{L_\infty(|\xi|\le ca_\nu)} \\
&\le e^{cM} M_\nu \left\| P_\nu(\xi) K_\nu\left(\frac{\xi-t}{da_\nu}\right) T_\nu(\xi) u(\xi)\right\|_{L_\infty(|\xi|\le ca_\nu)} \\
&\le \frac{e^{cM}\nu M_\nu}{a_\nu} \left\| P_\nu(\xi) K_\nu\left(\frac{\xi-t}{da_\nu}\right) T_\nu(\xi) u(\xi)\right\|_{L_1(\mathbf{R})} \\
&\le \frac{e^{cM}\nu M_\nu}{a_\nu} \left\| P_\nu(\xi) K_\nu\left(\frac{\xi-t}{da_\nu}\right) T_\nu(\xi) u(\xi)\right\|_{L_1(|\xi|\le ca_\nu)} \\
&\le \frac{e^{cM}\nu M_\nu}{a_\nu} \left\| P_\nu(\xi) K_\nu\left(\frac{\xi-t}{da_\nu}\right) w(\xi)\right\|_{L_1(|\xi|\le ca_\nu)}, \qquad x \in \mathbf{R}.
\end{aligned}$$

Substituting $t = x$, using $K_\nu(0) = 1$ and choosing $d = c > 1$ (i.e., the constant in the last inequality), by

$$\left\| K_\nu\left(\frac{u-x}{ca_\nu}\right)\right\|_{L_1(|u|\le ca_\nu)} = a_\nu \int_{-1-\frac{x}{ca_\nu}}^{1-\frac{x}{ca_\nu}} K_\nu(v)\,dv \le a_\nu \int_{-2a_\nu}^{2a_\nu} K_\nu(v)\,dv \le \frac{ca_\nu}{\nu},$$

$$|x| \le ca_\nu \tag{39}$$

(see (3.15)) we obtain, using the Cauchy–Schwarz inequality,

$$\begin{aligned}
|P_\nu'(x)|w(x) &\le \frac{e^{cM}\nu M_\nu}{a_\nu}\left\| P_\nu(\xi)K_\nu\left(\frac{\xi-x}{ca_\nu}\right)w(\xi)\right\|_{L_1(|\xi|\le ca_\nu)}\\
&\le \frac{e^{cM}\nu M_\nu}{a_\nu}\left(\int_{-ca_\nu}^{ca_\nu}[|P_\nu(\xi)|w(\xi)]^p K_\nu\left(\frac{\xi-x}{ca_\nu}\right)d\xi\right)^{1/p}\|\times\\
&\quad\times K_\nu(\xi - xca_\nu)\|_{L_1(|\xi|\le ca_\nu)}^{1-1/p} \le e^{cM}M_\nu\left(\frac{\nu}{a_\nu}\right)^{1/p}\\
&\quad\times\left(\int_{-ca_\nu}^{ca_\nu}[|P_\nu(\xi)|w(\xi)]^p K_\nu\left(\frac{\xi-x}{ca_\nu}\right)d\xi\right)^{1/p}, \qquad |x|\le ca_\nu.
\end{aligned}$$

Taking pth power on both sides, integrating, using Fubini's theorem and (3.16) again,

$$\int_{-a_\nu}^{a_\nu}[|P_\nu'(x)|w(x)]^p\,dx \le e^{cMp}M_\nu^p\int_{-ca_\nu}^{ca_\nu}[|P_\nu(\xi)|w(\xi)]^p\,d\xi.$$

With respect to Theorem 1, this proves Theorem 2 for $0 < p < \infty$. □

References

[1] P. Borwein and T. Erdélyi, *Polynomials and Polynomial Inequalities,* Springer Verlag (New York, 1995).

[2] G. Freud, *Orthogonal Polynomials,* Akadémiai Kiadó (Budapest, 1971).

[3] A.L. Levin and D.S. Lubinsky, *Orthogonal Polynomials for Exponential Weights,* Springer (2001).

[4] A.L. Levin and D.S. Lubinsky, *Weights on the real line that admit good relative polynomial approximation, with applications,* J. Approx. Theory, **49** (1987), 170–195.

[5] A.L. Levin and D.S. Lubinsky, *Christoffel functions, orthogonal polynomials, and Nevai's conjecture for Freud weights,* Constr. Approx., **8** (1992), 463–535.

Afréd Rényi Institute of Mathematics
H-1364 Budapest, P.O.B. 127
Email address: szabados@renyi.hu

Advanced Problems in Constructive Approximation
(Eds.) M.D. Buhmann and D.H. Mache
International Series of Numerical Mathematics Vol. 142, 237–253

Some Erdős-type Convergence Processes in Weighted Interpolation

László Szili and Péter Vértesi

Abstract

In 1943, P. Erdős [5] showed that if the interpolation point system $X_n \subset [-1,1]$ $(n \in \mathbf{N})$ such that the fundamental polynomials of Lagrange interpolation are uniformly bounded in $[-1,1]$ then for every $f \in C[-1,1]$ and $c > 0$ there exists a sequence of polynomials φ_n of degree $\leq n(1+c)$ $(n \in \mathbf{N})$ which interpolates f at the points X_n and it tends to f uniformly in $[-1,1]$. The weighted versions of this result were proved in [19] and [18] using Freud-type weights and exponential weights on $[-1,1]$. The aim of this paper is to show that analogue statements are true for weighted interpolation if we consider Erdős-type and some ultraspherical weights.

1 Introduction, notations, preliminaries

1.1

On an interval $I \subset \mathbf{R}$ one of the most natural approximating tools is the Lagrange interpolation. Namely, if $X_n := \{x_{n,n} < x_{n-1,n} < \cdots < x_{2,n} < x_{1,n}\} \subset I$ $(n \in \mathbf{N} := \{1, 2, \ldots\})$ is an interpolatory matrix on I and $f : I \to \mathbf{R}$ is a given function then we can construct the Lagrange interpolatory polynomials

$$L_n(f, X_n, x) := \sum_{k=1}^{n} f(x_{k,n})\ell_{k,n}(X_n, x) \quad (x \in I,\ n \in \mathbf{N}),$$

where

$$\ell_{k,n}(X_n, x) := \ell_{k,n}(x) := \frac{\omega_n(X_n, x)}{\omega_n'(X_n, x_{k,n})(x - x_{k,n})} \tag{1}$$

$$(x \in I,\ k = 1, 2, \ldots, n,\ n \in \mathbf{N})$$

(here $\omega_n(X_n, x) := c_n \prod_{k=1}^{n}(x - x_{k,n})$, $n \in \mathbf{N}$) are the fundamental polynomials of degree exactly $n-1$ ($\ell_{k,n}(X_n, \cdot) \in \Pi_{n-1} \setminus \Pi_{n-2}$ shortly) of Lagrange interpolation.

It was proved by G. Faber in 1914, *there is no point system* $X_n \subset [-1,1]$ ($n \in \mathbf{N}$) *for which the corresponding sequence of Lagrange interpolatory polynomials* $L_n(f, X_n, \cdot)$ ($n \in \mathbf{N}$) *would converge uniformly on* $[-1,1]$ *for every* $f \in C[-1,1]$. Two years later, relaxing the degree-restriction, L. Fejér proved as follows. *If*

$$X_n := T_n := \left\{c_{k,n} := \cos \frac{2k-1}{2n}\pi \mid k = 1, 2, \ldots, n\right\} \qquad (n \in \mathbf{N})$$

then for all $f \in C[-1,1]$ *the sequence* $H_n(f, T_n, x)$ ($x \in [-1,1]$, $n \in \mathbf{N}$) *tends to* f *uniformly on* $[-1,1]$. Here, the so-called Hermite-Fejér polynomials H_n of degree $\leq 2n-1$ satisfy

$$H_n^{(i)}(f, T_n, c_{j,n}) = f(c_{j,n})\delta_{i,0} \qquad (j = 1, 2, \ldots, n,\ n \in \mathbf{N},\ i = 0, 1).$$

A far reaching generalization was proved in 1943 by P. Erdős [5].

Theorem 1.1 *Let the point group* $X_n \subset [-1,1]$ ($n \in \mathbf{N}$) *be such that the fundamental polynomials of Lagrange interpolation* $\ell_{k,n}$ ($k = 1, 2, \ldots, n$, $n \in \mathbf{N}$) *are uniformly bounded in* $[-1,1]$. *Then to every* $f \in C[-1,1]$ *and* $c > 0$ *there exists a sequence of polynomials* $\varphi_n(x) := \varphi_n(f, c, x)$ ($x \in [-1,1]$, $n \in \mathbf{N}$) *such that*

(i) *the degree of* φ_n *is* $\leq n(1+c)$,

(ii) $\varphi_n(x_{j,n}) = f(x_{j,n})$ ($j = 1, 2, \ldots, n$, $n \in \mathbf{N}$),

(iii) (φ_n) *tends to* f *uniformly in* $[-1,1]$.

1.2

The goal of this paper is to investigate the corresponding results for weighted interpolation using some weights. The weighted version of Theorem 1.1 were proved in [19] and [18] using Freud-type weights and exponential weights on $[-1,1]$. In this paper we prove the corresponding results for Erdős-type weights and certain ultraspherical Jacobi weights, too. As it turns out, the above mentioned four cases can be treated in a unified way. That means, sometimes we only sketch the proofs, referring the corresponding parts in [19] or [18].

However, to do this, we have to prove a new Markov-Bernstein inequality (see (23)) if $w \in \mathcal{E}(\mathbf{R})$ (see Definition 1.5). Moreover, we show the root distribution of the orthonormal polynomials with the weight $w \in \mathcal{E}(\mathbf{R})$.

Giving up the unified treatment and applying ad hoc considerations, we hope to get similar Erdős-type theorems for other weights, too. This far-reaching program is the topic of some forthcoming papers.

1.3 Some classes of weight functions

In the sequel $I = (-1,1)$ or $I = \mathbf{R}$. We suppose that the weight function $w : I \to \mathbf{R}$ is even, continuous, $w(x) > 0$ $(x \in I)$ and it vanishes at the end points of I, i.e., $w(x) \to 0$ if $|x| \to \sup I$.

Definition 1.2 *The class of the ultraspherical weights*

$$w(x) := w^{(\alpha)}(x) := (1-x^2)^\alpha = \exp(\alpha \log(1-x^2)) =: \exp(-Q(x))$$
$$(x \in (-1,1),\ \alpha > 0)$$

will be denoted by $\mathcal{U}[-1,1]$.

Definition 1.3 *Let* $w := \exp(-Q)$, *where* $Q : (-1,1) \to \mathbf{R}$ *is even and twice continuously differentiable in* $(-1,1)$. *Assume moreover, that* $Q' \geq 0$, $Q'' \geq 0$ *in* $(0,1)$ *and* $\lim_{x \to 1-0} Q(x) = +\infty$. *We also suppose that the function*

$$T(x) := 1 + x\frac{Q''(x)}{Q'(x)} \qquad (x \in [0,1))$$

is increasing in $[0,1)$, *moreover*

(i) $T(0+) > 1$,
(ii) $T(x) \sim Q'(x)/Q(x)$, *x close enough to* 1,
(iii) *for some* $A > 2$

$$T(x) \geq \frac{A}{1-x^2}, \quad x \text{ close enough to } 1.$$

Then w is called an exponential weight on $[-1,1]$ *and we write* $w \in \mathcal{EXP}[-1,1]$.

The principle examples of $w \in \mathcal{EXP}[-1,1]$ are

$$w(x) = w^{k,\alpha}(x) := \exp(-\exp_k(1-x^2)^{-\alpha})$$
$$(x \in (-1,1),\ \alpha > 0,\ k \in \mathbf{N}_0 := \{0,1,\ldots\}),$$

where $\exp_0(x) := x$ $(x \in \mathbf{R})$ and for $k \in \mathbf{N}$ $\exp_k := \exp(\exp_{k-1})$. These are "strongly vanishing" weights at ± 1.

A routine calculation shows that the ultraspherical weights $w^{(\alpha)}$ do not belong to $\mathcal{EXP}[-1,1]$.

Definition 1.4 *A weight* $w := \exp(-Q)$ *is a Freud-type weight* ($w \in \mathcal{F}(\mathbf{R})$, *shortly*) *iff* $Q : \mathbf{R} \to \mathbf{R}$ *is even, continuous in* $\mathbf{R}$, Q'' *is continuous in* $(0,+\infty)$, $Q' > 0$ *on* $(0,+\infty)$ *and for some* $1 < c \leq C$

$$c \leq T(x) := 1 + x\frac{Q''(x)}{Q'(x)} \leq C \qquad (x \in (0,+\infty)). \tag{2}$$

The simplest cases are the so-called *Freud weights*

$$w_\alpha(x) := \exp(-|x|^\alpha) \qquad (x \in \mathbf{R},\ \alpha > 1).$$

From (2) it follows that Q is polynomial growth at $+\infty$. We also consider a class of Erdős weights, for which the exponent Q grows faster than any polynomial.

Definition 1.5 *We say that $w \in \mathcal{E}(\mathbf{R})$ (w is an Erdős-type weight on $\mathbf{R}$) iff $Q : \mathbf{R} \to \mathbf{R}$ is even and differentiable on $\mathbf{R}$, $Q' > 0$ and $Q'' > 0$ in $(0, +\infty)$ and the function*

$$T(x) = 1 + x\frac{Q''(x)}{Q'(x)} \qquad (x \in (0, +\infty))$$

is increasing in $(0, +\infty)$ with

$$\lim_{x \to +\infty} T(x) = +\infty, \qquad T(0+) := \lim_{x \to 0+} T(x) > 1.$$

Moreover we assume that for some $c_1, c_2, c_3 > 0$

$$c_1 \le T(x)\frac{Q(x)}{xQ'(x)} \le c_2 \quad \text{if} \quad x \ge c_3.$$

The prototype of $w \in \mathcal{E}(\mathbf{R})$ is the case when

$$Q(x) := Q_{k,\alpha}(x) := \exp_k(|x|^\alpha) \qquad (x \in \mathbf{R},\ k \ge 1,\ \alpha > 1).$$

The corresponding w will be denoted by $w_{k,\alpha}$.

Let us introduce the following notation:

$$\mathcal{W}(I) := \mathcal{U}[-1,1] \cup \mathcal{EXP}[-1,1] \cup \mathcal{F}(\mathbf{R}) \cup \mathcal{E}(\mathbf{R}).$$

By definitions, if $w \in \mathcal{W}(I)$ then $w^2 \in \mathcal{W}(I)$, too. If $w \in \mathcal{W}(I)$ and $u > 0$ then as it is well known ([13, p. 76]) there is exactly one positive root of the equation

$$u = \frac{2}{\pi}\int_0^1 \frac{\lambda t Q'(\lambda t)}{\sqrt{1 - x^2}}dt \qquad (\lambda \in \mathbf{R}). \tag{3}$$

It is called the *Mhaskar-Rahmanov-Saff number* (of w); we denote it by $a_u(w) =: a_u$. An important property of $a_n(w)$ is the following: For any $r_n \not\equiv 0, r_n \in \Pi_n$ (the set of polynomials of degree at most n)

$$\|r_n w\| := \max_{x \in I} |r_n(x)w(x)| = \max_{|x| \le a_n(w)} |r_n(x)w(x)|$$
$$\|r_n w\| > |r_n(x)w(x)| \quad \text{for } |x| > a_n(w) \tag{4}$$

and asymptotically as $n \to +\infty$, $a_n(w)$ is the smallest such number.

For a weight function $w = \exp(-Q) \in \mathcal{W}(I)$ let

$$T_u := T_u(w) := T(a_u(w)), \quad \delta_u := \delta_u(w) := \left(\frac{1}{uT_u}\right)^{2/3} \qquad (u > 0) \tag{5}$$

(as above, $T(x) = 1 + xQ''(x)/Q'(x)$ for $x \in \mathbf{R}^+ \cap I$) and with $L > 0$ let us define

$$x_{0,n} := x_{0,n}(w, L) := a_n(1 + 2L\delta_n) \qquad (n \in \mathbf{N}). \tag{6}$$

Remark 1.6 *If $I = \mathbf{R}$ (i.e., $w \in \mathcal{F}(\mathbf{R}) \cup \mathcal{E}(\mathbf{R})$) then $x_{0,n} \in I$ for any $L > 0$ and $n \in \mathbf{N}$. For $I = (-1, 1)$ it is not true in general. If $w \in \mathcal{U}[-1, 1]$ then one can choose $L_0 > 0$ and $n_0 \in \mathbf{N}$ such that $x_{0,n}(w, L) < 1$ and $1 - x_{0,n}(w, L) \sim 1 - a_n$ $(0 < L < L_0, \quad n > n_0)$.*

If $w \in \mathcal{EXP}[-1, 1]$ then $a_n\delta_n/(1 - a_n) \to 0$ if $n \to +\infty$ (see [6, (3.8) and p.30]) *therefore for every $L > 0$ there exists $n_0 \in \mathbf{N}$ such that $x_{0,n} < 1$ $(n > n_0)$. In the following we suppose that the points $x_{0,n}$ satisfy the above requirements.*

The functions

$$\Psi_n(x) := \max\left\{\sqrt{1 - \frac{|x|}{a_n} + 2L\delta_n}, \frac{1}{T_n\sqrt{1 - \frac{|x|}{a_n} + 2L\delta_n}}\right\} \tag{7}$$

$$(|x| < x_{0,n}, \ n \in \mathbf{N}).$$

play a fundamental role in the asymptotic formulae with respect to the orthogonal polynomials. We set $x =: x_{0,n} \cos\vartheta$ $(\vartheta \in (0, \pi))$ and define

$$\psi_n(\vartheta) := \Psi_n(x_{0,n} \cos\vartheta) \qquad (\vartheta \in (0, \pi)).$$

A routine calculation shows that for every fixed $0 < c < \pi\sqrt{T_1}/2$, $\vartheta \in (0, \pi/2)$ and $n \in \mathbf{N}$

$$\psi_n(\vartheta) \sim \max\left\{\sin\vartheta, \frac{1}{T_n \sin\vartheta}\right\}$$

$$\sim \left\{\begin{array}{ll} \dfrac{1}{\vartheta T_n}, & \text{if } 0 < \vartheta < cT_n^{-1/2} \\ \vartheta, & \text{if } \vartheta > cT_n^{-1/2} \end{array}\right\} \sim \frac{(\vartheta + T_n^{-1/2})^2}{\vartheta}. \tag{8}$$

Here and later $A_n \sim B_n$ means that $0 < c_1 \le A_n/B_n \le c_2$, where c_1 and c_2 do not depend on n (but may depend on other, previously fixed parameters).

1.4 Some classes of functions

Let $C(I)$ represent the linear space of real valued continuous functions defined on the interval $I \subset \mathbf{R}$. Fix a weight $w \in \mathcal{W}(I)$ and let

$$C_w(I) := \{f \in C(I) \mid \lim_{|x| \to \sup I} (fw)(x) = 0\}.$$

One can show that $\|f\|_w := \|fw\| := \max_{x \in I} |(fw)(x)|$ $(f \in C_w)$ is a norm on C_w and $(C_w, \|\cdot\|_w)$ is a Banach space.

1.5 Weighted Lagrange interpolation

If $X_n \subset I$ is an interpolatory matrix, $w \in \mathcal{W}(I)$ then for $f : I \to \mathbf{R}$ the *weighted Lagrange interpolation* is defined by

$$L_n(f, w, X_n, x) := \sum_{k=1}^{n} (fw)(x_{k,n}) q_{k,n}(w, X_n, x) \quad (x \in I,\ n \in \mathbf{N}),$$

where

$$q_{k,n}(w, X_n, x) := \frac{(w\omega_n)(x)}{(w\omega_n)'(x_{k,n})(x - x_{k,n})} = \frac{w(x)}{w(x_{k,n})} \ell_{k,n}(X_n, x) \tag{9}$$

$$(x \in I,\ k = 1, 2, \ldots, n,\ n \in \mathbf{N})$$

are the *fundamental functions of the weighted Lagrange interpolation.*

It is known that the *weighted Lebesgue functions*

$$\lambda_n(w, X_n, x) := \sum_{k=1}^{n} |q_{k,n}(w, X_n, x)| \quad (x \in I,\ n \in \mathbf{N})$$

and the *weighted Lebesgue constants*

$$\Lambda_n(w, X_n) := \max_{x \in I} \lambda_n(w, X_n, x) \qquad (n \in \mathbf{N})$$

play a fundamental role in the convergence-divergence behavior of sequences of weighted Lagrange interpolation polynomials.

Special cases of P. Vértesi [20], [21] result that if $w \in \mathcal{W}(I)$ and $X_n \subset I$ is an *arbitrary* interpolatory point system then $\Lambda_n(w, X_n) \geq c \log n$ $(n \in \mathbf{N})$. From these relations one can easily get a *Faber type result for the corresponding weighted Lagrange interpolation.*

1.6 Orthogonal polynomials

If $w \in \mathcal{W}(I)$ then we can define the orthogonal polynomials

$$p_n(x) := p_n(w^2, x) := \gamma_n(w^2)x^n + \cdots \qquad (\gamma_n(w^2) > 0,\ n \in \mathbf{N}_0)$$

for the weight w^2, so that $\int_I p_n(w^2, x) p_m(w^2, x) w^2(x) dx = \delta_{m,n}$ $(m, n \in \mathbf{N}_0)$.

Let us denote by

$$U_n := U_n(w^2) := \{y_{k,n} := y_{k,n}(w^2) \mid k = 1, 2, \ldots, n\} \qquad (n \in \mathbf{N}) \tag{10}$$

the n different roots of $p_n(w^2, \cdot)$. We index them as

$$-\infty < y_{n,n}(w^2) < y_{n-1,n}(w^2) < \cdots < y_{2,n}(w^2) < y_{1,n}(w^2) < \infty.$$

There is a close relation between $y_{1,n}(w^2)$ and $a_n(w)$. Namely, if $w \in \mathcal{W}(I)$ then there are a positive number C_0 and an index $n_1 \in \mathbf{N}$ such that

$$\left|y_{1,n}(w^2) - a_n(w)\right| \le C_0 a_n(w)\delta_n(w) \qquad (n \ge n_1). \tag{11}$$

If $w \in \mathcal{W}(I)\backslash\mathcal{U}[-1,1]$ then (11) is [7, (1.17)], [6, (1.33)] and [9, (1.23)], respectively. If $w \in \mathcal{U}[-1,1]$ then (11) comes from the relations

$$1 - y_{1,n}(w^2) \sim \frac{1}{n^2}, \quad 1 - a_n(w) \sim \frac{1}{T_n} \sim \frac{1}{n^2} \qquad (n \in \mathbf{N}) \tag{12}$$

(see [16, (8.9.1)] and [10, p. 553]).

The fundamental functions of the weighted Lagrange interpolation with respect to $U_n(w^2)$ will be denoted by

$$u_{k,n}(w, U_n(w^2), x) := \frac{p_n(w^2,x)w(x)}{p_n'(w^2, y_{k,n})w(y_{k,n})(x - y_{k,n})} \quad (k = 1, 2, \ldots, n, \ n \in \mathbf{N}). \tag{13}$$

2 New developments

To get the analogue of Theorem 1.1, we use some preliminaries being interesting in themselves.

2.1

For $w \in \mathcal{W}(I)$ and $L > C_0$ (see (11) and Remark 1.6) let

$$y_{0,n}(w^2) := y_{0,n}(w^2, L) := x_{0,n}(w^2, L) =: -y_{n+1,n}(w^2, L) =: -y_{n+1,n}(w^2),$$

where $x_{0,n}(w^2, L)$ is defined by (6). Then $a_n(w) \sim y_{0,n}(w^2)$ $(n \in \mathbf{N})$ and

$$y_{k,n}(w^2) \in \left(-y_{0,n}(w^2), y_{0,n}(w^2)\right) \qquad (k = 1, 2, \ldots, n, \quad n \ge n_1).$$

Let us also introduce the following notations:

$$\begin{aligned} y_{k,n} := y_{k,n}(w^2) &=: y_{0,n}(w^2)t_{k,n}(w^2, L) =: y_{0,n}(w^2)t_{k,n} \\ &=: y_{0,n}(w^2)\cos\vartheta_{k,n}(w^2, L) =: y_{0,n}(w^2)\cos\vartheta_{k,n}. \end{aligned}$$

For the distribution of the roots of $p_n(w^2, \cdot)$ the following holds.

Proposition 2.1 *For $w \in \mathcal{W}(I)$, $L > C_0$ (see (1.11) and Remark 1.6), $0 < c < \sqrt{T_1}/2$ and $n \in \mathbf{N}$ we have*

$$\vartheta_{k,n} \sim \begin{cases} \left(\dfrac{k}{nT_n}\right)^{1/3}, & \text{if } 1 \le k \le cnT_n^{-1/2} \\ \dfrac{k}{n}, & \text{if } cnT_n^{-1/2} \le k \le \left[\frac{n+1}{2}\right]; \end{cases} \tag{14}$$

$$\vartheta_{k+1,n} - \vartheta_{k,n} \sim \frac{\vartheta_{k,n}}{k} \sim \begin{cases} \frac{1}{(nT_n)^{1/3}k^{2/3}}, & \text{if } 1 \le k \le cnT_n^{-1/2} \\ \frac{1}{n}, & \text{if } cnT_n^{-1/2} \le k \le [\frac{n+1}{2}]; \end{cases} \tag{15}$$

$$t_{k,n} - t_{k+1,n} \sim \frac{\Psi_n(y_{k,n})}{n} \sim \frac{\vartheta_{k,n}^2}{k}$$

$$\sim \begin{cases} \frac{1}{(nT_n)^{2/3}k^{1/3}}, & \text{if } 1 \le k \le cnT_n^{-1/2} \\ \frac{k}{n^2}, & \text{if } cnT_n^{-1/2} \le k \le [\frac{n+1}{2}]. \end{cases} \tag{16}$$

If $w \in \mathcal{U}[-1,1]$ then $T_n \sim n^2$ $(n \in \mathbf{N})$, whence for the ultraspherical weights the *second* relations hold for *every* $1 \le k \le [(n+1)/2]$ in the formulae (14), (15) and (16).

If w is a Freud-type weight then $T_n \sim 1$ $(n \in \mathbf{N})$ and for $0 < c < \sqrt{T_1}/2$ we have $cnT_n^{-1/2} \le [(n+1)/2]$. In this case the *first* relations hold for *every* $1 \le k \le [(n+1)/2]$ in (14)–(16). For $w \in \mathcal{E}(\mathbf{R})$ we know that $T_n \to +\infty$ $(n \to +\infty)$ and for any $\varepsilon > 0$ there exists $c > 0$ independent of n such that $T_n = O(n^\varepsilon)$ $(n \in \mathbf{N})$ (see [2, (2.7)]). From these facts it follows that $nT_n^{-1/2}$ tends to $+\infty$ slowly than n if $n \to +\infty$. This is also true for $w \in \mathcal{EXP}[-1,1]$ since for some $\varepsilon > 0$ we have $T_n = O(n^{2-\varepsilon})$ $(n \to +\infty)$ (see [6, (3.8)]).

2.2

Using Proposition 2.1, we conclude some useful estimates. For $x \in I$ denote by $y_{j,n}$ (one of) the closest node(s) to x (shortly $x \approx y_{j,n}$) from $\{y_{k,n} \mid k = 1,2,\dots,n\}$ (see Part 1.6).

Proposition 2.2 *Let $w \in \mathcal{W}(I)$. Then*

$$H_n(w,x) := \sum_{k=1}^{n} u_{k,n}^2(w, U_n(w^2), x) \le c \quad (x \in I,\ n \in \mathbf{N}) \tag{17}$$

with some constant $c > 0$. Moreover there exists $c > 0$ such that for every fixed $s = 1,2,\dots,[(n+1)/2]$ and $K \ge 1$ we have

$$G_n(w) := \sum_{\substack{i=1 \\ |i-s| \ge K}}^{[(n+1)/2]} u_{s,n}^2(w, U_n(w^2), \tau_{i,n}) \le c\Big(\frac{1}{K} + \frac{1}{s}\Big) \tag{18}$$

for all $n \in \mathbf{N}$, where $\tau_{i,n} \in [y_{i,n}(w^2), y_{i-1,n}(w^2)]$ $(i = 1,2,\dots,[\frac{n+1}{2}])$.

2.3

For a given weight $w \in \mathcal{W}(I)$ we shall consider such point systems X_n $(n \in \mathbf{N})$ for which the fundamental functions $q_{k,n}$ of the weighted Lagrange interpolation (see (9)) are uniformly bounded on I, i.e., there exists a constant $A > 0$ such that

$$|q_{k,n}(w, X_n, x)| \le A \qquad (x \in I,\ k = 1, 2, \ldots, n,\ n \in \mathbf{N}). \tag{19}$$

They are the $E(w)$*-systems* (the letter E reminds of Erdős).

Proposition 2.3 *If $w \in \mathcal{W}(I)$ and $X_n = \{x_{k,n}\} \subset I$ $(n \in \mathbf{N})$ is an $E(w)$-system, then there exists a constant $S > 0$ such that*

$$|x_{k,n}| \le a_n(w)(1 + S\delta_n(w)) \qquad (1 \le k \le n,\ n \in \mathbf{N}). \tag{20}$$

Moreover, for $w \in \mathcal{U}[-1, 1]$ the constant S can be chosen such that

$$a_n(w)(1 + S\delta_n(w)) < 1, \quad 1 - a_n(w)(1 + S\delta_n(w)) \sim a_n(w)\delta_n(w) \quad (n \ge n_1)$$

also hold. (Actually, here $1 - a_n(w) \sim \delta_n(w) \sim n^{-2}$.)

2.4

A significant observation shows that the node-distribution of an $E(w)$-system *is similar to the root-distribution of the orthogonal polynomials $p_n(w^2, \cdot)$* $(n \in \mathbf{N})$. To formulate the exact result we have to introduce some notations. Let $X_n = \{x_{k,n}\}$ be an $E(w)$-system and $L > \max\{S, C_0\}$ a fixed constant (see (11) and (20)). Let

$$\begin{gathered} x_{0,n} := x_{0,n}(w, L) := a_n(w)(1 + 2L\delta_n(w)) =: -x_{n+1,n}(w, L) =: -x_{n+1,n} \\ x_{k,n} =: x_{0,n} r_{k,n}(w, L) =: x_{0,n} r_{k,n} =: x_{0,n} \cos\xi_{k,n}(w, L) =: x_{0,n} \cos\xi_{k,n} \\ (k = 0, 1, \ldots, n+1,\ n \in \mathbf{N}). \end{gathered} \tag{21}$$

Proposition 2.4 *If $w \in \mathcal{W}(I)$, $X_n = \{x_{k,n}\}$ $(n \in \mathbf{N})$ is an $E(w)$-system and $x_{0,n}, x_{n+1,n}$ are given by (21), then for every fixed $c \ge 0$ we have*

$$x_{k,n} - x_{k+1,n} \sim \frac{a_n}{n}\Psi_n(x_{t,n}) \tag{22}$$

$$(k = 0, 1, \ldots, n,\ t = 1, 2, \ldots, n-1,\ |t - k| \le c,\ n \in \mathbf{N}).$$

Moreover $\{r_{k,n}\}$ and $\{\xi_{k,n}\}$ satisfy formulae analogous (14)–(16).

To prove this result we need a Markov-Bernstein inequality.

Proposition 2.5 *Fix a weight $w \in \mathcal{W}(I)$, the number $L > 0$ and consider the point $x_{0,n}$ (see (6)). Then there exist $c > 0$ and $n_1 \in \mathbf{N}$ such that for $p \in \Pi_n$, $|x| \le a_n(1 + L\delta_n)$ and $n > n_1$*

$$\left|(pw)'(x)\right| \le c\frac{n}{a_n}\frac{1}{\Psi_n(x)}\|pw\|, \tag{23}$$

where Ψ_n is given by (7).

2.5

Now we formulate our main result which is analogue to the Theorem 1.1. First a definition. If $w \in \mathcal{W}(I)$ and $f \in C_w(I)$, then let

$$E_n(f,w) := \min_{P \in \Pi_n} \|w(f-P)\| \qquad (n \in \mathbf{N}).$$

(As it is well known, $E_n(f,w) \to 0$ for $n \to +\infty$.)

Theorem 2.6 *Let $w \in \mathcal{W}(I)$ and suppose that $X_n = \{x_{k,n}\}$ $(n \in \mathbf{N})$ is an $E(w)$-system. Then for every fixed $\varepsilon > 0$, to every $f \in C_{w^{1+\varepsilon}}(I)$ there exists a sequence of polynomials $\varphi_\Delta \in \Pi_\Delta$ such that*

(i) $\Delta \le n\big(1+\varepsilon + c\varepsilon(T_n/n^2)^{1/3}\big)$,
(ii) $\varphi_\Delta(x_{k,n}) = f(x_{k,n})$ $(k = 1,2,\dots,n,\ n \in \mathbf{N})$,
(iii) $\|w^{1+\varepsilon}(f-\varphi_\Delta)\| \le cE_\Delta(f, w^{1+\varepsilon})$.

3 Proofs

3.1 Proof of Proposition 2.1

If $w \in \mathcal{W}(I)$ then for $k = 1,2,\dots,n$ and $n \in \mathbf{N}$

$$y_{k,n} - y_{k+1,n} \sim \frac{a_n}{n}\Psi_n(y_{k,n}) \quad \text{and} \quad \Psi_n(y_{k,n}) \sim \Psi_n(y_{k+1,n}). \tag{24}$$

For $w \in \mathcal{F}(\mathbf{R})$, this follows from [1, Lemma 4.4 and (4.21)]; for $w \in \mathcal{E}(\mathbf{R})$ this follows from [2, (2.2)], [9, p. 205] and [3, (2.10)]; for $w \in \mathcal{EXP}[-1,1]$ this follows from [6, (1.35), (2.19) and (10.12)]; for $w \in \mathcal{U}[-1,1]$, see [15, p. 282].

In [19, Part 3.1] (see also [18, Part 4.1]) it was shown how to get (14)–(16) from (11) and (24).

3.2 Proof of Proposition 2.2

The proofs of this result for $w \in \mathcal{F}(\mathbf{R}) \cup \mathcal{EXP}[-1,1]$ (see [19, Part 3.10] and [18, Part 4.3]) show that beside the formulae (14)–(16) we used the following properties of w:

$$(p_n w)'(y_{k,n}) \sim \frac{n}{a_n^{3/2}\Psi_n(y_{k,n})} \cdot \frac{1}{\sqrt[4]{1 - \frac{|y_{k,n}|}{a_n} + 2L\delta_n}} \quad (k = 1,2,\dots,n,\ n \in \mathbf{N}); \tag{25}$$

$$|u_{k,n}(x)| := |u_{k,n}(w, U_n(w^2), x)| \le c \quad (x \in I,\ k = 1,2,\dots,n,\ n \in \mathbf{N}); \tag{26}$$

$$u_{k,n}(x) + u_{k+1,n}(x) \ge 1 \quad (x \in [y_{k+1,n}, y_{k,n}],\ k = 1,2,\dots,n,\ n \in \mathbf{N}); \tag{27}$$

$$|(p_n w)(x)| \sim |(p_n w)'(y_{j,n})| \cdot |x - y_{j,n}| \quad (x \approx y_{j,n} \in [-y_{0,n}, y_{0,n}],\ n \in \mathbf{N}). \tag{28}$$

For $w \in \mathcal{E}(\mathbf{R})$ (25) is (1.27) of [9], (26) is (2.14) of [3] (a more simple proof can be given using (14)–(16)), (27) is (2.15) of [3]. Finally, (28) can be proved in the same way as P. Vértesi did it for Freud-type weights (see [19, Part 3.2]). For $w \in \mathcal{U}[-1,1]$ (25) and (28) come from [15, (9.11) and (9.10)]; (26) can be obtained by a simple computation using (13), while (27) is a special case of [12].

3.3 Proof of Proposition 2.3

If $w \in \mathcal{F}(\mathbf{R}) \cup \mathcal{E}(\mathbf{R}) \cup \mathcal{EXP}[-1,1]$ then (20) was shown in [17]. (For $w \in \mathcal{F}(\mathbf{R}) \cup \mathcal{EXP}[-1,1]$ different proofs can be found in [19] and [18].)

Now, assume that $w \in \mathcal{U}[-1,1]$. Let $z_{1,n} \in [-1,1]$ be a point such that $\|\ell_{1,n}\| = |\ell_{1,n}(z_{1,n})|$ (see (1)) and let $z_{2,n} := z_{1,n} - \frac{1}{2n^2}$ if $z_{1,n} \geq 0$ and $z_{2,n} := z_{1,n} + \frac{1}{2n^2}$ if $z_{1,n} < 0$. Then $1 - z_{2,n}^2 = (1 - z_{2,n})(1 + z_{2,n}) \geq \frac{1}{2n^2}$. By the Markov-Bernstein inequality we have

$$|\ell_{1,n}(z_{1,n}) - \ell_{1,n}(z_{2,n})| = \left|\int_{z_{1,n}}^{z_{2,n}} \ell'_{1,n}(t)dt\right| \leq |z_{1,n} - z_{2,n}|n^2\|\ell_{1,n}\| \leq \frac{1}{2}|\ell_{1,n}(z_{1,n})|.$$

Therefore

$$|\ell_{1,n}(z_{2,n})| \geq |\ell_{1,n}(z_{1,n})| - |\ell_{1,n}(z_{1,n}) - \ell_{1,n}(z_{2,n})| \geq \frac{1}{2}|\ell_{1,n}(z_{1,n})| \geq \frac{1}{2}.$$

Let $1 - x_{1,n} := \frac{\varepsilon_n}{n^2}$ $(n \in \mathbf{N})$. We may assume that $x_{1,n} \geq 0$. Then by (19)

$$A \geq |q_{1,n}(z_{2,n})| = \left|\frac{w(z_{2,n})}{w(z_{1,n})}\ell_{1,n}(z_{2,n})\right| \geq \frac{(1 - z_{2,n}^2)^\alpha}{2(1 - x_{1,n}^2)^\alpha} \geq \frac{(\frac{1}{2n^2})^\alpha}{2^{\alpha+1}(\frac{\varepsilon_n}{n^2})^\alpha},$$

i.e., $\varepsilon_n \geq \left(4(2A)\right)^{-1/\alpha}$ $(n \in \mathbf{N})$. From it follows that there exists $c_1 > 0$ such that $0 \leq x_{1,n} \leq 1 - \frac{c_1}{n^2}$ $(n \in \mathbf{N})$. Similar argument shows that $1 + x_{n,n} \geq \frac{c_2}{n^2}$ $(n \in \mathbf{N})$.

It is known that (see [10, p. 553]) $1 - a_n(w) \sim n^{-2}$, $T_n(w) \sim n^2$ and $\delta_n(w) \sim n^{-2}$ $(n \in \mathbf{N})$ which completes the proof of the Proposition 2.3 for $w \in \mathcal{U}[-1,1]$.

3.4 Proof of Proposition 2.5

If $w \in \mathcal{F}(\mathbf{R})$ then there exists $c > 0$ such that for $p \in \Pi_n$, $n \in \mathbf{N}$ and $x \in \mathbf{R}$

$$|(pw)'(x)| \leq c\frac{n}{a_n}\left(\max\{1 - |x|/a_n, n^{-2/3}\}\right)^{1/2}\|pw\|$$

(see [1, (4.16)]). Therefore (23) follows from

$$\left(\max\{1 - |x|/a_n, n^{-2/3}\}\right)^{1/2} \sim \left(\Psi_n(x)\right)^{-1} \quad (|x| \leq a_n(1 + L\delta_n),\ n \in \mathbf{N}).$$

For $w \in \mathcal{EXP}[-1,1]$, (23) were proved in Part 4.12 of [18].

Now, suppose that $w \in \mathcal{E}(\mathbf{R})$. Fix $\eta \in (0,1)$ and $r > 0$. Let us consider three cases.

Case 1: $|x| \in [a_n(1-r\delta_n), a_n(1+L\delta_n)]$. Then for $p \in \Pi_n$ we have

$$|(pw)'(x)| \le c(a_n\delta_n)^{-1}\|pw\| \qquad (n \ge 1)$$

(see [9, Lemma 10.10]). Since

$$\frac{1}{a_n\delta_n} = \frac{1}{a_n}(nT_n)^{2/3} = \frac{n}{a_n}\Big(\frac{T_n^2}{n}\Big)^{1/3} = \frac{n}{a_n}T_n\sqrt{\delta_n} \qquad (n \ge 1)$$

and

$$\Psi_n(x) \sim \frac{1}{T_n\sqrt{\delta_n}} \qquad (|x| \in [a_n(1-r\delta_n), a_n(1+L\delta_n)],\ n \ge 1)$$

therefore (23) follows from the above relations.

Case 2: $|x| \le \eta a_n$. Then for $n > n_1$ and $p \in \Pi_n$ $|(pw)'(x)| \le c\frac{n}{a_n}\|pw\|$ (see [11, Lemma 4.2]) from which we obtain (23) using $\Psi_n(x) \sim 1$ $(|x| \le \eta a_n,\ n \ge 1)$.

Case 3: $|x| \in [\eta a_n, a_n(1-r\delta_n)] =: J_n$. It was shown in [11, (1.25)] and [9, (10.37)] that for $n > n_1$, $p \in \Pi_n$ and $|x| \in J_n$

$$|(pw)'(x)| \le c\Big(\frac{1}{1-\frac{|x|}{a_n}}\int_{\frac{|x|}{a_n}}^1 \kappa_n(t)\sqrt{1-t}\,dt\Big)\|pw\|, \tag{29}$$

where with a fixed $\xi > 0$

$$\kappa_n(t) := \int_{\xi/a_n}^1 \frac{a_ntQ'(a_nt) - a_nsQ'(a_ns)}{a_nt - a_ns}\frac{1}{\sqrt{1-s}}ds \quad (t \in [0,1]).$$

For $n = 1, 2, \dots$ and $t \in (-1,1)$, let

$$\mu_n(t) := \frac{2}{\pi^2}\int_0^1 \frac{a_nsQ'(a_ns) - a_ntQ'(a_nt)}{n(s^2-t^2)}\frac{\sqrt{1-t^2}}{\sqrt{1-s^2}}ds.$$

Then μ_n is even, $\mu_n(t) \ge 0$ a.e. in $(-1,1)$, $\int_{-1}^1 \mu_n(t)dt = 1$ (see [11, Lemma 3.1]) and for n large enough

$$\kappa_n(t) \sim \frac{n}{a_n}\frac{1}{\sqrt{1-|t|}}\mu_n(t) \quad \text{uniformly } \eta \le |t| < 1, \tag{30}$$

where $\eta \in (0,1)$ is a fixed number (see [11, (3.26)]). From (29) and (30) we get

$$|(pw)'(x)| \le c\Big(\frac{1}{1-\frac{|x|}{a_n}}\int_{\frac{|x|}{a_n}}^1 \mu_n(t)dt\Big)\frac{n}{a_n}\|pw\| \quad (x \in J_n,\ p \in \Pi_n,\ n \ge 1).$$

It is easy to show that the relations

$$|(pw)'(x)| \le c\Big(\frac{1}{1-\frac{|x|}{x_{0,n}}}\int_{\frac{|x|}{x_{0,n}}}^1 \mu_n(t)dt\Big)\frac{n}{a_n}\|pw\| \quad (x \in J_n,\ p \in \Pi_n,\ n \ge 1) \tag{31}$$

also hold.

Theorem 3.1 in [9] states that uniformly for $n \geq 1$ and $|t| < 1$ we have $\mu_n(t) \sim \min\{(1-t^2)^{-1/2}, T_n(1-t^2)^{1/2}\}$. Let $x =: x_{0,n}\cos\vartheta$, $t =: \cos\tau$ $(\tau, \vartheta \in (0, \pi/2))$. Then for $x \in J_n$ and $n > n_1$

$$\frac{1}{1-\frac{|x|}{x_{0,n}}}\int_{\frac{|x|}{x_{0,n}}}^{1}\mu_n(t)dt = \frac{1}{1-\cos\vartheta}\int_0^{\vartheta}\mu_n(\cos\tau)\sin\tau d\tau$$

$$\sim \frac{1}{\vartheta^2}\int_0^{\vartheta}\min\{\frac{1}{\tau}, T_n\tau\}\tau d\tau \sim \begin{cases} T_n\vartheta, & \text{if } 0 < \vartheta < \frac{1}{\sqrt{T_n}} \\ \frac{1}{\vartheta}, & \text{if } \frac{1}{\sqrt{T_n}} < \vartheta < \frac{\pi}{2} \end{cases} \sim \frac{1}{\Psi_n(x)}. \tag{32}$$

For $x \in J_n$, we obtain (23), using (29) and (32) which completes the proof of (23) for $w \in \mathcal{E}(\mathbf{R})$.

Now, suppose that $w \in \mathcal{U}[-1,1]$. Let $0 < L < L_0$ (see Remark 1.6), $n \in \mathbf{N}$, $x_{0,n} := a_n(1+2L\delta_n)$ and $x_n^* := x_{0,n} - a_nT_n^{-1}$. Then $\Psi_n(x) = \left(1-|x|/a_n+2L\delta_n\right)^{1/2}$ if $|x| \leq x_n^*$ and $\Psi_n(x) = T_n^{-1}\left(1-|x|/a_n+2L\delta_n\right)^{-1/2}$ if $x_n^* \leq |x| < x_{0,n}$. Since $1-a_n \sim 1/T_n \sim \delta_n \sim n^{-2}$ $(n \in \mathbf{N})$ thus

$$\Psi_n(x) \sim \Psi_n(a_n) \sim \frac{1}{T_n\sqrt{\delta_n}} \sim \frac{1}{n} \qquad (x_n^* \leq |x| < a_n(1+L\delta_n), \quad n \in \mathbf{N}) \tag{33}$$

and

$$\Psi_n(x) \sim \sqrt{1-x^2} \qquad (|x| \leq x_n^*, \quad n \in \mathbf{N}). \tag{34}$$

If $p \in \Pi_n$ then

$$(pw)'(x) = w(x)p'(x) - \frac{2\alpha x}{1-x^2}w(x)p(x) \qquad (x \in (-1,1)).$$

Let $x_n^* \leq x \leq a_n(1+L\delta_n)$. Using (33), $1-x^2 \geq cn^{-2}$ and $\|wp'\| \leq c_1n^2\|pw\|$ (see [10, (1.21)]) we have

$$(pw)'(x) \leq c_1n^2\|pw\| + c_2n^2\|pw\| \leq c_3\frac{n}{a_n}\frac{1}{\Psi_n(x)}\|pw\|.$$

If $0 \leq x \leq x_n^*$ then $|w(x)p'(x)| \leq c_1n\|wp\|/\sqrt{1-x^2}$ (see [4, (8.1.3)]). Therefore by $\frac{1}{1-x^2} \leq c_2\frac{n}{\sqrt{1-x^2}}$ we obtain

$$\left|(pw)'(x)\right| \leq c_3\frac{n}{\sqrt{1-x^2}}\|pw\| \qquad (0 \leq x \leq x_n^*)$$

whence we get (23) using $a_n \sim 1$ $(n \in \mathbf{N})$ and (34).

3.5 Proof of Proposition 2.4

First we prove the following statement. *Fix a weight $w \in \mathcal{W}(I)$. Let $X_n = \{x_{k,n}\}$ be an $E(w)$-system and $x_{0,n}$, $x_{n+1,n}$ $(n \in \mathbf{N})$ are given by (21). Then*

$$x_{k,n} - x_{k+1,n} \sim \frac{a_n}{n}\min\{\Psi_n(x_{k,n}), \Psi_n(x_{k+1,n})\} \; (k = 0,1,\dots,n, \; n \in \mathbf{N}). \tag{35}$$

The "$\geq$" part of (35) is a simple consequence of the Markov-Bernstein inequality (23) (see [19, Part 3.4] or [18, Part 4.12]). The proof of the "$\leq$" relation in (35) is more cumbersome. In [19, Part 3.5] and [18, 4.13] this statement were proved indirectly for $w \in \mathcal{F}(\mathbf{R}) \cup \mathcal{EXP}[-1,1]$ using Proposition 2.2. However, one can check that this argument works for $w \in \mathcal{E}(\mathbf{R}) \cup \mathcal{U}[-1,1]$, too.

From (35) it follows that $\Psi_n(x_{k,n}) \sim \Psi_n(x_{k+1,n})$, $k = 1,2,\ldots,n$, $n \in \mathbf{N}$. Moreover, $r_{k,n} := x_{k,n}/x_{0,n}$ and $\xi_{k,n}$ ($\cos\xi_{k,n} := r_{k,n}$) satisfy formulae analogous (14)–(16).

3.6 Proof of Theorem 2.6

Let $w \in \mathcal{W}(I)$, $n \in \mathbf{N}$, $\varepsilon > 0$,

$$m := [\frac{\varepsilon N}{2}] + 1 \quad \text{and} \quad N := [n(1 + d(\frac{T_n}{n^2})^{1/3}], \tag{36}$$

where $[x]$ denotes the integer part of $x \in \mathbf{R}$ and the constant $d > 0$ will be fixed later (see Part 3.7).

One of the crucial steps of the proof of Theorem 2.6 is the following result which we shall prove in the next part.

Proposition 3.1 *Let $w \in \mathcal{W}(I)$, m be given by (36) and suppose that $X_n = \{x_{k,n}\}$ ($n \in \mathbf{N}$) is an $E(w)$-system and $\varepsilon > 0$. Then there exist w^ε-weighted polynomials $h_{k,m}$ ($k = 1,2,\ldots,n$, $n \in \mathbf{N}$) such that*

(a) *the degree of the polynomial $h_{k,m}/w^\varepsilon$ is $\leq 2m$ ($k = 1,2,\ldots,n$, $n \in \mathbf{N}$);*
(b) *$h_{k,m}(x_{k,n}) = 1$ ($k = 1,2,\ldots,n$, $n \in \mathbf{N}$);*
(c) *$\sum_{k=1}^{n} |h_{k,m}(x)| \leq D$ ($x \in I$, $n \in \mathbf{N}$) with some constant $D > 0$.*

Now the proof of the Theorem 2.6 can be finished as follows. Let $\triangle := n - 1 + 2(m-1)$ (i.e., $\triangle \leq n(1 + \varepsilon + d\varepsilon(T_n/n^2)^{1/3})$) and define the polynomial

$$\varphi_\triangle(x) := P_\triangle(f,x) + \sum_{k=1}^{n} \delta_{k,n}(1+\varepsilon)\frac{q_{k,n}(w, X_n, x) h_{k,m}(x)}{w^{1+\varepsilon}(x)},$$

where $P_\triangle(f,\cdot) \in \Pi_\triangle$ is the polynomial for which

$$\|w^{1+\varepsilon}(f - P_\triangle(f,\cdot))\| = E_\triangle(f, w^{1+\varepsilon}),$$

$$\delta_{k,n}(1+\varepsilon) := \big(f(x_{k,n}) - P_\triangle(f, x_{k,n})\big) w^{1+\varepsilon}(x_{k,n}) \quad (k = 1,2,\ldots,n, \; n \in \mathbf{N})$$

and $q_{k,n}$ is given by (9).

Obviously $\varphi_\triangle$ interpolates f at the points of X_n. Moreover, by (19) and (c) we have

$$\big|w^{1+\varepsilon}(x)\big(\varphi_\triangle(x) - f(x)\big)\big| \leq \big|w^{1+\varepsilon}(x)\big(P_\triangle(f,x) - f(x)\big)\big|$$
$$+ \big|\sum_{k=1}^{n} \delta_{k,n}(1+\varepsilon) q_{k,n}(x) h_{k,m}(x)\big| \leq (1 + AD) E_\triangle(f, w^{1+\varepsilon})$$

for all $x \in I$, as it was stated.

3.7 Proof of Proposition 3.1

For $\varepsilon > 0$ let $p_m(x) := p_m(w^\varepsilon, x)$ be the mth orthonormal polynomial with respect to $w^\varepsilon \in \mathcal{W}(I)$. Its roots will be denoted by $y_{k,m} := y_{k,m}(w^\varepsilon)$ ($k = 1, 2, \ldots, m$, $m \in \mathbf{N}$). Then the following is true. *If* $\{x_{k,n}\}$ *is an* $E(w)$*-system then the constant* d *in* (36) *can be chosen such that*

$$x_{k,n} \in \left[y_{m,m}(w^\varepsilon), y_{1,m}(w^\varepsilon)\right] \quad (k = 1, 2, \ldots, n,\ n > n_0). \tag{37}$$

For $w \in \mathcal{F}(\mathbf{R}) \cup \mathcal{EXP}[-1,1]$ this statement were proved in [19, Part 3.9] and [18, Part 4.16]. For $w \in \mathcal{E}(\mathbf{R})$ one can prove (37) in the same way using (11), $a_u(w^v) = a_{u/v}(w)$ for all $u, v > 0$ (see (3)), $T_{\alpha u} \sim T_{\beta u}$ uniformly for $u \geq c$, where $0 < \alpha < \beta$ (see [9, (2.8)]) and

$$\left|\frac{a_u}{a_v} - 1\right| \sim \left|\frac{u}{v} - 1\right| \frac{1}{T_n} \quad (u \in (c, +\infty), v \in [u/2, 2u])$$

(see [9, (2.12)]). If $w \in \mathcal{U}[-1,1]$ then (37) is given by a simple calculation from $x_{1,n} = 1 - c_1 n^{-2}$ and $y_{1,m} = 1 - c_2 n^{-2}$.

Let us define the following weighted polynomials

$$H_{i,m}(x) := \left(\frac{\Omega_m(x)}{\Omega'_m(y_{i,m})(x - y_{i,m})}\right)^2 \quad (x \in I,\ i = 1, 2, \ldots, m,\ m \in \mathbf{N}),$$

where $\Omega_m(x) := w^{\varepsilon/2}(x) p_m(w^\varepsilon, x)$. Using (17) we have

$$\sum_{i=1}^{m} H_{i,m}(x) \leq d_1 \qquad (x \in I,\ m \in \mathbf{N}) \tag{38}$$

with a proper constant $d_1 > 0$. Now let us define the indices $c(k) = c(k, n)$ by

$$\min_{1 \leq i \leq n} |x_{k,n} - y_{i,m}| = |x_{k,n} - y_{c(k),m}| \qquad (k = 1, 2, \ldots, n,\ n \in \mathbf{N}).$$

(If there exist for a fixed k and n at least two $c(k,n)$, we can choose any of them.) Using Propositions 2.1 and 2.4 we have as follows. If

$$c(r, n) = c(s, n) \quad \Longrightarrow \quad |r - s| \leq d_0 \quad (r, s = 1, 2, \ldots, n,\ n \in \mathbf{N}) \tag{39}$$

with a constant $d_0 > 0$.

Now consider the weighted polynomials

$$h_{k,m}(x) := \frac{H_{c(k),m}(x)}{H_{c(k),m}(x_{k,n})} \qquad (x \in I,\ k = 1, 2, \ldots, n,\ n \in \mathbf{N}). \tag{40}$$

It is clear that $h_{k,m}/w^\varepsilon \in \Pi_{2m}$ and $h_{k,m}(x_{k,n}) = 1$ $k = 1, 2, \ldots, n$, $n \in \mathbf{N}$). Moreover by (28) and (37) we have $H_{c(k),m}(x_{k,n}) \sim 1$ ($k = 1, 2, \ldots, n$, $n \in \mathbf{N}$). Therefore from (38) and (39) it follows that there exists $D > 0$ such that $\sum_{k=1}^{n} |h_{k,m}(x)| \leq D$ for all $x \in I$ and $n \in \mathbf{N}$ which proves the Proposition 3.1.

References

[1] G. Criscuolo, B. Della Vecchia, D. Lubinsky and G. Mastroianni, *Functions of second kind for Freud weights*, J. Math. Anal. Appl., **189** (1994), 256–296.

[2] S.B. Damelin, *The Lebesgue functions and Lebesgue constant of Lagrange interpolation for Erdős weights*, J. Approx. Theory, **94** (1998), 235–262.

[3] S.B. Damelin and D.S. Lubinsky, *Necessary and sufficient conditions for mean convergence of Lagrange interpolation for Erdős weights*, Can. J. Math., **48** (1996), 710–736.

[4] Z. Ditzian and V. Totik, *Moduli of Smoothness*, Springer-Verlag, New York (1987).

[5] P. Erdős, *On some convergence properties of the interpolation polynomials*, Ann. of Math., **44** (1943), 330–337.

[6] A.L. Levin and D.S. Lubinsky, *Christoffel Functions and Orthogonal Polynomials for Exponential Weights on* $[-1,1]$, Mem. Amer. Math. Soc. 535, Vol. 111 (1994).

[7] A.L. Levin and D.S. Lubinsky, *Christoffel functions, orthogonal polynomials, and Nevai's conjecture for Freud weights*, Constr. Approx., **8** (1992), 463–535.

[8] A.L. Levin and D.S. Lubinsky, *Orthogonal Polynomials for Exponential Weights*, Springer-Verlag, New York-Berlin-Heidelberg (2001).

[9] A.L. Levin, D.S. Lubinsky and T.Z. Mthembu, *Christoffel functions and orthogonal polynomials for Erdős weights on* $(-\infty,\infty)$, Rendiconti di Matematica, Serie VII, **14**, Roma (1994), 199–289.

[10] D.S. Lubinsky and E.B. Saff, *Markov-Bernstein and Nikolskii inequalities, and Christoffel functions for exponential weights on* $(-1,1)$, SIAM J. Math. Anal., **24(2)** (1993), 528–556.

[11] D.S. Lubinsky, L_∞ *Markov and Bernstein inequalities for Erdős weights*, J. Approx. Theory, **60** (1990), 188–230.

[12] D.S. Lubinsky, *An extension of the Erdős-Turán inequality for the sum of succesive fundamental polynomials*, Ann. of Numer. Math., **2** (1995), 305–309.

[13] H.N. Mhaskar and E.B. Saff, *Where does the sup norm of a weighted polynomial live?* Constr. Approx., **1** (1985), 71–79.

[14] J. Szabados, *Weighted Lagrange and Hermite-Fejér interpolation on the real line*, J. of Inequal. and Appl., **1** (1997), 99–112.

[15] J. Szabados and P. Vértesi, *Interpolation of Functions*, World Sci. Publ., Singapore-New Jersy-London-Hong Kong (1990).

[16] G. Szegő, *Orthogonal Polynomials*, AMS Coll. Publ., Vol. 23, Providence, 1978.

[17] L. Szili, *On the "good" nodes of weighted Lagrange interpolation for exponential weights*, Annales Univ. Sci. Budapest., 2001, (to appear).

[18] L. Szili and P. Vértesi, *An Erdős-type convergence process in weighted interpolation. II (Exponential weights on $[-1,1]$)*, Acta Math. Hungar., **98(1–2)** (2003).

[19] P. Vértesi, *An Erdős-type convergence process in weighted interpolation. I (Freud-type weights)*, Acta Math. Hungar., **91 (3)** (2001), 195–215.

[20] P. Vértesi, *On the Lebesgue function of weighted Lagrange interpolation. II*, J. Austral. Math. Soc. (Series A), **65** (1998), 145–162.

[21] P. Vértesi, *On the Lebesgue function of weighted Lagrange interpolation. I. (Freud-type weights)*, Constr. Approx., **15** (1999), 355–367.

László Szili
Department of Numerical Analysis
Loránd Eötvös University
Budapest, Pázmány P. sétány I/C
H-1117, Hungary
Email address: szili@ludens.elte.hu

Péter Vértesi
Rényi Mathematical Institute of the
Hungarian Academy of Sciences
Budapest, Reáltanoda u. 13–15
H–1053, Hungary
Email address: vertesi@renyi.hu

Research supported by the Hungarian National Scientific Research Foundation (OTKA) under Grant Ns. T032719, T32872, T37299 and T22943.

Advanced Problems in Constructive Approximation
(Eds.) M.D. Buhmann and D.H. Mache
International Series of Numerical Mathematics Vol. 142, 255–262

Absolute Continuity of Spectral Measure for Certain Unbounded Jacobi Matrices

Ryszard Szwarc

Abstract

Spectral properties of unbounded symmetric Jacobi matrices are studied. Under mild assumptions on the coefficients absolute continuity of spectral measure is proved. Only operator theoretic proofs are provided. Some open problems of Ifantis are solved.

1 Introduction

Let J be a Jacobi matrix of the form

$$J = \begin{pmatrix} \beta_0 & \lambda_1 & 0 & 0 & \cdots \\ \lambda_1 & \beta_1 & \lambda_2 & 0 & \cdots \\ 0 & \lambda_2 & \beta_2 & \lambda_3 & \ddots \\ 0 & 0 & \lambda_3 & \beta_3 & \ddots \\ \vdots & \vdots & \ddots & \ddots & \ddots \end{pmatrix}, \tag{1}$$

where $\lambda_n > 0$, for $n \geq 1$, and $\beta_n \in \mathbf{R}$, for $n \geq 0$. The matrix J gives rise to a symmetric operator on the Hilbert space $\ell^2(\mathbf{N})$ of square summable complex sequences $a = \{a_n\}_{n=0}^{\infty}$, with the domain $D(J)$ consisting of sequences with finitely many nonzero terms. This operator acts by the rule

$$(Ja)_n = \lambda_{n+1}a_{n+1} + \beta_n a_n + \lambda_n a_{n-1},$$

for $n \geq 0$, with the convention that $a_{-1} = \lambda_0 = 0$. It is well known that this operator admits selfadjoint extensions (see [1]). In case the extension is unique the operator is called essentially selfadjoint. Then there exists a unique probability measure μ on $\mathbf{R}$, with finite moments, such that

$$(J^n\delta_0, \delta_0)_{\ell^2(\mathbf{N})} = \int_{\mathbf{R}} x^n \, d\mu(x),$$

where $\delta_0 = (1, 0, 0, \ldots)$. This measure is called the spectral measure of the operator J, because it can be shown that the operator J is unitarily equivalent to the operator M_x acting on $L^2(\mathbf{R}, \mu)$ by the rule

$$M_x f(x) = x f(x).$$

This unitary equivalence is defined as follows. Let $p_n(x)$ be a system of polynomials orthonormal with respect to the inner product in $L^2(\mathbf{R}, \mu)$. Then the operator $U\delta_n = p_n$ extends to an isometry from $\ell^2(\mathbf{N})$ onto $L^2(\mathbf{R}, \mu)$, where δ_n denotes the sequence whose nth term is equal to 1, and all other terms are equal to 0. Since

$$J\delta_n = \lambda_{n+1}\delta_{n+1} + \beta_n\delta_n + \lambda_{n-1}\delta_{n-1},$$

we have

$$xp_n = \lambda_{n+1}p_{n+1} + \beta_n p_n + \lambda_{n-1}p_{n-1}.$$

In this paper we will be dealing with special unbounded Jacobi matrices such that $\lambda_n \to +\infty$ and

$$\frac{\lambda_n^2}{\beta_n\beta_{n-1}} \xrightarrow{n} \alpha.$$

It is known that if J is essentially selfadjoint and $\alpha < \frac{1}{4}$ the measure μ is discrete (see [2]). In [5] Ifantis stated a problem of studying the spectra of operators for which $\alpha > \frac{1}{4}$. In this note we are going to show that the spectra of such operators cover the whole real line and, under some mild conditions on the coefficients, the spectral measure is absolutely continuous. We will also provide an operator theoretic proof for the case $\alpha < \frac{1}{4}$, which was also one of the problems stated by Ifantis.

2 Main results

Our considerations will rely heavily on the following generalization of a result of Maté and Nevai. We will state it in a form which will be useful for our considerations. We will also provide a proof different from the one in [8], and based on ideas from [4].

Theorem 2.1 (Maté, Nevai) *Let $\Lambda_n(x)$ be a positive valued sequence whose terms depend continuously on $x \in [a, b]$. Let $a_n(x)$ be a real valued sequence of continuous functions satisfying*

$$\Lambda_{n+1}(x)a_{n+1}(x) + Ba_n(x) + \Lambda_n(x)a_{n-1}(x) = 0,$$

for $n \geq N$. Assume the sequence $\Lambda_n(x)$ has bounded variation and $\Lambda_n(x) \to \frac{1}{2}$ for $x \in [a,b]$. Let $|B| < 1$. Then there is a positive function $f(x)$ continuous on $[a,b]$ such that

$$a_n^2(x) - a_{n-1}(x)a_{n+1}(x) \xrightarrow{n} f(x)$$

uniformly for $x \in [a,b]$. Moreover there is a constant c such that

$$|a_n(x)| \leq c$$

for $n \geq 0$ and $x \in [a,b]$.

Proof. Let

$$\Delta_n(x) = a_n^2(x) - a_{n-1}(x)a_{n+1}(x),$$

for $n \geq N$. By using the recurrence relation one can show that

$$\Delta_{n+1} - \Delta_n = \left(1 - \frac{\Lambda_n}{\Lambda_{n-1}}\right) a_{n+1}^2 + \left(1 - \frac{\Lambda_n}{\Lambda_{n+1}}\right) a_n^2 + B\left(\frac{1}{\Lambda_{n+1}} - \frac{1}{\Lambda_{n-1}}\right) a_n a_{n+1}.$$

Hence

$$|\Delta_{n+1} - \Delta_n| \leq c(|\Lambda_{n-1} - \Lambda_n| + |\Lambda_n - \Lambda_{n+1}|)(a_n^2 + a_{n+1}^2). \tag{2}$$

On the other hand

$$\begin{aligned}
\Delta_n &= a_n^2 + \frac{\Lambda_n}{\Lambda_{n-1}} a_{n+1}^2 + \frac{B}{\Lambda_{n-1}} a_n a_{n+1} \\
&= \left(a_n + \frac{B}{2\Lambda_{n-1}}\right)^2 + \left(\frac{\Lambda_n}{\Lambda_{n-1}} - \frac{B^2}{4\Lambda_{n-1}^2}\right) a_{n+1}^2 \\
&= \frac{\Lambda_n}{\Lambda_{n-1}} \left(a_{n+1} + \frac{B}{2\Lambda_n}\right)^2 + \left(1 - \frac{B^2}{4\Lambda_{n-1}\Lambda_n}\right) a_n^2.
\end{aligned}$$

Since $\Lambda_n \xrightarrow{n} \frac{1}{2}$, uniformly for $x \in [a,b]$, and $|B| < 1$ we have

$$a_n^2 + a_{n+1}^2 \leq 2c'\Delta_n, \qquad \text{where} \quad (c')^{-1} = \frac{1}{2} - \frac{B^2}{2}, \tag{3}$$

for n sufficiently large. Combining this with (2) gives

$$|\Delta_{n+1} - \Delta_n| \leq 2cc'(|\Lambda_{n-1} - \Lambda_n| + |\Lambda_n - \Lambda_{n+1}|)\Delta_n.$$

Let

$$\varepsilon_n = 2cc'(|\Lambda_{n-1} - \Lambda_n| + |\Lambda_n - \Lambda_{n+1}|).$$

Then

$$(1 - \varepsilon_n)\Delta_n \leq \Delta_{n+1} \leq (1 + \varepsilon_n)\Delta_n,$$

for n sufficiently large. Thus the sequence Δ_n is convergent uniformly to a positive function $f(x)$ for $x \in [a,b]$. Moreover by (3) we obtain the second part of the statement. □

The main result of this note is following.

Theorem 2.2 *Assume the sequences* λ_n *and* β_n *satisfy* $\lambda_n \to +\infty$, $|\beta_n| \xrightarrow{n} +\infty$, $\beta_n/\beta_{n-1} \xrightarrow{n} 1$ *and*

$$\frac{\lambda_n^2}{\beta_{n-1}\beta_n} \xrightarrow{n} \frac{1}{4B^2} > \frac{1}{4}.$$

Let the sequences

$$\frac{\lambda_n^2}{\beta_{n-1}\beta_n}, \qquad \frac{\beta_{n-1}+\beta_n}{\lambda_n^2}, \qquad \frac{1}{\lambda_n^2}$$

have bounded variation. Then the corresponding Jacobi matrix J *is essentially selfadjoint if and only if* $\sum \lambda_n^{-1} = \infty$. *In that case the spectrum of* J *coincides with the whole real line and the spectral measure is absolutely continuous.*

Proof. We may assume that $\beta_n \xrightarrow{n} +\infty$. Assume that J is essentially selfadjoint. Let μ denote the spectral measure of J. Fix a real number x. Consider the difference equation

$$xy_n = \lambda_{n+1}y_{n+1} + \beta_n y_n + \lambda_{n-1}y_{n-1}, \tag{4}$$

for $n \geq 1$. By [7] the measure μ is absolutely continuous on the set of those x for which the ratio

$$\frac{\sum_{k=1}^{n} |u_k|^2}{\sum_{k=1}^{n} |v_k|^2} \tag{5}$$

remains bounded above for any n, for any fixed solutions u_n and v_n of (4). We are going to show that this ratio is always bounded. Let a_n satisfy (4). Let N be large enough so that $\beta_n > x$ for $n \geq N$. Set

$$a_n(x) = y_n\sqrt{\beta_n - x}, \quad \text{for } n \geq N. \tag{6}$$

The equation (4) can be transformed into the following.

$$\Lambda_{n+1}(x)a_{n+1}(x) + Ba_n(x) + \Lambda_n a_{n-1}(x) = 0, \tag{7}$$

for $n \geq N$, where

$$\Lambda_n(x) = B\frac{\lambda_n}{(\beta_{n-1}-x)(\beta_n - x)}. \tag{8}$$

By assumptions we have $\Lambda_n \xrightarrow{n} \frac{1}{2}$ and $|B| < 1$. Moreover $\Lambda_n(x)$ has bounded variation if and only if $\Lambda_n^{-2}(x)$ has bounded variation. But

$$\Lambda_n^{-2}(x) = \frac{\beta_{n-1}\beta_n}{\lambda_n^2} - \frac{\beta_{n-1}+\beta_n}{\lambda_n^2}x + \frac{1}{\lambda_n^2}x^2.$$

Theorem 2.1 implies

$$a_n^2(x) - a_{n-1}(x)a_{n+1}(x) \xrightarrow{n} C > 0$$

and $a_n(x)$ is a bounded sequence. Using (6), the boundedness of $a_n(x)$ and the assumptions on β_n we obtain that

$$\beta_n(y_n^2 - y_{n-1}y_{n+1}) \xrightarrow{n} C$$

and $\beta_n y_n^2$ is bounded. Therefore there exist positive constants c and M such that

$$\begin{aligned} \beta_n y_n^2 &\leq c \\ \beta_n(y_n^2 - y_{n-1}y_{n+1}) &\geq c^{-1} \end{aligned}$$

for $n \geq M$. If J is essentially selfadjoint there exists a solution y_n of (4) which is not square summable. Thus $\sum \beta_n^{-1} = +\infty$. Hence $\sum \lambda_n^{-1} = +\infty$.

We have

$$c \leq \beta_n(y_n^2 - y_{n-1}y_{n+1}) \leq \beta_n(y_{n-1}^2 + y_n^2 + y_{n+1}^2) \leq c'.$$

for $n \geq M$. Now if u_n and v_n are arbitrary nonzero solutions of (4) we have

$$\frac{u_{n-1}^2 + u_n^2 + u_{n+1}^2}{v_{n-1}^2 + v_n^2 + v_{n+1}^2} \leq \frac{c'}{c}.$$

This implies the ratio in (5) is bounded. □

Remark 2.3 Let p_n be the polynomials satisfying

$$xp_n = \lambda_{n+1}p_{n+1} + \beta_n p_n + \lambda_n p_{n-1}.$$

By the proof of Theorem 2.2 we get that

$$\beta_n[p_n^2(x) - p_{n-1}(x)p_{n+1}(x)] \xrightarrow{n} f(x) > 0,$$

uniformly on any bounded interval. and

$$\beta_n p_n^2(x) \leq c$$

on any bounded interval. In the case of bounded λ_n and β_n Maté and Nevai showed that the limit $f(x) = \lim_n[p_n^2(x) - p_{n-1}(x)p_{n+1}(x)]$ is closely related with the density of the spectral measure of J, which coincides with the orthogonality measure for the polynomials p_n. Namely they showed that if $\lambda_n \xrightarrow{n} 1/2$ and $\beta_n \xrightarrow{n} 0$ then the orthogonality measure μ is absolutely continuous in the interval $(-1, 1)$ and its density is given by

$$\frac{2\sqrt{1-x^2}}{\pi f(x)}, \quad -1 < x < 1.$$

Remark 2.4 A similar result has been obtained recently by Janas and Moszyski [6] under stronger assumptions that the sequences

$$\frac{\lambda_{n-1}}{\lambda_n}, \quad \frac{1}{\lambda_n}, \quad \frac{\beta_{n-1}}{\lambda_n}$$

have all bounded variation. It can be verified easily that these assumptions imply the assumptions of Theorem 2.2. Moreover there are examples showing that our assumptions are actually weaker. Indeed, let

$$\beta_n = n + 1 + (-1)^n, \qquad \lambda_n = \sqrt{\beta_{n-1}\beta_n}.$$

One can verify that β_n/λ_n does not have bounded variation while the assumptions of Theorem 2.2 are satisfied.

Example 2.5 Let $\lambda_n = n^\kappa$ and $\beta_n = \beta n^\kappa$, where $|\beta| < 1$ and $0 < \kappa \le 1$ (see [5])). By the Carleman criterion the corresponding Jacobi matrix is essentially selfadjoint. Moreover all the assumptions of Theorem 2.2 are satisfied. Hence the spectrum of J cover the whole real line and the spectral measure is absolutely continuous. Also we have that the corresponding orthonormal polynomials satisfy

$$\begin{aligned} &n^\kappa \left[p_n^2(x) - p_{n-1}(x)p_{n+1}(x)\right] \to f(x) > 0, \\ &n^\kappa |p_n(x)| \le c, \end{aligned}$$

uniformly with respect to x from any bounded interval $[a, b]$.

The next theorem is known (see [3]). We give an operator theoretic proof. Finding such a proof was one of the open problems stated in [5].

Theorem 2.6 (Chihara) *Let J be a Jacobi matrix given by* (1) *and satisfying*

$$\frac{\lambda_n^2}{\beta_{n-1}\beta_n} \to \frac{1}{4B^2} < \frac{1}{4}.$$

Let $\lambda_n \to +\infty$ and $\beta_n \to \infty$. Assume J is essentially selfadjoint. Then the spectrum of J is discrete and consists of a sequence of points convergent to $+\infty$.

Proof. It suffices to show that for every real number M there are only finitely many points in the spectrum $\sigma(J)$ which are less than M. Fix M. By assumptions there is N such that $\beta_{n+N-1} > M$ and

$$\frac{\lambda_{n+N}^2}{(\beta_{n+N-1} - M)(\beta_{n+N} - M)} \le \frac{1}{4}, \tag{9}$$

for $n \ge 0$. Let J_N be the Jacobi matrix defined as

$$J_N = \begin{pmatrix} \beta_{N-1} & \lambda_N & 0 & 0 & \cdots \\ \lambda_N & \beta_N & \lambda_{N+1} & 0 & \cdots \\ 0 & \lambda_{N+1} & \beta_{N+1} & \lambda_{N+2} & \ddots \\ 0 & 0 & \lambda_{N+2} & \beta_{N+2} & \ddots \\ \vdots & \vdots & \ddots & \ddots & \ddots \end{pmatrix}.$$

We will show that

$$\sigma(J_N) \subseteq [M, +\infty)$$

by estimating the quadratic form $(J_N x, x)_{\ell^2(\mathbf{N})}$ from below by $M(x,x)_{\ell^2(\mathbf{N})}$. Let x be a real valued sequence. Set $\beta'_n = \beta_{n+N-1} - M$ and $\lambda'_n = \lambda_{n+N}$. Then by (9) we have

$$\begin{aligned}
(J_N x, x)_{\ell^2(\mathbf{N})} - M(x,x)_{\ell^2(\mathbf{N})} &= \sum_{n=0}^{\infty} \beta'_n x_n^2 + 2\sum_{n=0}^{\infty} \lambda'_n x_n x_{n+1} \\
&\geq \sum_{n=0}^{\infty} \beta'_n x_n^2 - 2\sum_{n=0}^{\infty} \lambda'_n |x_n||x_{n+1}| \\
&\geq \sum_{n=0}^{\infty} \beta'_n x_n^2 - \sum_{n=0}^{\infty} \sqrt{\beta'_n}\sqrt{\beta'_{n+1}}|x_n||x_{n+1}| \\
&\geq \sum_{n=0}^{\infty} \beta'_n x_n^2 - \frac{1}{2}\sum_{n=0}^{\infty} \beta'_n x_n^2 - \frac{1}{2}\sum_{n=1}^{\infty} \beta'_n x_n^2 \\
&= \frac{1}{2}\beta'_0 x_0^2 \geq 0.
\end{aligned}$$

Hence

$$(J_N x, x)_{\ell^2(\mathbf{N})} \geq M(x,x)_{\ell^2(\mathbf{N})},$$

and consequently $\sigma(J_N) \subseteq [M, +\infty)$. Let 0_N denote the $N \times N$ matrix with all entries equal to zero. Observe that the Jacobi matrix J can be written in the form

$$J = J_0 + (0_N \oplus J_N),$$

where J_0 is a finite-dimensional Jacobi matrix. We have

$$\sigma(0_N \oplus J_N) = \{0\} \cup \sigma(J_N).$$

By the Weyl perturbation theorem the spectra of J and $0_N \oplus J_N$ may differ by at most N points. Hence $\sigma(J)$ can have at most $N+1$ points to the left of M. □

References

[1] Akhiezer, N.I., *The Classical Moment Problem,* Hafner Publ. Co., New York, 1965.

[2] T. Chihara, *An Introduction to Orthogonal Polynomials,* Mathematics and Its Applications, Vol. **13**, Gordon and Breach, New York, London, Paris, 1978.

[3] T. Chihara, *Chain sequences and orthogonal polynomials,* Trans. Amer. Math. Soc., **104** (1962), 1–16.

[4] J. Dombrowski and P. Nevai, *Orthogonal polynomials, measures and recurrence relations,* SIAM J. Math. Anal., **17** (1986), 752–759.

[5] E.K. Ifantis, *On the spectral measure of a class of orthogonal polynomials,* J. Comp. Appl. Math., **133** (2001), 688–689.

[6] J. Janas and M. Moszyński, *Alternative approaches to the absolute continuity of Jacobi matrices with monotonic weights,* Int. Eq. Oper. Theory, (to appear).

[7] S. Khan and D.B. Pearson, *Subordinacy and spectral theory for inifinite matrices,* Helv. Phys. Acta **65** (1992), 505–527.

[8] A. Maté and P. Nevai, *Orthogonal polynomials and absolutely continuous measures,* In C.K. Chui et al., editor, Approximation Theory IV, Vol. **103**, Academic Press, New York, (1983), 611–617.

[9] B. Simon, *The classical moment problem as a self-adjoint finite difference operator,* Advances Math. **137** (1998), 82–203.

Institute of Mathematics
Wrocław University
pl. Grunwaldzki 2/4
50–384 Wrocław, Poland
Email address: szwarc@math.uni.wroc.pl

This work was partially supported by KBN (Poland) under grant 5 P03A 034 20.

Advanced Problems in Constructive Approximation
(Eds.) M.D. Buhmann and D.H. Mache
International Series of Numerical Mathematics Vol. 142, 263–274

Approximation on Compact Subsets of R

Vilmos Totik

Dedicated to Béla Csákány on his 70-th birthday

Abstract

The role polynomial approximation of the $|x|$ function in approximation theory, as well as some recent developments on the subject is discussed. In particular, we show how a strengthening of the classical approximation of $|x|$ can lead to the Jackson theory.

1 The role of approximation of $|x|$

Weierstrass' theorem on polynomial approximation is the most fundamental results in approximation, and there are many different approaches to it, see the excellent article [9].

One of the possible approaches, due to Lebesgue [8] (cf. [9]) is via approximating $|x|$ on $[-1,1]$. In fact, if we can approximate $|x|$ on $[-1,1]$ by polynomials with any accuracy, then by translation we get the same fact for any $|x-a|$, $a \in [0,1]$ on the interval $[0,1]$. Now every piecewise linear continuous function ("broken line") is a linear combination of such functions, therefore every "broken line" is approximable arbitrarily well by polynomials. But it is clear that to any continuous function we can put arbitrarily closely a piecewise linear continuous function, hence we can approximate any continuous function on $[0,1]$ by polynomials with any error.

Thus, the Weierstrass theorem follows from its special case for the function $|x|$, which can be achieved e.g. by taking an appropriate section of the expansion

$$|x| = [1-(1-x^2)]^{1/2} = \sum_{k=0}^{\infty} \binom{1/2}{k} (-1)^k (1-x^2)^k,$$

which can be shown to be uniformly convergent for $x \in [-1,1]$.

A refinement of the Weierstrass theory is Jackson's theory, which connects the rate of approximation to smoothness. It is a natural question if the above argument can be extended so as to obtain the Jackson theorems. To answer this question we need to know how well $|x|$ can be approximated by polynomials of a given degree. Let for $f \in F$

$$E_n(f, F) = \inf_{\deg(P_n)} \|f - P_n\|_F$$

denote the error of best approximation of f on the set $F \subset \mathbb{R}$ by polynomials of degree at most n. In 1914 S.N. Bernstein [1] proved that

$$\lim_{n\to\infty} nE_n(|x|, [-1, 1])$$

exists, it is finite and positive, which implies among others, that $E_n(|x|, [-1, 1])$ decreases to zero with rate n^{-1}. Thus, there are polynomials Q_n of degree n such that

$$||x| - Q_n(x)| \le \frac{C}{n}, \qquad x \in [-1, 1],$$

but this does not seem to be strong enough to deduce Jackson's theorem. To do that we need a refinement of the preceding inequality. This refinement is related to the fact that with $|x|$ there is trouble only at the origin, otherwise it is a nice (analytic) function. This suggests, that although on the whole interval $[-1, 1]$ we cannot approximate $|x|$ by polynomials of degree at most n better than C/n, perhaps this order can be improved as we move away from the origin. There are several papers on improved local approximation of functions away from singular points (see e.g. [7], [13]), and the main technique of those papers is to use fast decreasing polynomials. We shall illustrate this approach for the function $|x|$, and then we shall deduce Jackson's theorem from the improved approximation rate.

2 Fast decreasing polynomials

Many applications in mathematical analysis require polynomials that take the value 1 at the origin and are fast decreasing on $[-1, 1]$ as we move away from the origin. Thus, these fast decreasing polynomials (Paul Erdős used the term pin polynomials) imitate the Dirac delta as close as possible. Thus, we are looking for polynomials with the property

$$P(0) = 1, \quad |P(x)| \le e^{-\varphi(x)}, \quad x \in [-1, 1], \tag{1}$$

where φ is a given even function that typically involves the degree of P, as well.

As we shall see, the integrals of such polynomials provide good approximation to the signum function and thereby they serve as the building blocks for well-localized "polynomial partition of unity". Once we have such well-localized "partition of unity", the connection between smoothness and order of polynomial approximation becomes a fairly simple thing, for the "partition of unity" allows us to use local approximation.

The problem can be formulated in two different ways: find the fastest decreasing polynomials of a given degree, or alternatively, find the smallest possible degree for the polynomial P in (1). This degree is denoted by n_φ.

The order of n_φ can be estimated by an explicitly computable quantity as follows (see [6]):

Let φ be an even function, right continuous and increasing on $[0,1]$. Then

$$\frac{1}{6}N_\varphi \le n_\varphi \le 12N_\varphi$$

where $N_\varphi = 0$ if $\varphi(1) \le 0$ and

$$N_\varphi = 2 \sup_{\varphi^{-1}(0)\le x<b} \sqrt{\frac{\varphi(x)}{x^2}} + \int_b^{1/2} \frac{\varphi(x)}{x^2}dx + \sup_{1/2\le x<1} \frac{\varphi(x)}{-\log(1-x)} + 1,$$

$b = \min(\varphi^{-1}(1), 1/2)$, otherwise.

Here, for $u \ge 0$,

$$\varphi^{-1}(u) = \sup\{\tau \,|\, \tau \in [0,1],\ \ \varphi(\tau) \le u\}.$$

If $N_\varphi = \infty$, then the statement of the theorem means that there are no polynomials whatsoever with the stated properties.

As special cases the following hold. Let φ be an even and on $[0,\infty)$ increasing function with $\varphi(0) = \varphi(0+0) = 0$ and $\varphi(x) \le C\varphi(x/2)$ for $x \in [0,1]$. Then there are polynomials P_n of degree at most n satisfying

$$P_n(0) = 1, \quad |P_n(x)| \le D\exp(-dn\varphi(x)), \quad x \in [-1,1], \quad n = 0,1,\dots \tag{2}$$

for some constants $D > 0$ and $d > 0$ if and only if

$$\int_0^1 \frac{\varphi(u)}{u^2}du < \infty. \tag{3}$$

For example,

$$|P(0)| = 1, \quad |P(x)| \le C_1\exp(-n|x|^\beta)$$

with $\beta > 1$ can be achieved by polynomials of degree $\le Cn$, but for $\varphi(x) = |x|$ we get that the minimal degree n_φ of the polynomials P satisfying

$$|P(0)| = 1, \quad |P(x)| \le C_1e^{-n|x|}, \quad x \in [-1,1]$$

satisfies

$$\frac{1}{C}n\log n \le n_\varphi \le Cn\log n.$$

If we want instead of (2) the property

$$P_n(0)=1,\quad |P_n(x)|\le D\exp(-d\varphi(nx)),\quad x\in[-1,1],\quad n=0,1,\dots,$$

then for this the necessary and sufficient condition is that

$$\int_1^\infty \frac{\varphi(u)}{u^2}du<\infty.$$

E.g.

$$P_n(0)=1,\quad |P_n(x)|\le D\exp(-d|nx|^\beta),\quad x\in[-1,1],\quad n=0,1,\dots,\tag{4}$$

is possible if and only if $\beta<1$.

The substitution $x\to x^2$ easily changes the problem into fast decreasing polynomials on $[0,1]$. E.g. in this case (2) takes the form

$$P_n(0)=1,\qquad |P_n(x)|\le De^{-dn\varphi(x)}\quad x\in[0,1],\tag{5}$$

and for this (3) will change into the necessary and sufficient condition

$$\int_0^1 \frac{\varphi(t)}{t^{3/2}}dt<\infty.\tag{6}$$

3 Improved approximation of $|x|$ and Jackson's theorem

What we use from (4) is that if $\beta<1$, then there are P_n with

$$P_n(0)=1,\qquad |P_n(x)|\le De^{-d(n|x|)^\beta},\quad x\in[-1,1].$$

Actually, we need to use this for a fixed β, so let us put here $\beta=1/2$. Thus, there are even polynomials $P_n\ge 0$ with

$$P_n(0)=1,\qquad |P_n(x)|\le De^{-d\sqrt{n|x|}},\quad x\in[-1,1].$$

Consider

$$Q_n(x)=\frac{2}{\gamma_n}\int_{-1}^x P_n(t)dt-1$$

where

$$\gamma_n=\int_{-1}^1 P_n(t)dt\sim\frac{1}{n}.$$

If $x\in[-1,0]$, then

$$0\le Q_n(x)+1\le Cn\int_{-1}^x e^{-d\sqrt{n|t|}}dt\le C\int_{-\infty}^{nx} e^{-d\sqrt{|u|}}du\le Ce^{-d\sqrt{n|x|}/2}.$$

On the other hand if $x \in [0,1]$, then

$$0 \le 1 - Q_n(x) = \frac{2}{\gamma_n}\int_x^1 P_n(t)dt$$

$$\le Cn\int_x^1 e^{-d\sqrt{n|t|}}dt \le \int_{nx}^{\infty} e^{-d\sqrt{u}}du \le Ce^{-d\sqrt{n|x|}/2}$$

Thus, we have shown that

$$|\text{sign}(x) - Q_n(x)| \le Ce^{-d\sqrt{n|x|}/2},$$

i.e., the signum function can be approximated in this sense. On multiplying by x we obtain

$$\begin{aligned} ||x| - xQ_n(x)| &\le C|x|e^{-d\sqrt{n|x|}/2} \le \frac{C}{n}(n|x|)e^{-d\sqrt{n|x|}/2} \\ &\le \frac{C}{n}e^{-d\sqrt{n|x|}/4}, \end{aligned} \tag{7}$$

which is the improvement over the standard $O(1/n)$ approximation that one needs to deduce Jackson's theorem.

Indeed, (7) yields for the function $x_+ = (x + |x|)/2$ and for the polynomial $Q_n^*(x) = (x + Q_n(x))/2$ the estimate

$$|x_+ - Q_n^*(x)| \le \frac{C}{n}e^{-d\sqrt{n|x|}/4}, \quad x \in [-1,1].$$

Now let for example $f \in \text{Lip}\,\alpha$ on $[0,1]$. Set $f_n(k/n) = f(k/n)$ $k = 0,1,\dots,n$ and let f_n be linear on each $(k/n,(k+1)/n)$. Then clearly $|f - f_n| \le Cn^{-\alpha}$. We can write f_n it in the form

$$f_n(x) = f(0) + \sum_{k=0}^{n} B_{k,n}(x - k/n)_+,$$

where

$$|B_{k,n}| = \left|\frac{f((k+1)/n) - f(k/n)}{1/n}\right| \le Cn^{1-\alpha}.$$

But then for

$$P_n(x) = f(0) + \sum_{k=0}^{n} B_{k,n}Q_n^*(x - k/n)$$

we have

$$\begin{aligned} |f(x) - P_n(x)| &\le |f(x) - f_n(x)| + |f_n(x) - P_n(x)| \\ &\le Cn^{-\alpha} + \sum_{k=0}^{n}|B_{k,n}|\frac{C}{n}e^{-d\sqrt{n|x-k/n|}/4} \\ &\le Cn^{-\alpha} + \frac{C}{n}n^{1-\alpha}\sum_j e^{-d\sqrt{j}/4} \le \frac{C}{n^\alpha}, \end{aligned}$$

which shows that f can be approximated in the order $O(n^{-\alpha})$ by polynomials of degree at most n.

If we want the Jackson estimate not just for Lipshitz α function, and if

$$\omega(f,\delta) = \sup_{|x-x'|\le\delta} |f(x)-f(x')|$$

is the normal modulus of continuity of f, then all we have to do is to notice that $|f-f_n| \le \omega(f,1/n)$, $|B_{k,n}| \le n\omega(f,1/n)$, and

$$|f_n(x) - P_n(x)| \le \sum_{k=0}^{n} n\omega(f,1/n)\frac{C}{n}e^{-d\sqrt{n|x-k/n|/4}} \le C\omega(f,1/n),$$

and these give

$$|f(x) - P_n(x) \le C\omega(f,1/n),$$

i.e., we obtain Jackson's estimate.

It would be possible to obtain this way the Nikolskii-Dzjadyk-Timan-Gopengauz estimates as well as the ω_φ sharpening of the Jackson theorem, but to do that one needs nonsymmetric fast decreasing polynomials.

4 Order of approximation on compact sets

We have already mentioned that in 1914 S.N. Bernstein verified that the finite and positive limit

$$\lim_{n\to\infty} nE_n(|x|,[-1,1]) = \sigma$$

exists, and for the value of σ he obtained $0.278 < \sigma < 0.286$. That was a proof of about 50 pages. The exact value of σ is still unknown, though it was calculated up to 50 decimal digits in [14]. In the years 1938–46 Bernstein returned to the problem, and established that if $p > 0$ is not an even integer, then

$$\lim_{n\to\infty} n^p E_n(|x|^p,[-1,1]) = \sigma_p \tag{8}$$

exists and it is finite, and so is

$$\lim_{n\to\infty} n^p E_n(\text{sign}(x)|x|^p,[-1,1]) = \sigma_p^*.$$

He also showed that for $x_0 \in (-1,1)$

$$\lim_{n\to\infty} n^p E_n(|x-x_0|^p,[-1,1]) = (1-x_0^2)^{p/2}\sigma_p, \tag{9}$$

where this σ_p constant is the same as in (8).

It is a natural question what happens for more general sets. We need to measure somehow the density of F about a point x_0. Let this measure be

$$\Theta_F(t,x_0) = |[x_0-t,x_0+t]\setminus F|,$$

which is a variant of the density function in [5], but it also appears in a different form in the work of R.K. Vasiliev [15]. Now with this density function we have the following result.

Theorem 1 *If*

$$\int_0^1 \frac{\Theta_F(t,x_0)^2}{t^3}dt < \infty, \tag{10}$$

then for $p > 0$ *not an even integer we have*

$$\liminf_{n\to\infty} n^p E_n(|x-x_0|^p, F) > 0. \tag{11}$$

Conversely, if $0 \le \Theta(t) \le 2t$ *is an increasing function on* $[0,1]$ *with*

$$\int_0^1 \frac{\Theta(t)^2}{t^3}dt = \infty, \tag{12}$$

then there is a compact set $F \subset [-1,1]$ *such that* $\Theta_F(t,0) \le \Theta(t)$ *for all* t *and*

$$\lim_{n\to\infty} n^p E_n(|x|^p, F) = 0. \tag{13}$$

Let us mention that the same density condition as in the theorem appears in other problems, as well. For example, for the analogue of the Bernstein inequality

$$|P_n'(x)| \le \frac{n}{\sqrt{1-x^2}}\|P_n\|_{[-1,1]}$$

on compact sets we have the following:

Theorem 2 *If* (10) *holds, then*

$$|P_n'(x_0)| \le Cn\|P_n\|_F$$

for some $C > 0$.

Conversely, if $\Theta(t) \le 2t$, *is an increasing function such that* (12) *holds, then there is a set* $F \subset [-1,1]$ *such that* $\Theta_F(t,0) \le \Theta(t)$ *for all* $t > 0$, *but for some polynomials* P_n, $n = 1, 2, \ldots$

$$\lim_{n\to\infty} \frac{P_n'(0)}{n\|P_n\|_F} = \infty.$$

Or consider the analogue of (2) on fast decreasing polynomials. Suppose that $F \subset [-1,1]$ is a compact set, φ is an even, and on $[0,1]$ increasing function with $\varphi(0) = \varphi(0+0) = 0$, $\varphi(x) \le C\varphi(x/2)$ for $x \in [0,1]$. We consider the problem how fast a polynomial can decrease on F in the sense that we look for polynomials P_n of degree at most n with the property

$$P_n(0) = 1, \quad |P_n(x)| \le D\exp(-dn\varphi(x)), \quad x \in F, \quad n = 0, 1, \ldots. \tag{14}$$

It is clear that since F is not the whole $[-1,1]$ and we do not care how the polynomials behave outside F, there is more freedom in constructing such polynomials. Now we have the following result that tells us that (2) is true under the density condition (10).

Theorem 3 *If* (10) *holds, then there are polynomials with the property* (14) *if and only if*

$$\int_0^1 \frac{\varphi(t)}{t^2} dt < \infty. \tag{15}$$

Conversely, if $0 \le \Theta(t) \le 2t$ *is an increasing function on* $[0,1]$ *with the property* (12), *then there is a compact set* $F \subset [-1,1]$ *such that* $\Theta_F(t) \le \Theta(t)$ *for all* t *and there is a monotone* φ *with*

$$\int_0^1 \frac{\varphi(t)}{t^2} dt = \infty \tag{16}$$

for which there are polynomials P_n *with property* (14).

That the necessity of the condition in (10) have to be considered in the above sense is shown by the following surprising results, which show that the positive directions of the above theorem can be true for sets of measure 0. In fact, let F^* be symmetric onto the origin, and set $F = \{x^2 \mid x \in F^*\} \subseteq [0,1]$. Then

$$E_n(|x|^p, F^*) \sim n^{-p} \iff E_n(|x|^{p/2}, F) \sim n^{-p},$$

i.e., our approximation problem is equivalent to approximation of $x^{p/2}$ on a set $F \subset [0,1]$. Therefore, in what follows let $F \subseteq [0,1]$ be compact and $0 \in F$.

For $\varepsilon_1, \varepsilon_2, \ldots \in (0,1)$ do the Cantor set construction with this sequence, i.e., at the n-th step omit the middle ε_n part of each remaining interval. Let C_n be the set the we have after n steps. Then C_n consists of 2^n intervals of length,

$$\frac{1-\varepsilon_1}{2} \cdot \frac{1-\varepsilon_2}{2} \cdots \frac{1-\varepsilon_n}{2}$$

so their total length is

$$(1-\varepsilon_1)(1-\varepsilon_2)\cdots(1-\varepsilon_n).$$

This shows that $C = \cap_n C_n$ is of measure 0 if and only if $\sum_j \varepsilon_j = \infty$. Now for approximation on these Cantor sets the following is true.

Theorem 4 *Let* $p > 0$ *be not an integer. Then*

$$E_n(x^p, C) \ge \frac{c}{n^{2p}}, \quad n = 1, 2, \ldots$$

for some $c > 0$ *if and only if* $\sum_j \varepsilon_j^2 < \infty$.

With $\varepsilon_j = 1/(j+1)$ we get a set F of measure 0 such that $E_n(x^p, F) \geq c/n^{2p}$, $c > 0$, and by symmetrization a set $F \subseteq [-1,1]$ of measure 0 such that $E_n(|x|^p, F) \geq c/n^p$, $c > 0$. Note that for a set of measure 0 the function $\Theta_F(t, x_0)$ is identically $2t$, so the integral in (10) is infinite, i.e., (10) is not true. Thus, the strict necessity of (10) is not true, just in the sense that was discussed above.

The condition $\sum_i e_i^2 < \infty$ also appears in several other questions, for example the local Markoff inequality

$$|P_n'(0)| \leq Cn^2 \|P_n\|_C, \qquad \deg(P_n) \leq n, \quad n = 1, 2, \dots$$

holds if an only if $\sum_i e_i^2 < \infty$.

5 Vasiliev's results

R.K. Vasiliev in his paper [15] considered approximation of $|x|^p$ (or what is the same $|x - x_0|^p$) on compact sets. His approach was the following.

Let

$$F = [-1,1] \setminus \cup_{i=1}^{\infty} (\alpha_i, \beta_i),$$

and consider the sets

$$F_m = [-1,1] \setminus \cup_{i=1}^{m-1} (\alpha_i, \beta_i).$$

F_m consists of m intervals

$$F_m = \cup_{j=1}^{m} [a_j, b_j]$$

$a_1 < b_1 < a_2 < b_2 \cdots b_{m-1} < a_m < b_m$, and for it define

$$h_{F_m}(x) = \frac{\prod_{j=1}^{m-1} |x - \lambda_j|}{\sqrt{\prod_{j=1}^{m} |x - a_j||x - b_j|}},$$

where λ_j are chosen so that

$$\int_{b_k}^{a_{k+1}} \frac{\prod_{j=1}^{m-1} (t - \lambda_j)}{\sqrt{\prod_{j=1}^{m} |t - a_j||t - b_j|}} dt = 0$$

for all $k = 1, \dots, m-1$. Now set

$$h_F(x) = \lim_{n \to \infty} h_{F_n}(x) = \sup_n h_{F_n}(x),$$

where it can be shown that the limit exists (but it is not necessarily finite).

Now Vasiliev claims that

$$\lim_{n\to\infty} n^p E_n(|x-x_0|^p, F) = h_F(x_0)^{-p}\sigma_p, \tag{17}$$

$$\lim_{n\to\infty} n^p E_n(|x-x_0|^p, F) > 0 \Longleftrightarrow \int_0^1 \frac{\Theta_F(t,x_0)^2}{t^3}dt < \infty. \tag{18}$$

This second claim contradicts Theorem 4 (recall that for a set F of zero measure we have $\Theta_F(t,x_0) \equiv 2t$), and the correct relevant result is Theorem 1. Vasiliev's paper [15] is 166 pages long and it is solely dedicated to the proof of (17) and (18). Now it is difficult to say what is wrong in a 160 page proof. I do not know if the full (17) is correct. However, the following is certainly true, which shows that (17) is correct provided x_0 lies in the interior of F.

To formulate the result let us recall that the density of the equilibrium measure (cf. [11]) for a set

$$F = \cup_{j=1}^m [a_j, b_j]$$

$a_1 < b_1 < a_2 < b_2 \cdots b_{m-1} < a_m < b_m$ is given by

$$d\omega_F(x) = \frac{\prod_{j=1}^{m-1}|x-\lambda_j|}{\pi\sqrt{\prod_{j=1}^m |x-a_j||x-b_j|}}dx, \tag{19}$$

where λ_j are chosen so that

$$\int_{b_k}^{a_{k+1}} \frac{\prod_{j=1}^{m-1}(t-\lambda_j)}{\sqrt{\prod_{j=1}^m |t-a_j||t-b_j|}}dt = 0$$

for all $k = 1, \ldots, m-1$. Thus, Vasiliev's function is just $h_F(x) = \pi\omega_F(x)$ if F consists of a finite number of intervals, and this is also true if F is arbitrary compact, but x is in its interior. Now (17) for $x_0 \in \mathrm{Int}(F)$ takes the following form.

Theorem 5 (R.K. Vasiliev) *Let $F \subseteq \mathbb{R}$ be compact and let x_0 be a point in the interior of F. Then*

$$\lim_{n\to\infty} n^p E_n(|x-x_0|^p, F) = (\pi\omega_F(x_0))^{-p}\sigma_p, \tag{20}$$

where σ_p is the constant from Bernstein's theorem (8).

For example, if $F = [-1,1]$, then

$$\pi\omega_{[-1,1]}(x) = \frac{1}{\sqrt{1-x^2}},$$

and in this special case we recapture Bernstein's result (9).

Idea of the proof is to use polynomial inverse images of intervals.

Let T_N be a real polynomial of degree N. We call T_N admissible, if T_N has $(N-1)$ local extrema, and all the local extremal values of T_N are alternately ≥ 1 and ≤ -1. The inverse image of $[-1,1]$ under T_N is

$$T_N^{-1}([-1,1]) = \{x \,|\, T_N(x) \in [-1,1]\}.$$

Now it can be shown (see e.g. [4], [10], [12]) that if $F = \cup_{i=1}^{l}[a_j, b_j]$ consists of l intervals, and $\varepsilon > 0$, then there is an admissible T_N such that

$$T_N^{-1}([-1,1]) = \cup_{i=1}^{l}[a_i', b_i']$$

consists of the same number of intervals, and for all i we have

$$|a_i - a_i'| \leq \varepsilon, \quad |b_i - b_i'| \leq \varepsilon.$$

If $x_0 \in \operatorname{Int}(F)$, then it can also be achieved that $T_N(x_0) = 0$ (and also that $T_N^{-1}([-1,1]) \subset F$ or $F \subset T_N^{-1}([-1,1])$ as we wish).

Knowing this, the sketch of our proof for Vasiliev's theorem for $x_0 \in \operatorname{Int}(F)$ is as follows. Each of the following steps uses the result from the previous step.

- Apply Bernstein's original result, i.e., we have the statement for $F = [-1,1]$, $x_0 = 0$.
- If $F = T_N^{-1}([-1,1])$, $T_N(x_0) = 0$, then take polynomial inverse image, and transfer the result from the previous step. This is the step when we change from one interval to several intervals.
- If F consists of finitely many intervals, then approximate F by a polynomial inverse image set $T_N^{-1}([-1,1])$ with $T_N(x_0) = 0$. The approximation means that $\omega_F(x_0)$ and $\omega_{T_N^{-1}([-1,1])}(x_0)$ are close.
- If F is an arbitrary compact, then approximate it by a set of finitely many intervals.

References

[1] S.N. Bernstein, Sur la meilleure approximation de $|x|$ par des polynômes des degrés donnés, *Acta Math.* (Scandinavian), **37**(1914), 1–57.

[2] S.N. Bernstein, On the best approximation of $|x|^p$ by means of polynomials of extremely high degree, *Izv. Akad. Nauk SSSR,* Ser. Mat. **2**(1938), 160–180. Reprinted in S.N. Bernstein *Collected Works*, Vol. 2, pp. 262–272. Izdat. Nauk SSSR, Moscow, 1954 (Russian).

[3] S.N. Bernstein, On the best approximation of $|x - c|^p$, *Dokl. Akad. Nauk SSSR,* **18**(1938), 379–384. Reprinted in S.N. Bernstein *Collected Works*, Vol. 2, pp. 273–260. Izdat. Nauk SSSR, Moscow, 1954 (Russian).

[4] A.B. Bogatyrev, Effective computation of Chebyshev polynomials for several intervals, *Math. USSR Sb.*, **190**(1999), 1571–1605.

[5] T. Erdélyi, A. Kroó and J. Szabados, Markov-Bernstein type inequalities on compact subsets of **R**, *Analysis Math.*, **26**(2000), 17–34.

[6] K.G. Ivanov and V. Totik, Fast decreasing polynomials, *Constr. Approx.*, **6**(1990), 1–20.

[7] K.G. Ivanov, E.B. Saff and V. Totik, Approximation by polynomials with locally geometric rates, *Proc. Amer. Math. Soc.*, **106**(1989), 153–161.

[8] H. Lebesgue, Sur l'approximation des fonctions, *Bull. Sci. Math.* **22**(1898), 278–287.

[9] Allan Pinkus, Weierstrass and approximation theory, *J. Approx. Theory*, **107**(2000), 1–66.

[10] F. Peherstorfer, Deformation of minimizing polynomials and approximation of several intervals by an inverse polynomial mapping, *J. Approx. Theory*, **111**(2001), 180–195.

[11] E.B. Saff and V. Totik, *Logarithmic Potentials with External Fields*, Grundlehren der mathematischen Wissenschaften, **316**, Springer Verlag, Berlin, Heidelberg, 1997.

[12] V. Totik, Polynomial inverse images of intervals and polynomial inequalities, *Acta Math.* (Scandinavian), **187**(2001), 139–160.

[13] V. Totik, Polynomial approximation with locally geometric rates, *Approximation Theory*, Proc. Conf. in Kecskemét, Hungary, 1990, Coll. Math. Soc. János Bolyai, **58**, Akadémiai Kiadó/North-Holland, Budapest 1992, 663–671.

[14] R. Varga and A.J. Carpenter, On the Bernstein conjecture in approximation theory, *Constr. Approx.*, **1**(1985), 333–348.

[15] R.K. Vasiliev, *Chebyshev Polynomials and Approximation Theory on Compact Subsetc of the Real Axis*, Saratov University Publishing House, 1998.

Bolyai Institute
University of Szeged
Aradi v.tere 1
6720, Hungary and

Department of Mathematics
University of South Florida
Tampa, FL 33620, USA
Email address: totik@math.usf.edu

This work was partially supported by KBN (Poland) under grant 5 P03A 034 20.

MIX
Papier aus verantwortungsvollen Quellen
Paper from responsible sources
FSC® C105338

If you have any concerns about our products,
you can contact us on
ProductSafety@springernature.com

In case Publisher is established outside the EU,
the EU authorized representative is:
Springer Nature Customer Service Center GmbH
Europaplatz 3, 69115 Heidelberg, Germany

Printed by Libri Plureos GmbH
in Hamburg, Germany